Electrical Calculations and Guidelines for Generating Stations and Industrial Plants

Second Edition

Thomas E. Baker

CRC Press
Taylor & Francis Group
Boca Raton London New York

CRC Press is an imprint of the
Taylor & Francis Group, an **informa** business

CRC Press
Taylor & Francis Group
6000 Broken Sound Parkway NW, Suite 300
Boca Raton, FL 33487-2742

© 2018 by Taylor & Francis Group, LLC
CRC Press is an imprint of Taylor & Francis Group, an Informa business

No claim to original U.S. Government works

International Standard Book Number-13: 978-1-4987-6938-9 (Hardback)

This book contains information obtained from authentic and highly regarded sources. Reasonable efforts have been made to publish reliable data and information, but the author and publisher cannot assume responsibility for the validity of all materials or the consequences of their use. The authors and publishers have attempted to trace the copyright holders of all material reproduced in this publication and apologize to copyright holders if permission to publish in this form has not been obtained. If any copyright material has not been acknowledged, please write and let us know so we may rectify in any future reprint.

Except as permitted under U.S. Copyright Law, no part of this book may be reprinted, reproduced, transmitted, or utilized in any form by any electronic, mechanical, or other means, now known or hereafter invented, including photocopying, microfilming, and recording, or in any information storage or retrieval system, without written permission from the publishers.

For permission to photocopy or use material electronically from this work, please access www.copyright.com (http://www.copyright.com/) or contact the Copyright Clearance Center, Inc. (CCC), 222 Rosewood Drive, Danvers, MA 01923, 978-750-8400. CCC is a not-for-profit organization that provides licenses and registration for a variety of users. For organizations that have been granted a photocopy license by the CCC, a separate system of payment has been arranged.

Trademark Notice: Product or corporate names may be trademarks or registered trademarks, and are used only for identification and explanation without intent to infringe.

Visit the Taylor & Francis Web site at
http://www.taylorandfrancis.com

and the CRC Press Web site at
http://www.crcpress.com

Printed and bound in Great Britain by
TJ International Ltd, Padstow, Cornwall

Contents

Preface xiii
Caveat Emptor xvii
Acknowledgments xix
About the Author xxi

1 **The Basics** 1
   1.1   Three-Phase History 1
   1.2   Three-Phase System Advantages 2
   1.3   Theory 2
   1.4   Magnetism 3
   1.5   Voltage, Current, and Frequency 6
   1.6   Resistance 11
   1.7   Inductance 12
   1.8   Capacitance 13
   1.9   Circuits 13
   1.10 Time Constants 16
   1.11 Reactance 20
   1.12 Series Impedance 21
   1.13 Parallel Impedance 24
   1.14 Transformers 27
   1.15 Electrical Systems 30
   1.16 Generating Station Electrical Configurations 32
   1.17 Three-Phase Basics 34
   1.18 Power Transformer Connections 45
   1.19 Instrument Transformer Connections 49
   Bibliography 53

2 **Electrical Studies** 55
   2.1   Conversions 55
         2.1.1   Ohmic 55
         2.1.2   Megavolt-Amps (MVA) 56
   2.2   Transformer Tap Optimization 58
   2.3   Conductor Parameters 62
         2.3.1   Buses 63
         2.3.2   Insulated Cable 63
         2.3.3   Overhead Aluminum Conductor Steel-Reinforced
                 (ACSR) Cable 66
   2.4   Study Accuracy 70

2.5	Voltage Studies		70
	2.5.1	Bus Voltage Drop	70
	2.5.2	Line Voltage Drop	70
	2.5.3	Capacitive Voltage Rise	72
	2.5.4	Collapsing Delta	72
2.6	Power Transfer Calculations		72
2.7	Two-Generator System		74
2.8	Ohmic Short Circuit Calculations		75
	2.8.1	No Transformer	75
	2.8.2	Parallel Sources	76
2.9	The Per-Unit System		76
	2.9.1	Basic Formulas	77
	2.9.2	Corrected Voltage Base	78
	2.9.3	Converting Per-Unit Z to Amps	79
	2.9.4	Converting Amps to Per-Unit R and X	81
	2.9.5	New MVA Base	82
	2.9.6	Converting Per-Unit to Ohms	82
	2.9.7	Converting Amps to Per-Unit Z	82
2.10	Per-Unit Short Circuit Calculations		84
	2.10.1	Transformer Short Circuits	84
	2.10.2	Transformer Three-Phase and Phase-to-Phase Fault Procedures	84
	2.10.3	Three-Winding Transformer Short Circuits	86
	2.10.4	Transformer Ohmic Short Circuit Calculations	88
	2.10.5	Sequence Impedances	91
	2.10.6	Transformer Ground Fault Procedures	91
	2.10.7	Demystifying Ground Fault Calculations	93
	2.10.8	Generator Three-Phase Short Circuits	96
	2.10.9	Generator De-Excitation	97
	2.10.10	Motor Contribution	98
Bibliography			100

3	**Auxiliary System Protection**		**101**
3.1	Switchgear Overcurrent Coordination		103
3.2	Overcurrent Schematic		106
3.3	Current Transformer (CT)		107
	3.3.1	CT Safety Ground	107
	3.3.2	CT Open Circuit	108
	3.3.3	CT Reflected Ohms	108
	3.3.4	CT Burden	108
	3.3.5	CT Saturation	109
3.4	Motor Overcurrent		112
	3.4.1	Stator Overcurrent Protection (51)	112
	3.4.2	Rotor Overcurrent Protection (51)	114
	3.4.3	Short Circuit Protection (50)	116

		3.4.4 Digital Motor Protection 117
		3.4.5 Motor Overcurrent Oversights 117
	3.5	Motor Control Center (MCC) Source Overcurrent (50/51).......... 118
		3.5.1 MCC Source Feeder Protection Oversights....................... 121
	3.6	Bus Tie Overcurrent.. 121
		3.6.1 Roughly Estimating Bus Transfer Motor Currents 122
		3.6.2 Delta Bus Transfer Currents 122
		3.6.3 Bus Tie Overcurrent (51)... 123
		3.6.4 Bus Tie Protection Oversights 125
	3.7	Transformer Secondary Side Overcurrent (51) 125
		3.7.1 Transformer Secondary Side Protection Oversights........ 126
	3.8	Transformer Primary Side Overcurrent (50/51) 126
		3.8.1 Transformer Primary Side Protection Oversights............ 128
	3.9	Residual Ground Protection (51G)... 129
		3.9.1 Residual Ground Protection Oversights........................... 130
	3.10	High-Impedance Grounding.. 131
		3.10.1 Induced Voltages 132
		3.10.2 Transient Voltage Mitigation 134
		3.10.3 Primary to Secondary Capacitive Coupling 138
		3.10.4 Neutral Grounding (59G).. 138
		3.10.5 Grounded Wye–Broken Delta Grounding (59G).............. 142
		3.10.6 High-Impedance Ground Detection Oversights 146
	3.11	Transformer High-Speed, Sudden-Pressure Protection (63)....... 147
	3.12	Transformer Current Differential Protection (87) 148
	3.13	Bus Transfer Schemes (27R)... 151
		3.13.1 Bus Transfer Scheme Oversights 154
	Bibliography 155
4	**Generator Protection**... 157
	4.1	Generator Relay Data... 157
	4.2	High-Voltage Switchyard Configurations 161
	4.3	High-Voltage Switchyard Protection Concerns 163
	4.4	Generator Protective Functions 168
		4.4.1 Backup Impedance (21)... 168
		4.4.2 Volts/Hz (24) 172
		4.4.3 Sync Check (25) 178
		4.4.4 Reverse Power (32) 182
		4.4.5 Disconnect (33M)... 183
		4.4.6 Loss of Field (40) 183
		4.4.7 Generator Deexcitation (41).. 190
		4.4.8 Negative Phase Sequence (46) 191
		4.4.9 Inadvertent Energization (50/27)..................................... 194
		4.4.10 Breaker Failure (50BF) 197
		4.4.11 GSUT Instantaneous Neutral Overcurrent Ground Fault (50N).. 198

  4.4.12 GSUT Neutral Overcurrent Breaker Pole Flashover (50PF) 199
  4.4.13 Transformer GSUT Ground Bank Neutral Overcurrent (51GB) 201
  4.4.14 Overvoltage (59) 201
  4.4.15 Isolated Phase Bus Ground Detector (59BG) 202
  4.4.16 Loss of Potential (60) 205
  4.4.17 Stator Ground (64) 206
  4.4.18 Subharmonic (64S) 209
  4.4.19 Generator Third-Harmonic Monitoring (64T) 211
  4.4.20 Out of Step (78) 213
  4.4.21 Overfrequency and Underfrequency (81) 217
  4.4.22 Lockout Relay (86) 221
  4.4.23 Generator Differential (87) 222
 Bibliography 222

5 Electrical Apparatus Calculations 225
 5.1 Buses 226
 5.2 Cable 228
  5.2.1 Short Circuit Withstand Seconds 228
  5.2.2 Short Circuit Fusion Seconds 228
  5.2.3 Cable Line Loss 230
 5.3 Switchgear Circuit Breakers 231
  5.3.1 Alternating Current (AC) Hipot Testing 232
  5.3.2 Circuit Breaker Duty 233
 5.4 Generators 235
  5.4.1 Generator Stator Acceptance DC Hipot 235
  5.4.2 Generator Stator Routine DC Hipot 236
  5.4.3 Generator Rotor Acceptance DC Hipot 237
  5.4.4 Generator Rotor First-Year Warranty DC Hipot 238
  5.4.5 Rotor Routine Overhaul Insulation Testing 238
  5.4.6 Generator Temperature 240
  5.4.7 Cylindrical Rotor Shorted Turns 242
  5.4.8 X/R Ratio 244
 5.5 Metering 244
  5.5.1 Theory 247
  5.5.2 Watt Demand 248
  5.5.3 Watts 249
 5.6 Motors 250
  5.6.1 Motor Insulation Resistance 250
  5.6.2 Acceptance DC Hipot 252
  5.6.3 Routine DC Hipot 253
  5.6.4 Locked Rotor Amps 253
  5.6.5 Unbalanced Voltages 254

		5.6.6	X/R Ratio	255
		5.6.7	Switching Transients	256
		5.6.8	Reliability	258
		5.6.9	Voltage Drop	258
	5.7	Transformers		261
		5.7.1	Power Transformer Losses	261
		5.7.2	Power Transformer X/R Ratio	263
	Bibliography			264
6	**Electrical Operating Guidelines**			**265**
	6.1	Operation of Large Generators		265
		6.1.1	Purpose	265
		6.1.2	Startup Operation	266
		6.1.3	Shutdown Operation	267
		6.1.4	On-line Operation	268
		6.1.5	System Separation	269
		6.1.6	Field Grounds	269
		6.1.7	Voltage Regulators	270
		6.1.8	Moisture Intrusion	270
		6.1.9	Routine Operator Inspections	271
		6.1.10	Generator Protection	272
			6.1.10.1 Differential (87)	272
			6.1.10.2 Stator Ground (64) or (59G)	272
			6.1.10.3 Bus Ground Detectors (59BG)	273
			6.1.10.4 Loss of Excitation (40)	274
			6.1.10.5 Overexcitation (24)	274
			6.1.10.6 Reverse Power (32)	275
			6.1.10.7 Negative Phase Sequence (46)	275
			6.1.10.8 Backup Impedance (21) or Voltage Restraint Overcurrent (51V)	276
			6.1.10.9 Out of Step (78)	277
			6.1.10.10 Overfrequency and Underfrequency (81)	277
			6.1.10.11 Sync Check (25)	278
			6.1.10.12 Inadvertent Energization (50/27)	278
			6.1.10.13 Pole Flashover (50NF)	278
			6.1.10.14 Main and Auxiliary Transformer Differential (87)	279
			6.1.10.15 Feeder Differential (87)	279
			6.1.10.16 Overall Unit Differential (87)	279
			6.1.10.17 Unit Switchyard Disconnect Position Switch (33M)	280
			6.1.10.18 Auxiliary and Main Transformer Sudden Pressure (63)	280
			6.1.10.19 DC Low-Voltage (27DC)	281
			6.1.10.20 DC High-Voltage (59DC)	282

	6.2	Operation of Large Power Transformers 282	
		6.2.1 Purpose.. 282	
		6.2.2 Operator Inspections... 282	
		6.2.3 Sudden Pressure Relays 283	
		6.2.4 Transformer Differential or Sudden Pressure Relay Operations.. 284	
		6.2.5 Emergency Cooling and Loading.................................. 284	
		6.2.6 Oil Pump Operation 285	
	6.3	Operation of Large Electric Motors 285	
		6.3.1 Purpose.. 285	
		6.3.2 Operator Inspections... 285	
		6.3.3 Starting Duty 286	
		6.3.4 Heaters... 286	
		6.3.5 Protection 286	
		6.3.5.1 Instantaneous Phase Overcurrent Tripping 286	
		6.3.5.2 Time-Phase Overcurrent Tripping 287	
		6.3.5.3 Feeder Ground Tripping 288	
	6.4	Operation of Auxiliary System Switchgear..................................... 288	
		6.4.1 Purpose.. 288	
		6.4.2 Operator Inspections... 288	
		6.4.3 Protection 289	
		6.4.3.1 Load Feeder Overcurrent Protection 289	
		6.4.3.2 Load Feeder Ground Protection 290	
		6.4.3.3 Source and Tie Overcurrent Protection............ 290	
		6.4.3.4 High-Side Source Transformer Overcurrent Protection 290	
		6.4.3.5 Source and Tie Residual Ground Protection.... 290	
		6.4.3.6 Source Transformer Neutral Ground Protection 291	
		6.4.3.7 Alarm-Only Ground Schemes 291	
		6.4.4 Switchgear Bus Transfers... 292	
		6.4.4.1 Paralleling Two Sources 292	
		6.4.4.2 Drop Pickup Transfers 293	
		6.4.4.3 Automatic Bus Transfer Schemes 293	
	Bibliography 293		

7 Electrical Maintenance Guidelines.. 295
7.1 Generator Electrical Maintenance 295
7.1.1 Purpose... 295
7.1.2 Routine On-line Slip-Ring Brush-Rigging Inspections... 296
7.1.3 Inspection of Rotor Grounding Brushes and Bearing Insulation 298
7.1.4 Routine Unit Outages 298
7.1.5 Overhauls 299
7.1.6 Vibration 301

7.2	Transformer Electrical Maintenance		302
	7.2.1	Purpose	302
	7.2.2	Inspections	302
	7.2.3	Transformer Testing	304
	7.2.4	Avoiding Pyrolytic Growth in Tap Changers	304
	7.2.5	Internal Inspection	305
	7.2.6	Electrostatic Voltage Transfer	306
	7.2.7	DGA	306
	7.2.8	Dielectric Breakdown Test	307
	7.2.9	Insulators and Bushings	307
	7.2.10	Sudden-Pressure Relays	308
	7.2.11	Spare Transformer Maintenance	308
	7.2.12	Phasing Test	308
7.3	Motor Electrical Maintenance		309
	7.3.1	Purpose	309
	7.3.2	Electrical Protection	309
		7.3.2.1 Instantaneous Phase Overcurrent Tripping (50)	309
		7.3.2.2 Time-Phase Overcurrent Tripping (51)	310
		7.3.2.3 Feeder Ground Tripping (51G)	311
	7.3.3	Routine Testing	312
	7.3.4	Internal Inspections	312
	7.3.5	On-line and Off-line Routine Inspections	314
	7.3.6	Motor Monitoring and Diagnostics	314
7.4	Switchgear Circuit Breaker Maintenance		315
	7.4.1	Purpose	315
	7.4.2	General—Switchgear Circuit Breakers (200 V to 15 kV)	315
	7.4.3	Inspection and Testing Frequencies	316
	7.4.4	Mechanical Inspection	316
	7.4.5	Electrical Testing	317
	7.4.6	Operational Tests	320
	7.4.7	Cubicle Inspection	321
	7.4.8	Rack-In Inspection	321
	7.4.9	Generator DC Field Breakers	322
7.5	Insulation Testing of Electrical Apparatus		322
	7.5.1	Purpose	322
	7.5.2	Apparatus 460 V and Higher	322
	7.5.3	DC High Potential Testing	324
	7.5.4	Generator Rotor Winding Overvoltage Testing	328
	7.5.5	Generator-Neutral Buses or Cables	330
	7.5.6	Cables 5 kV and Higher	330
7.6	Bus and MCC Maintenance		330
	7.6.1	Purpose	330
	7.6.2	Bus Inspections	331
	7.6.3	Bus Testing	331

	7.6.4	MCC Position Inspections	332
	7.6.5	MCC Position Testing	332
7.7		Protective Relay Testing	332
	7.7.1	Purpose	332
	7.7.2	General	332
	7.7.3	Testing Schedule (440 V to 765 kV)	333
	7.7.4	Relay Routine Tests	333
	7.7.5	Primary Overall Test of CTs	334
	7.7.6	Documentation	334
	7.7.7	Multifunction Digital Relay Concerns	334
7.8		Battery Inspection and Maintenance	335
	7.8.1	Purpose	335
	7.8.2	General	335
	7.8.3	Floating Charges	336
	7.8.4	Inspection Schedules	336
	7.8.5	Safety Precautions	338
	7.8.6	Operation and Troubleshooting	339
7.9		Personnel Safety Grounds	340
	7.9.1	Purpose	340
	7.9.2	General	341
	7.9.3	Special Grounding Considerations	342
	7.9.4	Maintenance	344
	7.9.5	Electrical Testing	345
7.10		Generator Automatic Voltage Regulators and Power System Stabilizers	345
	7.10.1	Purpose	345
	7.10.2	AVRs	346
	7.10.3	PSSs	346
	7.10.4	Certification Tests	346
	7.10.5	Routine Tests	348
	7.10.6	Generating Station Responsibilities	348
	7.10.7	Excitation Engineering Responsibilities	349
		Bibliography	349

8 Managing a Technical Workforce 351
8.1	Strategies	351
8.2	Methodology	352
8.3	Area Maintenance	353
8.4	Safety	354
	Bibliography	354

Index 355

Preface

The objectives of the first and second editions of this book are to simplify the math, emphasize the theory, and consolidate the information needed by electrical engineers and technicians who support operations, maintenance, protective relay systems, and betterment projects for generating stations and industrial facilities. Perhaps, the following quotes from Albert Einstein express the author's intent best:

- "If you can't explain it simply, you don't understand it well enough."
- "Education is what remains after one has forgotten what one has learned in school."

The quotations seem to be applicable for theories that explain the reasons for observed electrical phenomenon in a simple manner that can be understood, retained, and utilized for future problem-solving. Complex math, on the other hand, seems to be more about memorizing a procedure that you may not inherently understand and then applying it in a logical manner or sequence. Complex math that is not periodically reinforced will not normally be retained.

Because generating stations and industrial plants are interconnected with utility transmission, subtransmission, and distributions systems, and since theory is theory regardless of voltage, much of the content in this book also relates directly to the transmission and distribution of electrical power.

This book is organized into the following eight chapters: "The Basics," "Electrical Studies," "Auxiliary System Protection," "Generator Protection," "Electrical Apparatus Calculations," "Electrical Operating Guidelines," "Electrical Maintenance Guidelines," and "Managing a Technical Workforce."

In the interest of being complete, and to further explain the mechanisms associated with electrical theory than is typically provided in other books or sources, Chapter 1, "The Basics," starts as a review of single-phase alternating current (AC) theory before delving into the more complex aspects of three-phase power systems. This chapter is particularly unique in that it provides Bohr model explanations for much of the phenomena associated with AC electricity.

Chapter 2, "Electrical Studies," discusses the conversions, data gathering, and calculation expressions for voltage studies, power transfer, and short circuit analysis. In the interest of simplicity, when a calculation does not involve transformation, a three-phase version of Ohm's Law is applied to simplify the procedure. When transformers are involved in the calculation (short circuit studies), the per-unit system is utilized; however, an ohmic method is also presented to compare the number of steps and as a proof for the per-unit system.

Additionally, ground fault calculations are typically performed with a somewhat complicated procedure that applies symmetrical component per-unit sequence impedances. Example calculations are provided for the symmetrical component method, as well as a more conventional approach that utilizes actual ohms, in order to demystify ground fault theory, simplify the process, and allow the reader to more fully understand the expressions. In both of the foregoing cases, three-phase and ground faults, the results of the ohmic comparisons seem to be contrary to conventional thinking on short circuit calculations for three-phase power systems and were not anticipated.

Chapter 3, "Auxiliary System Protection," covers switchgear overcurrent coordination, high-impedance ground detection schemes, transformer protection, and residual voltage bus transfer circuits. Detailed background information is provided on the time coordination of overcurrent relays, as well as on the benefits and electrical phenomena associated with the application of high-impedance (resistance) ground detection schemes. This chapter also provides a cookbook approach to developing protective relay settings for low- and medium-voltage switchgear for fossil (commonly used to cover oil, gas, and coal generating stations) and nuclear generating station auxiliary power systems that are configured according to typical practices.

Chapter 4, "Generator Protection," explains the gathering of data needed to set the various functions, the calculation procedures for actually setting the elements, and the math associated with the various types of impedance measurements. It also provides substantial details on the need for the protection functions and typical generator and turbine withstand times for associated abnormal operating conditions. The chapter also presents the negative impact that generating stations can have on utility bulk-power electrical systems and proposes the application of relatively low cost protective relay functions or elements and more refined settings that are not commonly applied in the interest of improving both the reliability of the high-voltage electrical system and plant availability. As in the previous chapter on low- and medium-voltage auxiliary power systems, it provides a cookbook approach for developing protective relay settings for 50 megavolt-amps (MVA) and larger two- and four-pole generators that are interconnected and designed according to common practices.

Chapter 5, "Electrical Apparatus Calculations," discusses practical apparatus calculations that were not included in the preceding chapters. Calculation procedures are provided for various aspects of buses, cable, circuit breakers, generators, meters, motors, and power transformers.

Chapters 6 and 7, "Electrical Operating Guidelines" and "Electrical Maintenance Guidelines," present the suggested electrical operating and maintenance guidelines for generating station and industrial facilities. In reality, guidelines are moving targets that require periodic revisions or updates to reflect site-specific experience and conditions, industry experience, manufacturers' recommendations, and the ever-changing regulatory requirements. Much of the content in these chapters is based on the author's

Preface xv

experience in overseeing and providing electrical engineering support for a large utility's gas-, oil-, and coal-generating assets for many years, followed by additional years of consulting work in the United States and other countries, and has more of a personal flavor to it.

Chapter 8, "Managing a Technical Workforce," discusses a method for optimizing and managing a technical workforce at large generating stations, including electricians, instrument technicians, and protective relay technicians.

> The continuum of life is a series of experiments; one only needs to analyze the results.
>
> **Thomas E. Baker**

Caveat Emptor

This book presents the author's experience over 50 years working with many different aspects of electrical power engineering. However, while it offers suggestions and recommendations, this book cannot possibly foresee all the on-site conditions or their ramifications, which could affect decision-making. Therefore, the final responsibility must rest with the individual engineer or technician, who has greater clarity and involvement with the actual situation. As time moves on, designs, materials, technology, regulations, and experience change, and it is up to the practicing engineer to keep current. I wish you all the best with your exciting careers in electrical power engineering.

Thomas E. Baker

Acknowledgments

In memory of my wife, Janet, who not only put up with the many hours it takes to complete an engineering book, but encouraged me to see it through to completion.

My son John Baker, BS in Computer Science, employed by CalTech in California, created the graphics and developed the programming code for the EE Helper Power Engineering software program. The software graphics are utilized throughout the text, and his assistance was extremely valuable in the development of this book.

I also want to thank my good friend and associate Dr. Isidor Kerszenbaum, Institute of Electrical and Electronics Engineers (IEEE) fellow, the author and coauthor of the more recent IEEE and CRC Press books on generators, a generator consultant, and a lecturer, for important suggestions that have been incorporated in the text to improve the presentation of the material.

Finally, I want to thank David Hughes, electrical engineer, consultant, and training instructor in the petroleum industry in Scotland, for his interest in understanding electrical theory and his unsolicited back-cover comments that were provided for an Amazon book review.

About the Author

Thomas E. Baker was working for Southern California Edison (SCE) as a protective relay technician performing new construction commissioning and overhaul testing of large generating station electrical systems and associated high-voltage switchyards when he was inducted into the Marine Corps. Following release from active duty in 1968, he returned to work as a protective relay technician and took advantage of the GI Bill, eventually completing a master's degree in electrical power engineering and management. He is a member of the Phi Kappa Phi honor society.

Following graduation, Baker held various electrical engineering positions with progressive responsibility as a metering engineer, protection engineer, distribution engineer, and apparatus engineer; and for the last 15 years of his career with SCE, he was the principal electrical engineer responsible for overseeing the electrical support of 12 multiunit fossil-fueled generating stations. A specific generating station usually does not have many significant electrical events, but overseeing 12 stations exposed him to many operating and maintenance errors, short circuits, electrical apparatus failures, and protective relay oversights.

After an early retirement from SCE due to deregulation in California, Baker worked as a consultant in the United States and overseas. He found it a little unsettling to be sent halfway around the world and show up with only a calculator and no reference material. His son, John, was completing a computer science degree at that time, and they cooperated in the design of the Sumatron EE Helper Power Engineering software program (originally 132 and currently being expanded to over 200 calculations) in order to facilitate Baker's consulting work. Arrangements for purchasing the software can be made at www.sumatron.com.

Each software calculation has an associated graphic that illustrates the circuitry, formulas, and references; these graphics are utilized throughout this book. Later, Baker realized that the graphics could also be used for electrical power engineering training, and as of this writing, he has presented 17 four day local California hotel seminars, and 27 additional on-site hotel seminars in the United States and Canada. The seminars are approved by IEEE for earning continuing educational credit.

Considering both his SCE experience and his consulting work, Baker has completed protection reviews for approximately 35,000 megawatts (MW) of gas, oil, coal, and nuclear generation. Consequently, he is well aware of common protective relay oversights that can affect station productivity and the reliability of the bulk-power electrical system.

Baker holds two patents in his name and had the originating ideas for two additional SCE patents or Company patents, all in the electrical power field. He has also written numerous articles and papers on electrical power engineering education and protective relaying that are also covered in the book.

1
The Basics

This chapter addresses the basic theories and conventions associated with three-phase, alternating current (AC) electrical power systems. Some of the theories contained in this chapter and throughout this book are not commonly presented, and the content is meant to enhance the reader's understanding of three-phase power. Although the main focus is generation, the interconnection of generation with transmission, as well as subtransmission and distribution, are also covered in this book. In addition, much of the material presented for generation can be directly applied to the transmission and distribution of electrical power as well.

1.1 Three-Phase History

In 1888, Nikola Tesla (1856–1943), a Serbian-American engineer and physicist, delivered a lecture on the advantages of polyphase AC power. The first transmission of three-phase power occurred in Germany during 1891 at approximately 25 kilovolts (kV). The line was more than 100 miles long and ran from Lauften to Frankfurt to provide power for the International Electro-Technical Conference exhibitions being held there. During the early twentieth century, three-phase, high-voltage transmission became commonplace in the more developed countries. Today in the United States, 115 kV, 138 kV, 230 kV, 345 kV, 525 kV, and 765 kV are commonly used in bulk power electrical systems to convey power several hundred miles to link large areas and multiple states into a single network.

Direct current (DC), high-voltage transmission has also been in use since the invention of high-voltage power inverters and rectifiers, and it offers advantages in situations where there is a need to control the direction of network power flow and where the inductive reactance of longer AC lines becomes too limiting.

1.2 Three-Phase System Advantages

Three-phase systems provide the following advantages:

- Three-phase power creates a rotating magnetic field in the stator windings of large induction and synchronous motors. This rotating magnetic field significantly simplifies the design and applies the required rotational force to the rotor, and then in turn to the driven equipment.
- A single-phase system requires two conductors: one to deliver current to the load, and the other to return the current to the source, thereby completing the circuit. By simply adding one more conductor, significant economies can be realized in the amount of power transfer capability. If the three phase-phase voltages are equal to the single-phase system voltage, the amount of power transferred can be increased by a factor of 1.732 and the line watt losses increase by only a factor of 1.5. The incremental cost to add one more conductor is small compared to the cost of towers or poles and the real estate for right-of-way purposes.

1.3 Theory

Before this book delves too deeply into three-phase systems, a cursory review of basic electrical theory concepts and conventions that also apply to three-phase power may be beneficial. As you will see, magnetism and the development of voltages and currents are particularly complex, especially when one looks at the Bohr model of the atom and the relationships of the various components. Three-phase motors and generators on the surface appear to be relatively simple machines, but in actuality, the electrical relationships are extremely complex and involve many three-dimensional magnetic and electrical parameters that present a spatial challenge.

Developing a simple understanding of magnetic and electrical parameters by relating them to the Bohr model of the atom is difficult because the vast majority of electrical engineering reference and textbooks address only the mathematical expressions and do not explain the theoretical mechanisms. Bohr model explanations are commonly provided for DC, vacuum tube, solid-state electronics, biology, chemistry, and physics, but basic AC theory may be the only field of science that does not rely on the Bohr model of the atom to explain the electrical phenomenon. Even the most basic AC questions, listed below, are not typically addressed:

- Why is a changing magnetic flux required to induce a voltage?
- How does AC current appear to flow at or near the speed of light?
- What causes hysteresis magnetic iron loss?
- What is the atomic mechanism for conductor skin effect?

Assuming that the Bohr model and known relationships are approximately correct, and by researching various writings on the subject, one can deduce simple concepts or explanations that may help the reader grasp and retain the theory. One must keep in mind, however, that no one has ever seen an atom, proton, or electron, and that a proton is approximately 1836 times larger in mass than an electron. Although Bohr model explanations may not be 100% accurate, they do provide students with a way of visualizing the theory. After all, if atomic theory were fully understood, the scientific community would not be spending large sums on particle accelerators and other related research.

1.4 Magnetism

There are two parameters that can affect electrical circuits at any location: stray magnetic flux and capacitance (covered later in this chapter). No material can prevent magnetic lines of force or flux from flowing. Consequently, a doughnut type of current transformer can be applied around shielded cable and function properly; any eddy current effects in the shield will be cancelled out by the circular nature of the current transformer core and cable shield. However, if the shield is grounded at both ends, longitudinal current flow in the shield will cause an error in the current measurement.

Magnetic flux shielding can be provided by enclosing the area of interest inside a magnetic material that will attract the flux away from the interior. A magnetic material is one that has low reluctivity or reluctance and facilitates or attracts magnetic flux. In the case of motors and generators, most of the reluctance (roughly 90%) is in the air gap between the rotor and stator windings. Another method for shielding against magnetic flux is to allow eddy currents to flow in a conducting shield (often used in generators and transformers), which produces an opposing flux that reduces the original flux, that would otherwise have a negative impact on a particular surface area.

Many of the magnetic concepts and mathematical expressions were developed during the first half of the nineteenth century. Some of the more notable pioneers during that period were Michael Faraday (English) for induced voltages; André-Marie Ampère, Jean-Baptiste Biot, and Victor Savart (French) for induced forces; Joseph Henry (American) for self-inductance; and Heinrich Lenz (Estonian) for opposing effects. During the second half of the nineteenth century, Hendrik Lorentz (Dutch) improved the expressions for opposing force calculations.

The following list presents some of the basic elements or relationships in magnetic theory:

- Reluctivity (nu or Greek letter ν)
  = 0.313 for nonmagnetic material
  = 0.007 to 0.00009 for magnetic material
- Reluctance (script $\mathcal{R}$) = V(L/A)
- Permeability (mu or Greek letter μ) = $1/V$ = B/H
- Ampere turns (**IT**) or magneto-motive force (mmf)
- Flux (phi or Greek letter Φ) = IT/$\mathcal{R}$
- Flux density per square inch (**B**) = Φ/A
- Voltage (**E**) = 10^8 flux lines per second per volt = $\Delta \Phi/(S10^8)$
- Magnetic force required to overcome length (**H**) = IT/L
- Lenz's Law: Induced currents will try to cancel out the originating cause
- Force (**F**) in pounds = $(8.85BIL)/10^8$
  A = cross-sectional area (inches)
  I = amps
  L = length of magnetic path or active length of conductor (inches)
  S = seconds

Figure 1.1 shows Fleming's right-hand rule for generators in vector form, without regard to magnitude. As you can see, the original flux (B) is shown as a reference at 0°, the initiating motion (M) or velocity at 90° (the cross section of a conductor moving upward and crossing the flux), a current (I) flowing three-dimensionally into the paper or figure at the point of origin (closed resistive circuit), an opposing force (F) from current flow at 270°, and an opposing flux from current flow at 180°.

The lower-left area of the figure denotes some of the conditions that will affect the placement of a particular vector—that is, time, phase, power angle, power factor, and coil construction. Coils can change the position of current-induced magnetic flux.

In a single conductor, the flux produced by current flow will occupy a position that is 90° from the current flow in a circular fashion around the conductor. However, in a coil, the circular flux from one turn is in opposition to the circular flux of an adjacent turn. Consequently, the flux flows in the interior of the coil and exits at the ends to enclose the coil in an inner and outer flux. However, it still links each coil turn by 90° as each turn is mechanically or physically displaced by 90°. It also amplifies the magnetic flux linking each turn since it now includes the flux from all turns.

The Basics

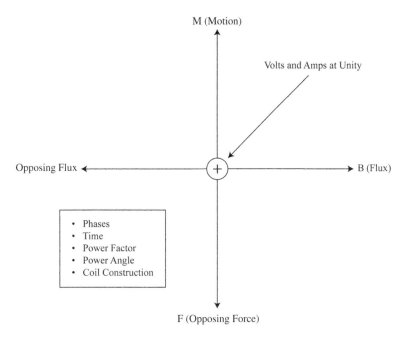

FIGURE 1.1
Magnetic phasors.

Figure 1.2 shows the Bohr model of a simple hydrogen atom, which consists of one positively charged proton and one negatively charged electron. The model was introduced by Niels Bohr (Danish) in 1913. Although the basic hydrogen atom does not have a neutron, one is shown in the figure for clarity because other elements have neutrons. The mass of neutrons are thought to provide a binding force for atoms that contain more than one proton in their nucleus (because like charges repel).

If a strong-enough magnetic flux links with atoms, the charge and mass of the protons reacts with it and aligns or orientates the nucleus and flips the atom

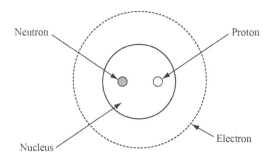

FIGURE 1.2
Bohr model of the atom.

accordingly. This realignment has an energy requirement before it can occur, and atoms in magnetic materials will align at much lower energy thresholds. The amount of energy required before the atom flips during each reversal of AC sine waves, causes a lag in alignment and a hysteresis loss in the core iron of transformers, generators, and motors. Hysteresis loss is the predominant loss of iron in AC electrical apparatuses and represents roughly two-thirds of the total iron loss. The remaining loss is associated with eddy currents in the iron laminations. Atoms can also flip in nonmagnetic materials, but the energy threshold is much greater. This loss can be reduced by adding silicon to the steel to reduce the flip energy. Different elements can also be identified by evaluating the reset time, or the time that the atoms take to return to their original position after the magnetic field is removed. In a permanent magnet (hard material), many of the atoms do not reset, and the material maintains a residual magnetism.

If enough atoms align, which readily happens in the more common ferromagnetic materials (such as iron, steel, nickel, and cobalt), a usable magnetic flux will be generated. According to conventional theory, the magnetic flux is generated at 90° from the charged spinning electrons and radiates outward. In relation to Figure 1.2, it would either be toward the reader or into the paper, depending on the relative polarities or orbit spin directions (clockwise or counter-clockwise).

1.5 Voltage, Current, and Frequency

In 1800, Alessandro Volta (Italian) invented the electric battery. The electrical unit for voltage, the volt (V), is named after him; the letter E is commonly used in calculation formulas. In this context, 1 V will produce a flow of 1 amp in a closed circuit that has 1 ohm. In 1827, George Ohm (Bavarian) developed the expression for Ohm's Law, which describes the mathematical relationship of volts, amps, and resistance, or ohms, in a closed circuit. The current can be determined by simply dividing the voltage by the resistance, or impedance, in an AC circuit.

Photons are packets of quanta of energy that are thought to be without mass and consist of self-propagating electrical and magnetic waves that travel at or near the speed of light, or approximately 186,000 miles per second, regardless of frequency. This electromagnetic propagation includes low frequencies in the communication range, infrared, the visible spectrum, ultraviolet, x-rays, and gamma and cosmic rays at the high end. In the visible spectrum (light), the various frequencies produce different colors. In 1900, Max Planck (German) developed a constant, h, for photon energy. Basically, he proposed that the energy in a photon was a function of the frequency; the higher the frequency, the higher the energy content; the number of photons striking a surface in a given time period increases at higher frequencies.

An orbiting electron's potential energy increases as the radius or distance from the nucleus increases. A photon striking an atom can increase the electron orbit radius, imparting an increase in potential energy. A fall in the orbit level is thought to initiate an electromagnetic propagation as the electron gives up its potential energy and settles into the original lower or closer orbit. The distance of the fall from one orbit level to another determines the frequency and wavelength of the propagation.

The generation of alternating voltage requires a continual changing magnetic flux. One explanation at the atomic level is that the magnetic flux sweeps the electron orbit, causing it to pick up potential energy as it becomes more elliptical and less symmetrical. It is well known that magnetic fields are used to control electron beams in cathode ray tubes, which may validate the concept that a sweeping magnetic flux could distort electron orbits. By analyzing the orbit distortion and remembering some basic physics, one can put together a simple theory for AC current flow. As mentioned previously, an electron's potential energy depends on the radius or distance from the nucleus; the greater the distance, the greater the potential energy is. When a sweeping magnetic flux distorts the electron orbits, they momentarily pick up potential energy as their radii increase, until the neighboring repelling electron forces and proton attractions drive them back to their original orbits, causing a loss of potential energy. This loss of potential energy may represent the voltage that is measured in an AC circuit.

If the circuit is complete or closed, a current will flow. Current flow or amperes (amps) are named in honor of André-Marie Ampère and are denoted by the symbol A; the letter I is often used in calculation expressions. Current flow involving AC has a different mechanism than DC. Conventional theory postulates that current flow and electron flow are the same. At the atomic level, this theory does not seem possible. Practically speaking, AC current in a conductor appears to flow at or near the speed of light. Electrons do have some mass, and consequently, they should not be capable of traveling at the speed of electromagnetic waves or photons that do not possess mass. Also, the orbiting speed of electrons is much slower than the speed of light, and electrons would need to accelerate significantly to attain the speed of AC current flow in a conductor. A magnetic field that sweeps or distorts the electron orbit causes it to gain potential energy, and conversely, the loss of potential energy as the orbit returns to its more original symmetrical position likely produces a weak, low-frequency photon wave.

As a matter of technical interest, electron orbits that are closer to the nucleus require the fastest orbit speed (approximately 1/100 the speed of light) since their attraction to protons is greater. Electrons orbit in order to acquire centripetal force, which prevents them from being sucked into the nucleus by the proton attraction. It appears to be a balancing act, and one might speculate that if the electron orbit speed increases or decreases enough, the electron may be flung free or move closer to the nucleus, respectively, by acquiring enough mass to maintain the required centripetal force (relativity). In Albert

Einstein's 1905 relativity theory, as an object or spaceship moves faster, the electron orbits in the direction of the motion would pick up speed and need to slow down to avoid being flung free. As their speed decreases, they will move closer to the nucleus and increase their mass in order to maintain the requisite centripetal force for the new, closer orbit position. Accordingly, the atoms will compress in the direction of motion and accumulate mass, and time will slow down as a result of the decrease in electron orbit speed.

As mentioned before, basically the same mechanism that launches a photon may be involved with AC current flow. Although this very-low-frequency electromagnetic wave would not have enough energy to propagate outside the conductor, it may have enough energy to flow if there is a complete or closed circuit, and it may account for the known effects of AC current flow in a conductor. Some sources indicate that the electromagnetic waves cause orbiting electrons to oscillate or vibrate as they travel through the conductor, which in turn creates a watt loss heating effect. This watt loss would prevent continued self-propagation because the energy is dissipated as heat. Another explanation that is commonly given in some introductory courses is to visualize it as a straw or tube full of touching marbles or balls; if you bump the first one, the last one on the other end will fall out almost instantaneously. Like charges repel; consequently, orbiting electrons should not be in contact with each other, and any electron jump speeds would need to occur at approximately the speed of light to be feasible within AC theory.

The electromagnetic wave also may have more of a tendency to flow on the outside surfaces (i.e., the skin effect) to reduce frictional watt losses from electron orbit vibrations by traveling near the insulation medium, where electrons are either less dense (in the case of air or other gases) or more tightly bound (with regard to insulation materials). The foregoing skin effect phenomenon may also explain why light bends when going by a planet or mass. As with AC current, photons that travel through atoms will vibrate their electron orbits, causing an energy loss. The light may be inclined to bend to avoid the corresponding losses. Although this and the previous discussion have been speculative in nature, they do seem to fit the Bohr model of the atom, seem to be consistent with the way that basic photon theory is presented in physic courses, and provide the nonphysicist or layperson with some insight into AC electricity and atomic behavior.

Alternating voltages and currents from generators and in three-phase power systems are sinusoidal or take the form of sine waves, as illustrated in Figure 1.3. In AC power systems, the voltage changes in polarity (positive to negative) and the current reverses direction during each half-cycle due to the change in magnetic field pole polarity (north to south) as the generator rotates. The changes in polarity during each consecutive half-cycle will cause vibration of twice the frequency (120 Hz) of the end-turn windings of generator and motor stator conductors because of the magnetic opposing forces created by the bidirectional current flow. One cycle represents both directions of current flow.

The Basics

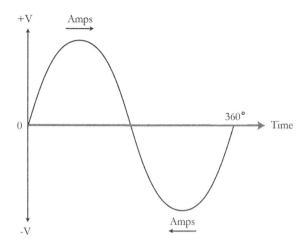

FIGURE 1.3
One cycle sine wave.

Frequency is expressed as hertz (Hz) in honor of Heinrich Hertz (German), for his work in electromagnetic fields during the last half of the 1800s, and this unit of measure is used to define cycles per second; the letter F is commonly used in calculation formulas to represent hertz. Figure 1.4 shows how to calculate the period, or time interval, for a specified base frequency for one cycle. Figure 1.5 provides an example of converting cycles to seconds, and

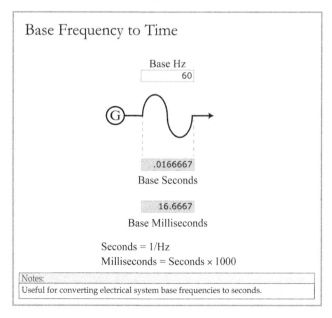

FIGURE 1.4
Base frequency to time.

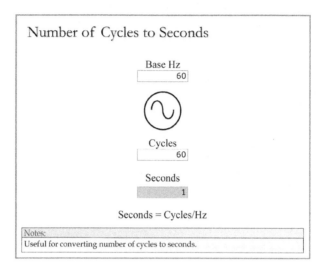

FIGURE 1.5
Number of cycles to seconds.

Figure 1.6 shows how to convert time or seconds to the number of cycles at a specified base frequency.

Graphic modules developed for the EE Helper Power Engineering software program are used in these figures and throughout the book. The answers or output boxes in the graphic modules are shaded and the input variables are clear or white.

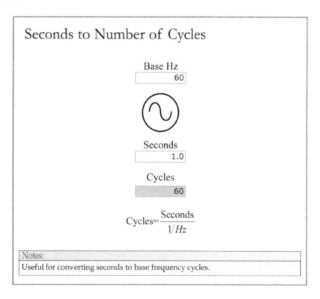

FIGURE 1.6
Number of seconds to cycles.

The frequency and number of poles affect the speed of generators and motors; 60-Hz, two-pole synchronous machines will rotate at 3600 RPM, while the speed of 50-Hz generators and motors will decrease to a factor of (50/60), or 3000 RPM. The inductive reactance will also be reduced by the same factor, which improves the power transfer capability of overhead transmission lines. Transformers and generators are smaller at 60 Hz because of the higher rate of change; conductor AC resistance will decrease a small amount at 50 Hz because of a reduction in skin effect and is dependent on the size of the conductor and the ratio of the skin effect area to the remaining conductor material. Most countries, including those in Europe, use 50 Hz, but North America, as well as some countries in Central and South America, utilize 60 Hz; Japan uses both. There is probably not a big advantage with either frequency, although Tesla thought that 60 Hz was more efficient.

The AC root mean square (RMS) values replicate the voltage and currents used to calculate watts in a DC circuit. The RMS values for both currents and voltages can be determined simply by multiplying the peak or maximum values by the sine of a 45° angle, or 0.7071. Unless specified otherwise, the RMS values will be used throughout this book for AC voltage and current magnitudes for both sine wave peaks and momentary or asymmetrical magnitudes.

1.6 Resistance

Electrical resistance impedes the flow of current in conductors and insulators and is affected by the temperature of the conductor or insulator. Resistance is measured in ohms in honor of George Ohm, and the Greek letter omega Ω is used to represent it; the letter R is used in calculation expressions. The resistance of a conductive material increases with an increase in temperature. This phenomenon is thought to be associated with the more chaotic motions of atoms at higher temperatures.

With insulating materials, the opposite occurs; the resistance decreases with an increase in temperature and may be associated with less tightly bound orbiting electrons. For both conductors and insulators, resistance values are normally given at specific temperatures that need to be corrected for any temperature differences; that is, buses at 80°C, insulated cable at 75°C, overhead aluminum conductor steel-reinforced (ACSR) at 50°C, and minimum megohms for insulation systems at 40°C.

The ohmic or resistance values of conductors with AC flowing is higher than with DC flowing. An AC resistance value is used that is congruent with the amps squared R (I^2R) heating from the current flow. According to conventional theory, the main difference is caused by skin effect in AC, where much of the current wants to flow near the outside surfaces instead of taking advantage of all the conducting material. There can also be additional effects from conduits, shielding, and proximity to other conductors.

Another likely contribution is that the subtractive magnetic flux from current flowing in other phases or neutral conductors has the most impact near the surface areas where they are stronger and lower the surface impedance path accordingly. Unless specified otherwise, resistance values mentioned in this book will be based on AC resistance, not DC resistance. The AC resistance values for conductors are readily available in manufacturers' data, the IEEE Standard 141, and other reference books.

1.7 Inductance

The inductance of conductors and coils impedes current flow in an AC circuit. It is caused by a self-magnetic flux that induces opposition voltages in conductors when current flows. The opposition voltage also tries to resist any change in current flow. As the source sine wave approaches zero, the opposing voltage will try to prevent the new direction of current flow and force the current to lag the source voltage. The electrical unit for inductance is the henry (symbolized by H), which is named after Joseph Henry; the letter L is commonly used in calculation formulas. When a rate of change in current of 1 amp per second results in an induced voltage of 1 V, the inductance of the circuit is said to be 1 H. When current flows and returns to conductors in proximity, the magnetic flux from the return conductor subtracts from the source conductor flux, which reduces the overall inductance of the conductors. If the wires are spread apart, the inductance will increase and the subtractive flux will weaken. If wound in a coil, the inductance will increase dramatically because the magnetic flux seen by each turn is the total flux from all the turns, and it will increase more dramatically if wound around a ferromagnetic material since the self-flux will increase due to the reduced reluctance of the magnetic path.

Inductance properties can be illustrated with a simple experiment. This particular example uses 27 feet of small gauge hookup wire, a clamp-on ammeter, and an inductance meter. From the highest measured millihenries to the lowest, the following applies:

- A coil with an average diameter of 3.5 inches with a clamp-on ammeter around the coil (for the laminated iron effect) = **0.183 millihenries**
- Without the clamp-on ammeter = **0.106 millihenries**
- Wire in an equilateral triangle arrangement = **0.02 millihenries**
- Parallel wires in close proximity or touching = **0.011 millihenries**

The higher the inductance, the greater the opposing voltage is. In the first example, the coil/iron produces the greatest opposing voltage and associated impedance to current flow. Removing the iron increases the magnetic reluctance, resulting in reduced flux and corresponding inductance. Arranging the wires in the shape of an equilateral triangle reduces the subtractive flux from the return path. Finally, with the wires in close proximity, the subtractive flux from the return path increases, which reduces the overall inductance or self-flux opposition voltage.

In this particular experiment, the inductance for the coil with iron was almost 17 times higher than when the source and return conductors are in close proximity. The differences can be increased further by introducing more iron, more coil turns, and having a smaller coil diameter, which also means more turns for a given length of hookup wire.

1.8 Capacitance

Capacitance can also impede current flow in an AC circuit. Capacitors store voltage or potential energy, perhaps by moving electrons to a higher orbit, thereby increasing their potential energy. The stored voltage tries to maintain and/or increase the circuit voltage. As the source sine wave approaches zero, the stored voltage pushes current in the new direction in advance of the source voltage, causing the current to lead the source voltage. The electrical unit for capacitance is the farad (symbolized by F), and is named in honor of Michael Faraday; the letter C is often used in calculation expressions. Here, 1 farad can store 1 coulomb (amp-second) of charge at 1 V. The coulomb (symbolized by Q) is named in honor of Charles-Augustin de Coulomb (French) for his work in electrostatic forces in the 1700s.

Capacitance is a function of a dielectric constant multiplied by the conductive plate areas and then divided by the distance between the plates. The dielectric (insulating materials or air/gas) could be the insulation on wires or cable, and the plates could be composed of the conductor on one side and another conductor or earth ground for the other plate. Consequently, stray capacitance is everywhere and can have an impact on every AC electrical circuit.

1.9 Circuits

Series circuits, as the name implies, denotes a single source with the devices connected in a consecutive manner, in which the same current flows through all the devices. With the exception of capacitors (in which farads are treated as if they were connected in parallel since a series connection effectively increases the distance between plates), the total value for like devices connected in series is simply the mathematical sum of each—that is, resistors, inductors, reactance,

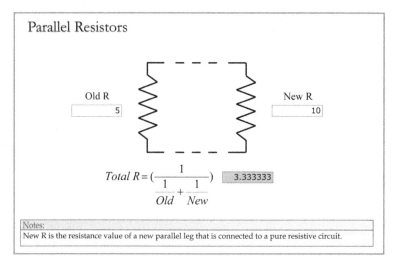

FIGURE 1.7
Parallel resistors.

and impedance (if the angles are the same). For example, three 2-ohm resistors connected in series would have a total ohmic value of 6.0.

Parallel circuits, as the name also indicates, denotes a single source with two or more legs, each with the same voltage but an independent current. Each leg internally can be handled as a series circuit, but the total seen by the source would be the reciprocal of the sum of the reciprocals for simple legs using the same or like components; that is, resistors, inductors, reactance, and impedance (if the angles are the same).

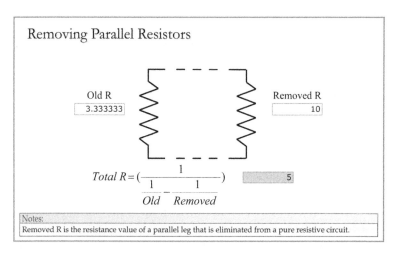

FIGURE 1.8
Removing parallel resistors.

The Basics

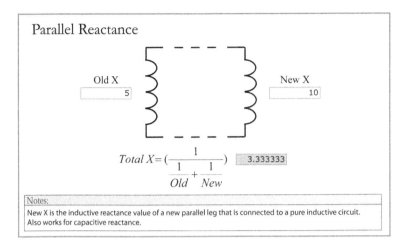

FIGURE 1.9
Parallel reactance or impedance.

Again, capacitors are an exception because the total farads of each parallel leg would simply be added together because it effectively increases the surface area of the plates. Figures 1.7 through 1.10 illustrate the expressions for adding and removing parallel resistance, inductive reactance, and impedance. The procedures for parallel capacitive reactances are the same as the one presented for inductive reactances; only farad values are handled differently. The calculations for series microfarads are illustrated in Figures 1.11 and 1.12.

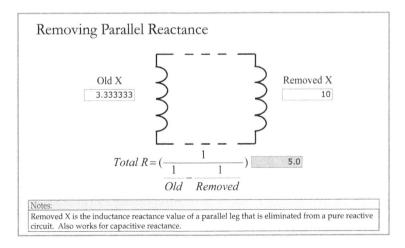

FIGURE 1.10
Removing parallel reactance or impedance.

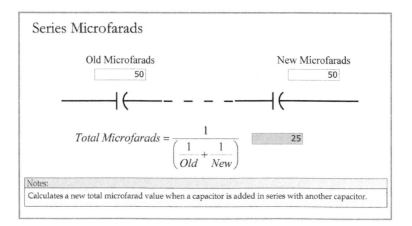

FIGURE 1.11
Series microfarads.

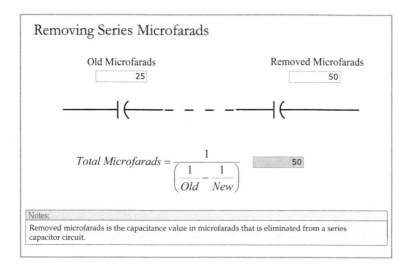

FIGURE 1.12
Removing series microfarads.

1.10 Time Constants

Figures 1.13 through 1.16 display time constant calculations and associated charge and discharge times for capacitors and inductors. Capacitors and inductors cannot charge or discharge instantaneously if resistance is in the circuit. The resistance limits the amount of current that can flow at a given time. At minimum, some resistance will always be present in the source and in the

The Basics   17

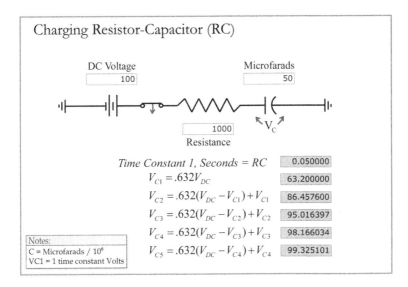

FIGURE 1.13
Charging resistor–capacitor (RC).

conductors that feed the capacitance or inductance. Although the figures show a DC voltage, resistance, and current, the same phenomenon occurs as the sine wave goes through peaks, valleys, and reversals in an AC circuit.

Figure 1.13 shows a resistor–capacitor (RC) charging circuit consisting of a 100-V DC source and a 1000-ohm resistor connected in series with a

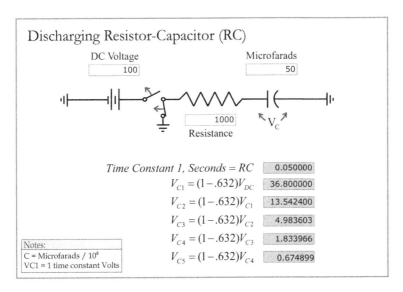

FIGURE 1.14
Discharging resistor–capacitor (RC).

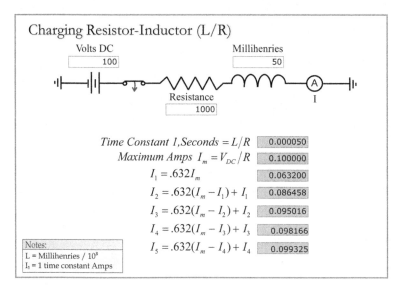

FIGURE 1.15
Charging resistor–inductor (L/R).

50-microfarad (µF) capacitor. One time constant (in seconds) is the ohmic value of the resistor multiplied by the capacitance value in farads. It takes approximately five time constants to fully charge a discharged capacitor. As you can see, the capacitor has a 63.2% charge in one time constant and is almost fully charged in five time constants. The discharging circuit is shown in Figure 1.14. The time constant is the same, and the capacitor will

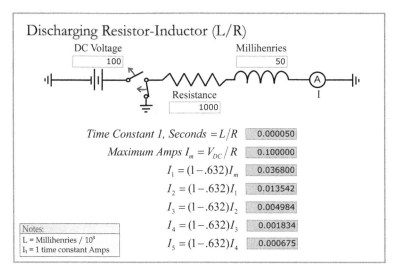

FIGURE 1.16
Discharging resistor–inductor (L/R).

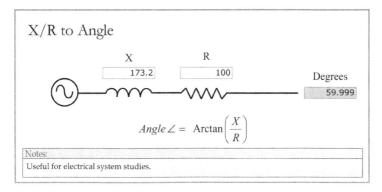

FIGURE 1.17
X/R to angle.

be discharged 63.2% in one time constant and, practically speaking, fully discharged in five time constants.

Figure 1.15 presents a charging circuit for inductance consisting of a 100-V DC source feeding a 1000-ohm resistor connected in series with a 50-millihenry (mH) inductance. The time constant (in seconds in this case) is inductance/resistance (L/R), and the measurement is current. As you can see, the current attains 63.2% of the maximum value in one time constant and reaches almost 100% in five time constants.

Figure 1.16 exhibits the discharging parameters for the same circuit. The time constant is the same, and the current is reduced by 63.2% in one time constant and almost to zero in five time constants.

As you will see in the following text and the rest of the chapters in this book, this L/R relationship is applied in AC power engineering as an X/R ratio to determine current decrements, PF, and short circuit angles. Figures 1.17 and 1.18 illustrate the process for converting X/R ratios to angles and the inverse calculation for AC circuits.

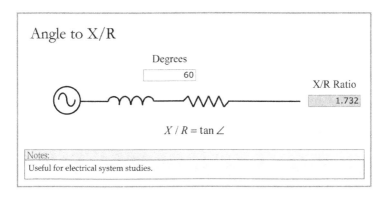

FIGURE 1.18
Angle to X/R.

1.11 Reactance

In order to solve electrical problems and perform studies, the capacitive and inductive parameters that also impede the flow of a changing alternating current need to be converted to a base that is similar to resistive ohms. Figure 1.19 illustrates the procedure for converting farads to a similar ohmic value, denoted by the symbol X_C; and Figure 1.20 shows a process for converting henries into X_L, which also has a similarity to resistive ohms. The capacitive and inductive ohms are called *capacitive reactance* and *inductive reactance*, respectively. Figures 1.21 and 1.22 illustrate the inverse calculations.

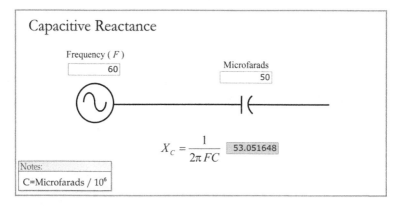

FIGURE 1.19
Capacitive reactance.

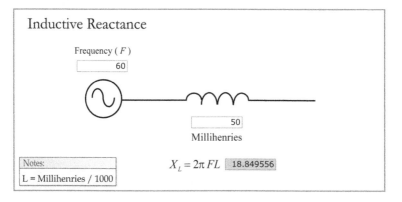

FIGURE 1.20
Inductive reactance.

The Basics

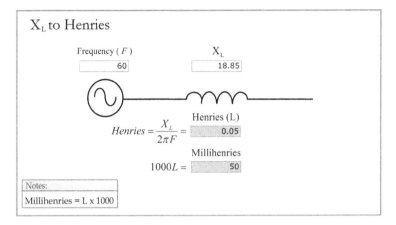

FIGURE 1.21
Inductive reactance to henries.

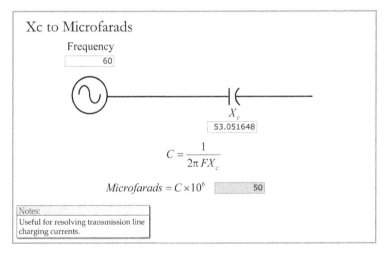

FIGURE 1.22
Capacitive reactance to microfarads.

1.12 Series Impedance

Figures 1.23 through 1.30 illustrate procedures for handling series impedance and their vector relationships. In this case, impedance represents the vector sum of resistive, inductive, and capacitive ohms. As mentioned earlier, capacitance causes the current to lead, inductance causes it to lag, and any resistive component will be in phase with the reference. Current is used as a reference in series circuits because it is the same in all parts of the circuit and the voltage drops are the vector quantities.

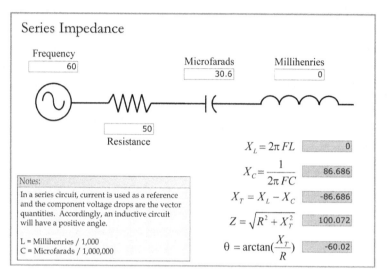

FIGURE 1.23
Series capacitive impedance.

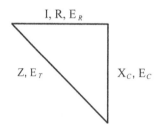

FIGURE 1.24
Series capacitive right triangle.

Figure 1.23 illustrates determining the impedance of a circuit that has resistance in series with capacitance, and Figure 1.24 shows the vector or phasor relationships. Although some in the industry prefer to use the term *phasors* to describe vectors associated with sinusoidal complex numbers, many in industry still use the term *vector*, and both terms will be used interchangeably in this book.

As you can see, it is a right triangle relationship, with the triangle intruding into the fourth quadrant. Because vector rotation is counterclockwise and current is the reference, I, R, and E_R (resistive voltage drop) are shown at 0° and X_C and E_C (capacitive voltage drop) at 270°, and the hypotenuse represents the total impedance Z and total circuit voltage E_T at an angle of 300°. The total current that flows is (E_T/Z).

Figure 1.25 shows the same procedure for determining the impedance of a circuit that has resistance in series with inductance, and Figure 1.26 shows the vector relationships. As you can see, this is also a right triangle relationship,

The Basics

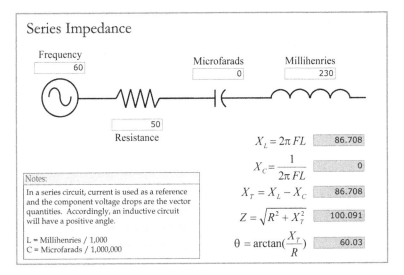

FIGURE 1.25
Series inductive impedance.

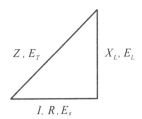

FIGURE 1.26
Series inductive right triangle.

with the triangle intruding into the first quadrant. In this case, I, R, and E_R are shown at 0° and X_L and E_L (inductive voltage drop) at 90°, and the hypotenuse represents the total impedance Z and total circuit voltage E_T at an angle of 60°. This right triangle illustration will be used throughout this book, as most power circuits are series circuit representations that are inductive in nature. Even grid transmission lines that are part of a network are represented in a series fashion when looking at the impedance components of the line itself.

A series circuit with all three components—resistance, capacitance, and inductance—is illustrated in Figure 1.27. As you can see, the capacitive reactance is equal to the inductive reactance, and the current is limited only by the resistive component. This is referred to as *resonance*. If the circuit resistance is low enough, very high or short circuit currents can flow. Figure 1.28 displays a procedure for calculating resonant frequencies, Figure 1.29 shows how to calculate the capacitive portion at a given frequency, and Figure 1.30 illustrates the inductive portion.

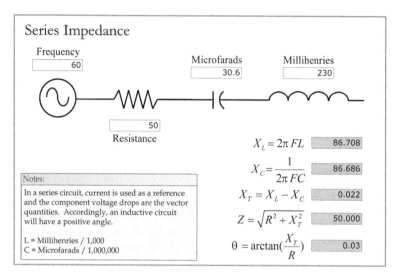

FIGURE 1.27
Series resonant impedance.

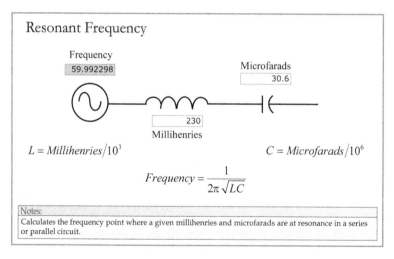

FIGURE 1.28
Resonant frequency.

1.13 Parallel Impedance

Figure 1.31 presents the procedure for calculating parallel impedance. In parallel circuits, the voltage is the same in each leg, and each leg current is independent of the others; consequently, the voltage is the reference and the leg currents are the vector quantities. The old impedance represents the

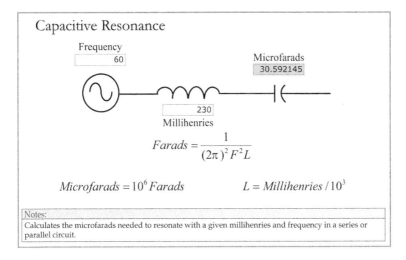

FIGURE 1.29
Capacitive resonance.

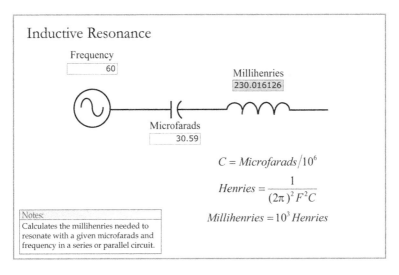

FIGURE 1.30
Inductive resonance.

capacitive impedance of Figure 1.23, except that the polarity of the angle is shown as positive instead of negative as a result of the change in reference from current to voltage. The second leg has a 5-ohm resistor in series with an 11.3-ohm inductance. As presented in Figure 1.7 on parallel resistors (reciprocal of the sum of the reciprocals for like components), a somewhat similar procedure will apply for parallel impedance. The reciprocal of a leg with pure resistive ohms is labeled *conductance* (G), pure reactance ohms as

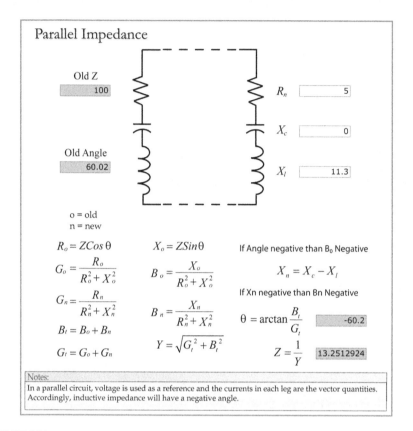

FIGURE 1.31
Parallel impedance.

susceptance (B), and total impedance ohms as *admittance* (Y). However, if the leg has mixed reactance and resistance, the expression for $G = R/(R^2 + X^2)$ and B becomes $X/(R^2 + X^2)$.

The old leg is reduced to the R and X components by using the cosine of the angle to determine R and the sine of the angle to find X. Because of the change in reference (voltage instead of current), inductive quantities are assigned a negative value. G and B are determined for each leg, and then the square root of the total G^2 plus the total B^2 is calculated to determine the circuit Y. The reciprocal of Y equals the total circuit impedance Z, and the angle can be determined by the arc-tangent of B/G. The right triangle relationships are shown in Figure 1.32 and intrude into the fourth quadrant to represent an inductive circuit. The final output of the calculation represents an equivalent simple inductive series circuit that now can be represented in the first quadrant if desired by using current as the reference.

The aforementioned parallel impedance procedure does not provide accurate results when the applied frequency is close to resonance. Unlike

The Basics

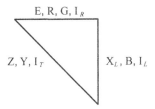

FIGURE 1.32
Parallel impedance right triangle.

a low-impedance, series-resonant circuit, parallel resonance blocks current flow from the source, causing high impedance. The series-resonant expressions presented in Figures 1.28 through 1.30 are also used to identify resonant conditions in parallel circuits. In a parallel circuit near resonance, the circuit can partially sustain oscillations between the capacitive leg and the inductive leg (the *flywheel effect* or *tank circuit*), significantly reducing the source current. However, any resistance in the tank circuit will cause watt losses that will need to be replenished by drawing some current from the source. Parallel resonant circuits (*wave traps*) are used by electric utilities to confine power line carrier communication frequencies to high-voltage transmission lines. The wave traps will pass 60 Hz, but they are tuned to block the much-higher power line communication frequencies from propagating through the electrical system and confine them to the line.

1.14 Transformers

The first closed iron core transformer was built by Westinghouse in 1886 based on a patent secured by William Stanley. Figure 1.33 represents a simple closed core transformer. The source side is called the *primary* and the load side is called the *secondary*. Without load, an excitation current flows in

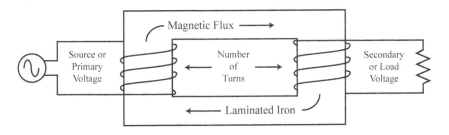

FIGURE 1.33
Simple transformer.

the primary, which is limited by the primary impedance. The excitation current is expressed as a percentage of full load amps and produces a magnetic flux (mutual flux) based on the ampere turns in the primary side, which also links with the secondary winding through the core iron. The magnetic flux induces a voltage in the secondary side proportional to the relative number of turns in the winding. The secondary voltage magnitude is the transformer ratio (primary turns/secondary turns) divided into the primary or source voltage. The secondary voltage can be higher or lower depending on the respective number of turns or winding ratios. An increase in the number of turns in the secondary winding increases the voltage.

When a transformer is loaded, the secondary winding current produces a magnetic flux that opposes the flux originating in the primary winding. This counter or opposing flux reduces the impedance of the primary winding, allowing additional current to flow from the source; in this way, the transformer is self-regulating. The secondary current flow magnitude is also dependent on the ratio (primary turns/secondary turns), and the secondary current can be determined by multiplying the primary current (excluding excitation current) by the turns ratio. The winding with the lower number of turns will have the higher current flow. The foregoing transformer voltage and current relationships are illustrated in Figure 1.34.

Hyphens or black dots are commonly used to represent primary/secondary polarity marks and provide the relative instantaneous direction of current

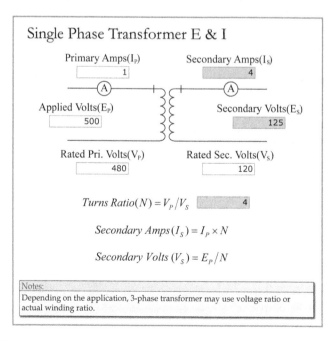

FIGURE 1.34
Transformer voltage and current relationships.

The Basics

flow. The general rule is that current flows into the polarity mark on the primary side will produce current flows out of the polarity mark on the secondary side (in on polarity, and out on polarity). For phasor or vector purposes, the direction of current flow is in agreement with the direction of power flow from the source to the load.

The short circuit ohms of a transformer can be determined by testing. A jumper or short circuit is applied across the secondary winding, and the primary voltage that allows full load amps to flow can be measured. This voltage is denoted as an impedance voltage drop. The primary impedance ohms of the transformer can then be determined by dividing the voltage drop by the full load amps. The same test process can be reversed to determine the impedance ohms on the secondary side. Because the primary and secondary side will have different ohmic values, impedance values are normally expressed as a percent impedance or percentage value that will work on both sides. The per-unit impedance is determined by dividing the impedance voltage drop by the rated voltage or mid-tap position for tapped windings. Multiplying the per-unit value by 100 yields the %Z of the transformer and will normally be provided on nameplates for power transformers. The impedance of the secondary or load can be reflected to the primary side by the following expression as illustrated in Figures 1.35 and 1.36:

$$Z \text{ primary} \times (\text{secondary turns})^2 = Z \text{ secondary} \times (\text{primary turns})^2$$

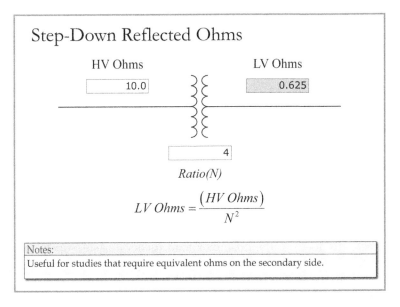

FIGURE 1.35
Reflected ohms HV to LV.

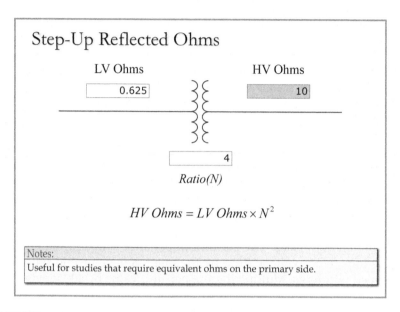

FIGURE 1.36
Reflected ohms LV to HV.

Because apparent power, volt-amps (VA), is a product of volts and amps, as the symbol indicates, the amount of current required for a given VA can be reduced by using a transformer to step up the voltage to a higher level. This is convenient for transferring power long distances to reduce losses and to permit smaller conductors for transmission purposes and also in the design of large motors and generators. Neglecting losses, a transformer's primary VA is equal to its secondary VA.

1.15 Electrical Systems

Besides increasing voltage, transformers are also used to step down transmission, subtransmission, and distribution voltages to more practical levels for the particular applications; that is, industrial, agricultural, commercial, and residential loads. Typically, there are at least four transformers involved before actually feeding the customer load (generator step-up to transmission, transmission to subtransmission, subtransmission to distribution, and distribution to the final customer service entrance voltage).

In addition to an increased insulation requirement for all electrical apparatus, with regard to transformers, there are some tradeoffs with increasing the voltage. The primary turns can be reduced, which increases the excitation current, or the number of secondary turns can be increased, which will cause additional load losses due to the increase in copper conductor resistance and increased var consumption due to the increase in inductance (offset by a lower current for the same MVA).

In the case of generators, increasing the voltage can require significant design changes. These could involve increasing the number of stator turns, which may require a larger bore diameter to accommodate additional stator slots; increasing the excitation field current and corresponding magnetic flux in the stator core iron, which likely requires a bigger stator core to handle the increase in magnetic flux density; and increasing the active length of the slot conductors, which would also require additional core iron. There are also design limitations associated with the rotor diameter due to increased centrifugal forces for a longer radius, and the overall length of the rotor due to flexing. Generators smaller than 150 MVA tend to run around 13.8 kV, and larger machines 16, 18, 20, 22, 24, and 26 kV or higher, depending on their MVA ratings and the manufacturers' design tradeoffs or customer preferences. In general, the higher the MVA, the higher the output voltage is, in order to reduce the amount of current in the stator conductors.

Probably, the most common voltage for generators 100 MVA and larger would be 18 kV (medium-voltage); for transmission, 230 kV (high-voltage); for subtransmission, 69 kV (medium-voltage); and for distribution, 12 kV (medium-voltage). For overhead conductors, the economics generally favor 1 kV per mile of distance. In other words, the foregoing voltages were probably initially selected for roughly transferring power 230, 69, and 12 miles, respectively. The cost of overhead conductors and support towers or poles for transmission, subtransmission, and distribution is roughly one-tenth the cost of undergrounding, depending on the distance, particular voltage, and construction details. Large industrial plants feed their loads with different voltages depending on the horsepower (HP) of connected motors; typical ranges for switchgear fed motors are 75–299 HP for 480-V (low-voltage) buses, 300–5000 HP for 4 kV (medium-voltage) buses, and 6.9 kV or 13.8 kV (medium-voltage) for motors greater than 5000 HP. Commercial facilities are commonly fed with 480 V, and residential customers with 220/110 V.

Figure 1.37 illustrates a simplified one-line (shows one line to represent all three phases) view of a typical utility system. The 230-kV high-voltage transmission system is part of a bulk power electrical network that has multiple geographically dispersed sources of generation and substations. It is used to link large areas together for reliability and economic reasons and may be composed of multiple utilities and states. Medium-voltage 69 kV subtransmission systems are normally fed from a single 230/69-kV substation that is operated by a single utility in an isolated mode and not paralleled with other subtransmission systems. In more recent years, with the advent of peaking units, distributed generation, cogeneration (combined heat and power, or CHP), battery storage, and renewables, it has become much more commonplace for subtransmission systems to also have other sources of power.

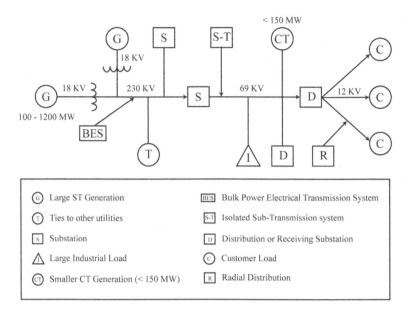

FIGURE 1.37
Typical utility configuration.

Subtransmission systems feed large industrial customers and multiple receiving or distribution substations over an isolated network that may cover a county or portion of a large county.

The 12-kV distribution substations serve much smaller areas, with the majority of the loads coming from smaller industrial, agriculture, commercial facilities, and residential homes. The 12-kV distribution lines radiate from their respective substation and consequently, are referred to as *radial systems*. The term *radial* is also used to describe non-network isolated electrical systems used for auxiliary power for large industrial plants and generating stations.

1.16 Generating Station Electrical Configurations

Figures 1.38 and 1.39 present the more common one-line electrical configurations for a large fossil steam turbine (ST) generating station connected to the bulk power electrical system (BES) at 230 kV and a smaller combustion turbine (CT) plant connected to subtransmission at 69 kV, respectively. In both cases, a generator step-up transformer (GSUT) increases the output voltage for connection to transmission or subtransmission systems, plant medium-voltage 4-kV auxiliary power buses are being fed by unit auxiliary transformers (UATs), and plant low-voltage 480-V buses are fed from the 4 kV buses through station service transformers (SSTs).

The Basics

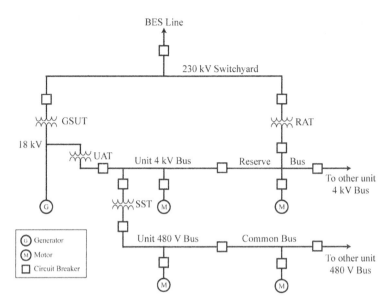

FIGURE 1.38
Steam turbine (ST) generating station simplified one-line.

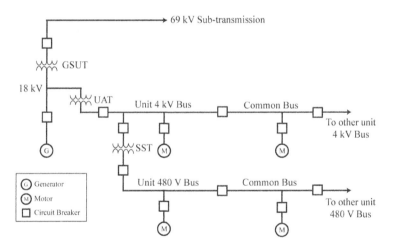

FIGURE 1.39
Combustion turbine (CT) plant simplified one-line.

In the past, providing a generator bus breaker for a large generating unit was cost-prohibitive because the short circuit current is very high at that location; today, with advances in circuit breaker–interrupting technology, it is much more affordable. The application of generator bus circuit breakers for fossil plants started to be more prominent in the 1970s with the advent of combustion turbine units that were smaller in size.

The large fossil unit is equipped with a reserve auxiliary transformer (RAT). This particular transformer is directly connected to the bulk power 230-kV local switchyard (although it could be connected to a 69-kV subtransmission system instead) and is provided for off-line operation and startup power purposes; once a parallel or synchronization operation connects the unit with the high-voltage switchyard, the load is transferred to the UAT. UAT loads are more critical to the operation of the unit and RAT, and common buses would normally be used to feed less critical balance of plant loads.

Depending on the size of the unit and design details involving costs, short circuit currents, voltage drops, and reliability, there may be multiple UAT, RAT, and SST transformers or windings that feed multiple isolated buses. Because the combustion turbine unit is equipped with a generator bus breaker that isolates the machine when the unit is off-line, startup power can be provided by simply backfeeding through the GSUT.

1.17 Three-Phase Basics

Figure 1.40 illustrates a wye-winding connection, a balanced wye-connected resistive load, and the related mathematical expressions and phasor relationships. The wye-winding connection could be associated with either a generator or transformer. The mathematical calculations assume a balanced load and, in that case, neither the dashed earth ground connection nor the dashed connection to the common point of the load can affect the results.

The voltage from each winding or phase (277 V) is displaced by 120°. The phasor wheel shows A0 hitting a peak at 90°, and B0 at 330°, followed by C0 at 210°. The outside or arrowhead of each phasor is the phase letter, and 0 represents the inside or common or neutral point. The phase sequence is ABC, with a counterclockwise rotation. Three phase-phase voltage combinations are presented: AC, BA, and CB, and each can be determined by following the circular arrow on the inside of the delta or equilateral triangle. The direction of the circular arrow can also be reversed to determine the three phase-phase combinations (AB, BC, and CA) in the opposite direction. Each phase-phase voltage is the vector sum of two winding voltages; for example, AC equals the vector sum of A0 + 0C. Because A0 and 0C are equal in magnitude and 60° apart, the vector sum equals the square root of 3 (1.732) multiplied by A0, or 480 V. Calculations for determining phase-phase and phase-neutral voltages are shown in Figures 1.41 and 1.42, respectively.

Balanced currents of 10 amps resistive flows in each line or phase that are in phase with their associated phase-neutral voltages. Each line current is derived from two phase-to-phase combinations, as shown by the dashed lines on the phasor wheel in Figure 1.40. In the case of the A-phase, an AC current is added to an AB current. Because the two currents are 60° apart,

The Basics

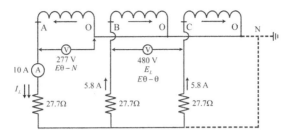

- $E_L = E\theta\text{-}\theta = E\theta\text{-}N \times \sqrt{3} = 277 \times 1.732 = 480\,V$
- $3\theta\ VA = 277 \times 10 \times 3 = 8310\,VA$ or $8.31\,kVA$
- Voltage drop $E_R = 10 \times 27.7 = 277\,V$
- $3\theta\ VA = \dfrac{E^2}{Z} \times 3 = \dfrac{277^2/27.7}{1000} \times 3 = 8.3\,kVA$
- $3\theta\ VA = I^2 Z \times 3 = \dfrac{100 \times 27.7}{1000} \times 3 = 8.3\,kVA$
- $3\theta\ W = \dfrac{E^2}{R} \times 3 = \dfrac{277^2/27.7}{1000} \times 3 = 8.3\,kW$
- $3\theta\ W = I^2 R \times 3 = \dfrac{100 \times 27.7}{1000} \times 3 = 8.3\,kW$

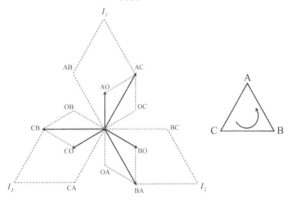

FIGURE 1.40
Wye configurations and relationships.

the total is one of the currents multiplied by 1.732 with balanced load. The arrows show how the A-phase winding current exits polarity at 10 amps and returns on polarity at the other two windings, with a magnitude of 5.8 amps each. This, of course, only shows the current for the A-phase, and a similar process can be applied to determine the magnitudes and phasor positions for the other two windings.

The total voltage drop per phase can be calculated by multiplying the per-phase impedance by the line current. The VA for the A-phase can be determined by simply multiplying the line current (which is the same as the winding current with a wye connection) by the phase-neutral voltage.

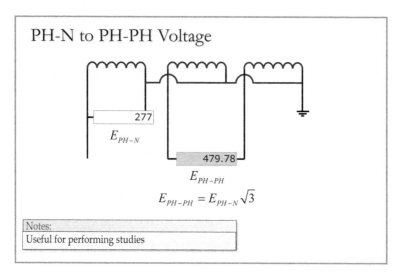

FIGURE 1.41
Phase-neutral to phase-phase.

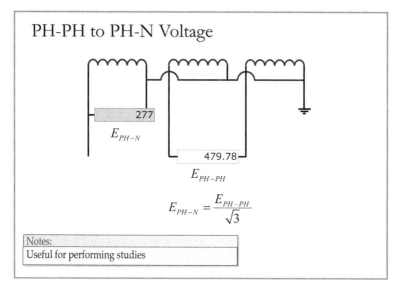

FIGURE 1.42
Phase-phase to phase-neutral.

Because the load is balanced, the VA for A-phase can be multiplied by 3 and then divided by 1000 for the total three-phase kilovolt-amps (kVA). The square of the phase to neutral voltage divided by the phase impedance or the square of the current times the line impedance can also be used to determine the per-phase VA and, with balanced loads, multiplied by 3 to determine the total VA, as presented in Figures 1.43 and 1.44. In balanced

The Basics

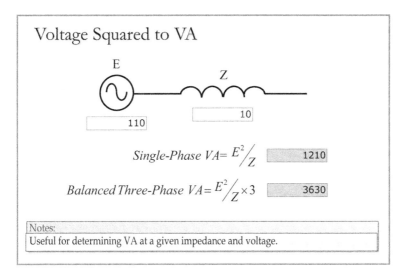

FIGURE 1.43
Voltage squared to VA.

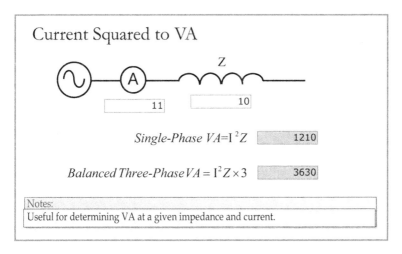

FIGURE 1.44
Current squared to VA.

three-phase power, only the line phase impedance for each conductor and associated load needs to be considered. The return conductor impedance (which involves the other phases) can be ignored.

Similar formulas are shown in Figures 1.45 and 1.46 for determining real power or watts (symbolized by W) in honor of James Watt (Scottish), who improved steam engine efficiencies in the late 1700s. A watt is 1 joule per second. A joule (symbolized by J) is 1 watt per second; it is named after James Joule (English) for his work in thermodynamics in the 1800s. In this example,

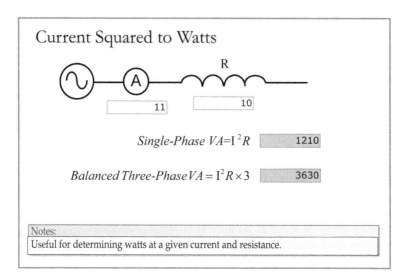

FIGURE 1.45
Current squared to watts.

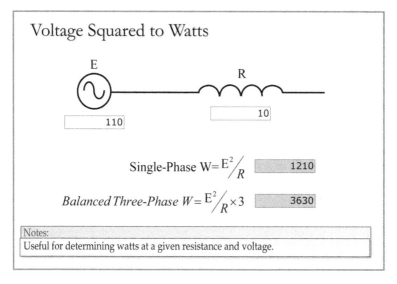

FIGURE 1.46
Voltage squared to watts.

because the load is resistive and there are no angular differences between the voltage and current, the real power equals the apparent power, and watts and volt-amps have the same magnitude.

Figure 1.47 presents a delta-winding connection, a balanced delta-connected resistive load, and the related mathematical expressions and phasor wheel

The Basics

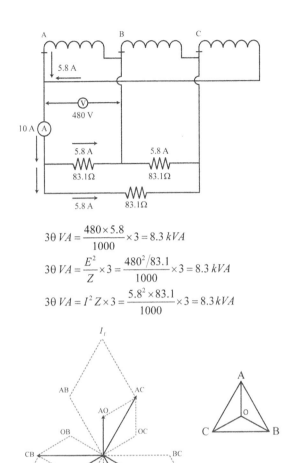

FIGURE 1.47
Delta configurations and relationships.

relationships. The delta-winding connection could be associated with either a transformer or generator, although almost all generators 10 megavolt-amps (MVA) and larger are normally wye-connected. As you can see by comparing Figures 1.40 and 1.47, the line currents, the line or phase-phase voltages, the total kilovolt-amp values, and the phasor wheels are all identical. In addition to the delta-winding and load configurations, there are differences in current flows that are internal to the windings and load, the load ohmic values, and how the kilovolt-amp magnitudes are developed in the figures.

For both the delta-connected transformer and delta-connected load, each phase line current has two possible paths. For balanced conditions, the

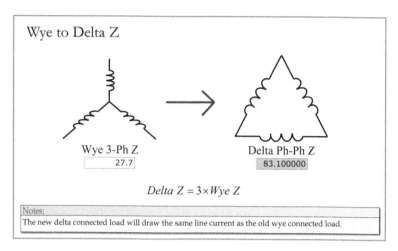

FIGURE 1.48
Wye–delta conversion.

current in each path, transformer, or load can be determined by dividing the total line current by the square root of 3. The actual load current for each leg of the delta can also be determined by dividing the line voltage by the impedance associated with that particular leg.

As previously mentioned, the phasors for both figures have the same alignment. A phase-neutral voltage is still represented, even though the transformer delta-winding configuration does not have a neutral. Figure 1.48 shows how to convert the balanced wye load impedance of Figure 1.40 to an equivalent balanced delta impedance. Conversely, the delta load impedance in Figure 1.47 can be divided by 3 to reestablish the original wye load impedance to help visualize the phase-neutral load voltage drop relationships. The line voltage drops from the delta load currents will also be in phase with the phase-neutral voltages, as shown on the phasor wheel.

Even without a load on delta transformer configurations, the stray insulation capacitance to ground for the three-phase conductors, which is normally distributed equally, will establish a phase to neutral voltage. Although with sufficiently long and horizontal 480-V busways, the capacitance for each phase will not be equal because the phase-phase capacitance of the outside bus bars will be different, which can affect the resulting phase-neutral voltage magnitudes and angles.

For high-impedance, alarm-only ground detector schemes, discussed in Chapter 2, it may be necessary to connect microfarad capacitors to the outside phases to balance out the capacitive reactance to ground to avoid nuisance alarms. The delta or equilateral triangle associated with Figure 1.47 represents a transposition that extrapolates the angular relationships of the phase-neutral voltages for the 480-V delta windings. With phasors, it is customary to use voltage as the reference and shift the currents to show angular

The Basics

differences between the two quantities from inductive or capacitive loading. One rationale for this is that the source normally feeds a number of different legs and, consequently, represents a parallel circuit.

The kilovolt-amp expressions for the delta and wye configurations are similar; both have the same calculated magnitudes, but are a little different because they are using current magnitudes inside the delta load instead of line or per-phase quantities. Both the E^2/Z and the I^2Z expressions can be used to determine the volt-amp for one leg of the delta. The results are divided by 1000 to convert volt-amps to kilovolt-amps and then multiplied by 3 for a total three-phase kilovolt-amp. Because the load is purely resistive, the phase-phase current will be in phase with the phase-phase voltage.

Figure 1.49 displays a more common approach for determining three-phase volt-amps, watts, and vars that is transparent to actual transformer winding and load configurations, so long as the loads are balanced and the measurements are confined to line quantities (i.e., phase-phase or line voltage and line phase current). Volt-amps-reactive (VARs) are shown at the opposite side of the right triangle relationship in the figure. Watts require output energy from generator prime movers and can be calculated using the cosine of the angle between the phase-neutral voltage and the line current; but vars do not supply real energy and only feed the reactive requirement of the load. They can be calculated using the sine of the angle as presented in the figure, or using an I^2X or (E phase-n)$^2/X$ expression multiplied by 3, as shown in Figures 1.50 and 1.51. Normally, the output power and load flows for generating station transformers (i.e., GSUT, UAT, RAT, and SST transformers) are reasonably balanced, and the calculations presented assume balanced conditions.

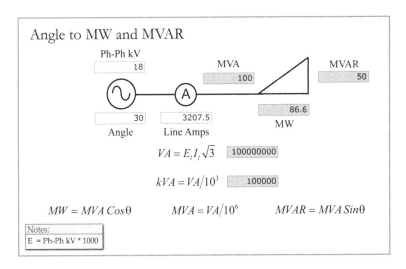

FIGURE 1.49
Angle to MVA and MW.

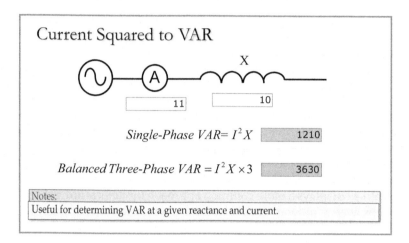

FIGURE 1.50
Amps to VAR.

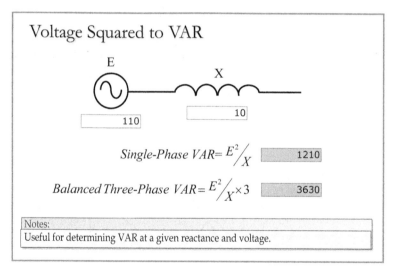

FIGURE 1.51
Volts to VAR.

The power factor (PF) is the cosine of the angle between the phase-neutral voltage and the line current; it can be used to determine watts. The percent PF is calculated by multiplying PF by 100. The loads shown in Figures 1.40 and 1.47 are purely resistive, and the currents and voltages are in phase; this is referred to as the *unity power factor* because the cosine of 0 is 1, or 100%. Usually, the angle is somewhere around 30°, or cosine 0.866, or 86.6% PF lagging due to the inductive nature of transformers and motor loads. Figure 1.52

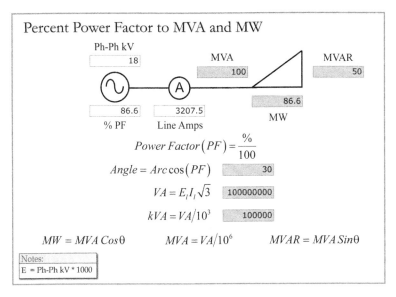

FIGURE 1.52
Percent power factor to MVA and MW.

is almost identical to Figure 1.49 except that the percent PF is used in the first part of the calculation sequence instead of the angle. The arc-cosine of (percent PF/100) will yield the angle, and the remaining calculations are the same.

In the United States today, there are two phase rotation sequences used by large utilities. Figure 1.53 shows both, as follows: (a) represents CBA or 321 rotation and (b) represents ABC or 123 rotation. In the case of (a), A and C phases displayed in (b) are swapped. With reference to (b), because vector rotation is counterclockwise, A-phase voltage peaks occur first, followed by B 120° later, and then by C 240° after A.

Although ABC is more prevalent in the industry, several large utilities do use the phase sequence rotation presented in (a). In the case of (a), motors

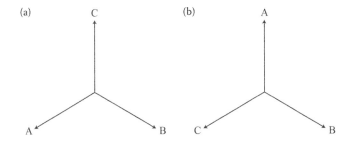

FIGURE 1.53
Phase rotation sequence.

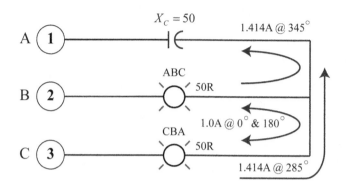

FIGURE 1.54
Two lamp phase rotation sequence indicator.

that are connected ABC to motor terminals 123, respectively, will operate in a reverse rotation mode, and care needs to be taken to ensure that protective relay functions are not affected by the sequence change. Since (b) is more prevalent, unless stated otherwise, ABC rotation will be assumed in the remainder of this book. The phase rotation sequence can be verified in the field with phase angle meters, electromechanical phase rotation indicators (basically a small, three-phase motors that will show the direction of rotation), and lamps or electronic instrumentation.

As a matter of technical interest, Figure 1.54 illustrates a two-lamp method for determining a phase rotation sequence with terminals 123 from top to bottom. Terminal 1 has a capacitive reactance of 50 ohms, and terminals 2 and 3 are each equipped with a 50-ohm resistive lamp. Accordingly, the current that flows between terminals 2 and 3 would be 100 V divided by 100 ohms or 1 amp at 0° and 180° respectively. The current flow for the other two combinations would be 1.414 amps apiece, at a 45° current leading angle. For an ABC, 123 rotation, the BA voltage is initially at 300° and the capacitive current advances the current phasor to 345°. As shown in the vector results in Figure 1.55, the sums of the phase-phase current combinations indicate that the terminal 2 lamp has a greater current of 2.4 amps for an ABC, 123 rotation and will glow brighter. The CBA lamp has a current of only 1.5 amps. If A and C are swapped for a CBA, 321 rotation, then the higher current would be in the terminal 3 lamp.

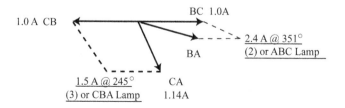

FIGURE 1.55
Vector results for ABC rotation.

The Basics

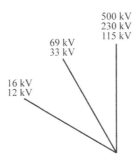

FIGURE 1.56
Phase to neutral voltage relationships.

1.18 Power Transformer Connections

Figure 1.56 displays the phase-to-neutral voltage relationships for a large utility. As you can see, their bulk power electrical system voltages (115, 230, and 500 kV) are in phase, the subtransmission voltages (33 and 69 kV) lead by 30°, and the distribution voltages (12 and 16 kV) lead by another 30°.

Figure 1.57 illustrates a delta-wye transformer leading configuration. A delta-wye connection is almost always used for the GSUT and UAT transformer applications presented in Figures 1.38 and 1.39. In both cases, the delta primary windings would be on the 18-kV or generator side to avoid ground fault currents that could damage the generator stator iron. The transformer windings do not cause a phase shift; the primary winding voltage is in phase with the secondary winding voltage.

In the case of the A-phase, the AB 18-kV voltage for the primary is across from and in phase with an A0 voltage on the secondary or 230-kV side. With regard to Figure 1.38, where generator parallels with the electrical system are made by closing a 230-kV breaker, those voltages could be used for synchronizing purposes. However, there is a phase shift between the phase-to-neutral voltages on each side of the transformer. A0 230 kV leads A0 18 kV by 30° and consequently represents a leading connection. The phase-phase 230-kV voltages are determined by taking the vector sum of a phase-neutral voltage from one phase and a reverse voltage from another phase. For example, AC current flow exits on polarity for the A-phase winding and enters on nonpolarity for the C-phase winding, which is the reason for reversing the C-phase vector. The phase-to-neutral 18-kV voltages are determined by drawing or transposing the phase-phase phasors into a delta form and then extrapolating the corresponding phase-neutral voltages, as shown in the Figure 1.57. For a wye–delta configuration, with 230 kV on top and 18 kV on the bottom, the process would be the same, as would be the outcome or phasor relationships, so long as the connections are not changed. The leading connection could be used to develop the 30° system voltage relationships between bulk

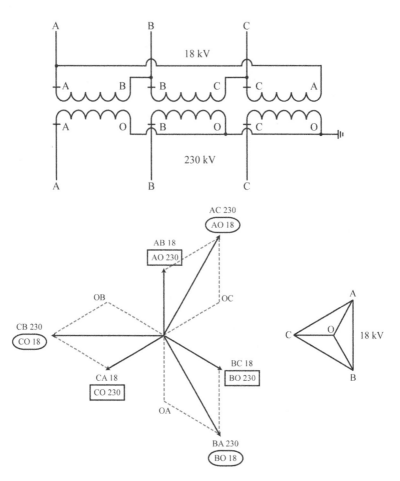

FIGURE 1.57
Delta–wye leading connection.

power transmission and subtransmission and between subtransmission and distribution, as presented in Figure 1.56.

A lagging connection is shown in Figure 1.58. The phasor or vector procedure is the same as just described, but A0 230 kV is now in phase with AC instead of AB, which causes A0 230 to lag A0 18 instead of lead. If a leading delta–wye connection is used for both the GSUT and the UAT, which is a common practice, the 4 kV would be in phase with the 230 kV, and a connection that does not provide a phase-neutral shift between the primary and secondary would be required for the RAT of Figure 1.38.

An in-phase connection could be provided by a delta–delta transformer configuration (Figure 1.59), a wye–wye transformer bank (Figure 1.60), or a delta–wye–zigzag transformer configuration (Figure 1.61). The delta–delta could be used if an independent ground bank is applied for ground tripping

The Basics   47

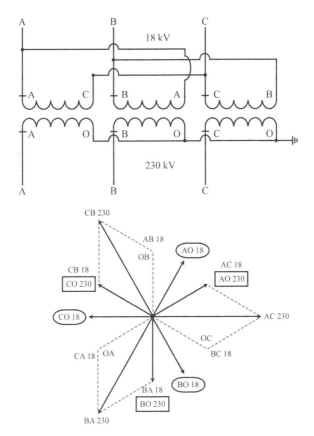

FIGURE 1.58
Delta–wye lagging connection.

or alarm purposes on the 4-kV side; the wye–wye for ground schemes (if both neutrals are grounded), and the delta–wye zigzag where there is concern over third harmonic levels with wye–wye connections. Because the neutral does not normally carry load in a generating station and ground fault currents are usually limited, concern about third-harmonic levels are reduced. Third-harmonic levels in wye–wye configurations can also be mitigated with delta-connected tertiary windings that circulate harmonics or through improved core iron designs that reduce the third harmonic levels.

The easiest way to vector the delta–wye–zigzag connection illustrated in Figure 1.61 is to start with the secondary phase connections. Using the A-phase as an example, the 1 end of the 1–2 winding is connected to the A0 phasor and provides a 30° leading voltage. By analyzing the phasors, it can be determined that the 12 end of the 11–12 winding provides a lagging voltage, which results in an overall secondary phase-neutral voltage that is in phase with the primary phase-to-neutral voltage. A similar process can be

48 Electrical Calculations and Guidelines

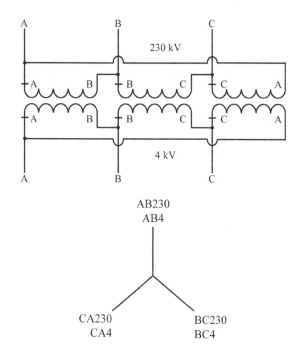

FIGURE 1.59
Delta–delta connection.

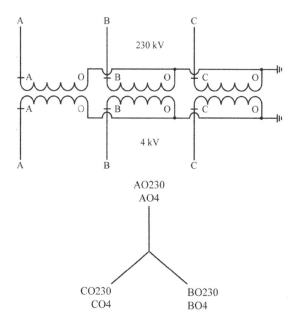

FIGURE 1.60
Wye–wye connection.

The Basics

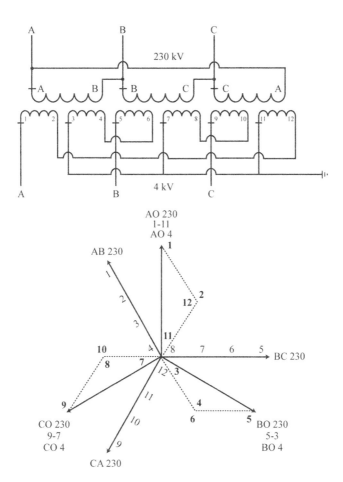

FIGURE 1.61
Delta–wye–zigzag connection.

applied to the other secondary phases. If 1, 5, and 9 represent the secondary phase connections, 12 and 2, 6 and 4, and 10 and 8 can be tied together, respectively. The neutral connection can then be made by connecting 3, 7, and 11.

1.19 Instrument Transformer Connections

Instrument transformers are required to reduce voltage and current levels to practical magnitudes for revenue metering, monitoring, and protective relay applications. The most common connections for potential transformers

(PTs) are open delta and wye–wye. Some in the industry prefer to use the more recent term, *voltage transformer (VT)*, but the term *PT* is also commonly used. As mentioned previously, there is no phase shift with the wye–wye connection.

Figure 1.62 shows a typical open delta connection. As the phasor wheel indicates, AB 18 kV is in phase with B1-2. Some utilities use B1 to indicate an A-phase secondary voltage or potential. For most applications, the primary phase-phase voltage is reduced to 115 or 120 V on the secondary side. The secondary arrows show a reverse current flow for the B3-1 potential. Accordingly, the phasors for both B2-3 and B1-2 are reversed to B3-2 and B2-1, respectively, and the vector sum results in the B3-1 voltage. Because they are 120° apart, the secondary voltage magnitude is not affected, but there is an angular change, as shown on the phasor wheel. Depending on their loading and associated PF, PT or VT instrument transformers will normally be specified with an accuracy class of 0.3, 0.6, or 1.2%, with 0.3% preferred for revenue metering applications.

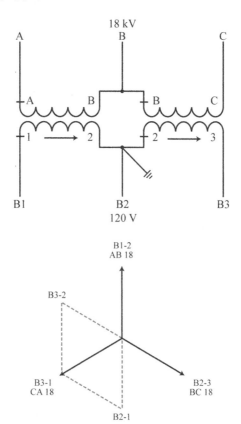

FIGURE 1.62
Open delta connection.

The Basics

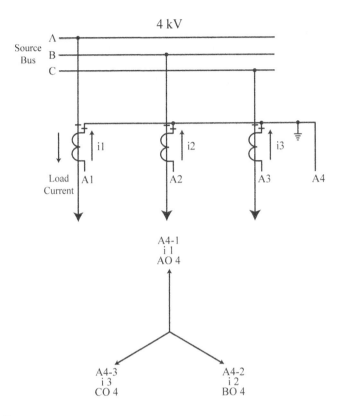

FIGURE 1.63
Current transformer wye connection.

The most common current transformer configuration is a wye connection, as illustrated in Figure 1.63. Some utilities use A1 to indicate an A-phase secondary current with A4, denoting the neutral or star point conductor. Current transformer ratios are usually shown with 5 amps on the secondary side; that is, 3000/5 for a turns ratio of 600/1, as indicated in Figure 1.64. The continuous current thermal rating of current transformers with rating factors of 1 is 5 amps. A current transformer with a rating factor of 3 would have a continuous secondary current rating of 15 amps.

Per ANSI/IEEE standards, current transformers used for protective relay applications are normally rated not to exceed a 10% error at 100 amps secondary at or below a specified external impedance burden, depending on the voltage class (amount of laminated iron) at 50% PF; and revenue metering current transformers are normally designed to be within 0.3% accuracy, with either a 0.1- or 0.2-ohm impedance secondary burden at 90% PF at 5 amps secondary.

Most current transformers resemble a doughnut, with the primary conductor going through the hole. It is basically a transformer with one turn

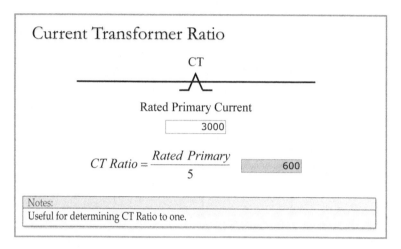

FIGURE 1.64
Current transformer ratio.

on the primary side and multiple turns on the secondary side. It will work in reverse if the primary is shorted and current flows in the secondary, but the driving voltage is low, and any impedance introduced in the primary would also need to be low. For phasor purposes, the current is referenced to the direction of load flow, and the load is assumed to be at unity PF where the current is in phase with the phase-neutral voltage. The polarity marks are shown, and primary currents into polarity develop secondary currents i1, i2, and i3 that exit on polarity, as shown on the phasor wheel. Accordingly, an A4-1 current would be in phase with an A0 primary voltage.

For delta–wye electromechanical transformer differential schemes, delta-connected current transformers are needed on the wye side of the transformer to bring the currents back into phase because of the 30° phase shift in the power transformer phase-neutral voltages. With newer digital relays, the phase angle differences can be accounted for in the software, and wye-connected current transformers can be used on both sides of the transformer.

Figure 1.65 shows the connections and phasors for both 30° lead and lag current transformer connections. In the case of the A-phase, the vector sum of i1 − i2 produces a current that leads the phase-to-neutral voltage by 30°, as illustrated under (a). Under (b), the vector sum of i1 − i3 develops a secondary current that lags the phase-to-neutral voltage by 30°. In both cases, the delta CT connection increases the magnitude of balanced secondary current flows in A1, A2, or A3 by the square root of 3. The secondary currents for each phase enter their current transformers through A1, A2, and A3, respectively.

The Basics

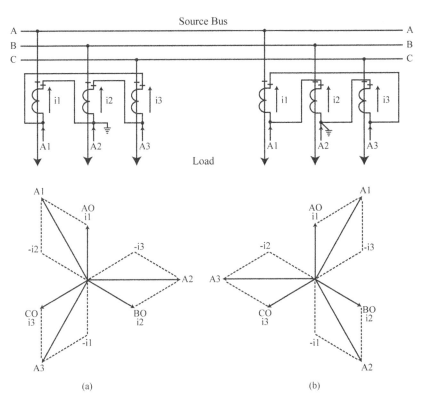

FIGURE 1.65
Current transformer delta lead and lag connections.

Bibliography

Baker, T., *EE Helper Power Engineering Software Program* (Laguna Niguel, CA: Sumatron, Inc.), 2002.

Calhoun, W.R., *Vector Analysis and AC Circuit Theory* (Alhambra, CA: Southern California Edison Company, Substation Training School), 1964.

Klempner, G. and Kerszenbaum, I., *Handbook of Large Turbo-Generator Operation and Maintenance*, 2nd ed. (Piscataway, NJ: IEEE Press), 2008.

2
Electrical Studies

Electrical engineers who support generating stations and large industrial facilities will periodically need to complete studies that address operational considerations and design issues for betterment projects. This chapter covers the theoretical basics of power system analysis and provides examples of voltage, power transfer, and short circuit studies.

2.1 Conversions

When performing electrical studies, a good starting point is to find out the short circuit duty at the source. In the case of a generating station, the source would be the high-voltage switchyard associated with the particular plant, and subsequent calculations would start from that location. Short circuit duty can be expressed as current (amps), impedance (ohms), or apparent power (megavolt-amps, or MVA). Accordingly, electrical power engineers need to be able to convert from one form to another as required for the specific calculation.

2.1.1 Ohmic

In Figure 2.1, balanced three-phase impedance is converted to amps, and the inverse calculation is shown in Figure 2.2. Both formulas are basically three-phase versions of Ohm's Law and are commonly used when performing studies. A typical three-phase short circuit current level for large generating station (auxiliary power system) 4 kV and for 480-V main unit buses is 30,000 amps. For the 4-kV bus, the total short circuit impedance, including the 230-kV system, the unit auxiliary transformer (UAT) or reserve auxiliary transformer (RAT) impedance, and the busway or cable impedance from the UAT or RAT to the bus would have a value of less than 0.1 ohms.

Balanced conditions are assumed for most studies, and in that case, three-phase Z is equal to Z phase-neutral (E-N/I), and both quantities are interchangeable. Three-phase Z is assumed to be in a wye configuration. Ohmic values for cables and busways are typically given for one line or phase and correspondingly are equivalent to phase-neutral values or wye connections for three-phase short circuit analyses. In Figure 2.3, a delta configuration is

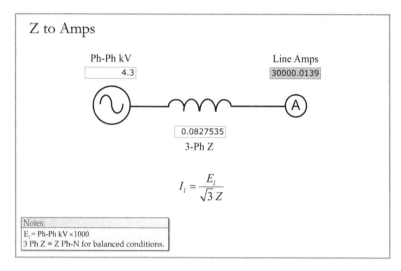

FIGURE 2.1
Z to amps conversion.

converted to a wye connection. Load and line currents for before and after the delta/wye conversion will have the same magnitude.

2.1.2 Megavolt-Amps (MVA)

By convention, when MVA is used to express short circuit duty, the three-phase balanced apparent power formula is used to determine current

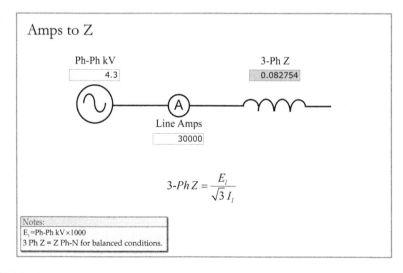

FIGURE 2.2
Amps to Z conversion.

Electrical Studies

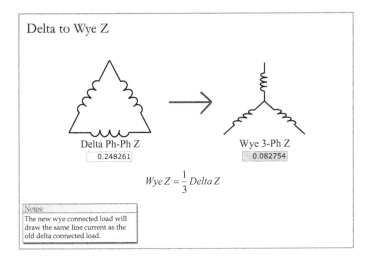

FIGURE 2.3
Delta to wye Z.

magnitudes for both balanced and unbalanced faults. Figures 2.4 through 2.6 cover balanced three-phase conversion calculations for MVA, amps, and ohms, respectively.

By analyzing these figures, you will find that 30,000 amps, 0.83 ohms, or 223 MVA can all be used independently to represent the short circuit duty and to calculate the missing parameters for 4.3-kV, three-phase, balanced short circuit conditions.

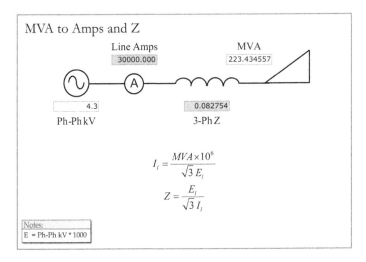

FIGURE 2.4
MVA to amps and Z.

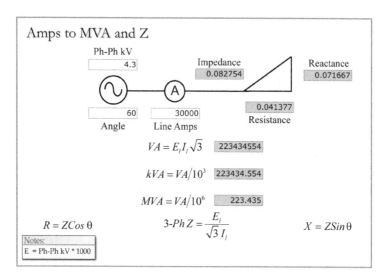

FIGURE 2.5
Amps to MVA and Z.

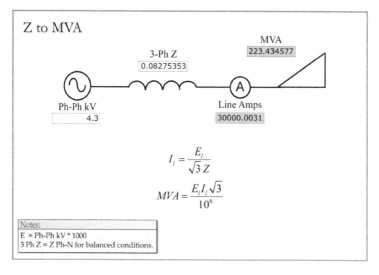

FIGURE 2.6
Z to MVA.

2.2 Transformer Tap Optimization

The second phase of electrical studies usually involves optimizing transformer winding tap positions, as any changes in ratio will affect the studies. If the annual peak primary voltage is known, an optimum tap can be selected for the transformer under study. If measured data is not available, a peak voltage of 105% of nominal (midpoint of the operating range) can be assumed; that is, 230 × 1.05 = 241.5 kV. As presented in Figure 2.7, transformers normally have five no-load tap positions: two above mid-tap and two below. Each step or tap change increases or reduces the mid-tap voltage ratio by 2.5%.

Although it is hard to say anything absolute about motors because there are numerous design variations, most motors will run more efficiently at higher voltages. In many cases, the motors will require fewer stator amps to produce the same horsepower (HP), which reduces the corresponding watt and var losses (amps squared × R and × X, respectively). The output of a generating station unit could be fan or pump limited; operation at higher voltages may increase the available HP as the motor slip diminishes and correspondingly can increase the output of a limited fan or pump, so long as there are no thermal limitations. Also, operating at higher voltages provides additional margins for system low-voltage conditions. The limiting factor is normally the continuous voltage rating of motor loads on the secondary side of transformers. The optimum tap setting provides the highest voltage that will not reduce the life of motors during worst-case operating conditions, when the primary voltage is at its highest and the transformer load is low (minimizing the voltage drops in the source transformer and conductors).

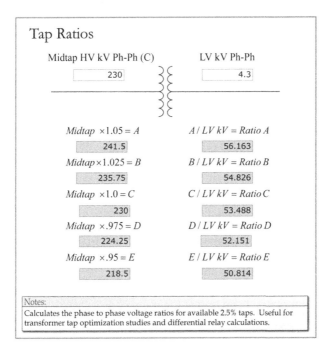

FIGURE 2.7
Transformer winding tap ratios.

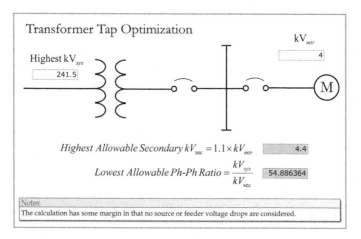

FIGURE 2.8
Transformer winding tap optimization.

The procedure illustrated in Figure 2.8 provides the lowest allowable phase–phase voltage ratio that can be applied without exceeding the continuous voltage rating of the motors. The lower the winding ratio, the higher the voltage is on the secondary side. Although motors have a continuous voltage rating of 110% of the nameplate, the nameplate value is normally below the nominal bus voltage; that is, 480-V motors normally have nameplate ratings of 460 volts and 4-kV motors either 4 kV or 4.16 kV, with a nominal bus voltage of 4.3 kV. The author prefers 4.16-kV motor nameplate values for improved operating margins and less likelihood that the 480-volt levels will be limited by the 4-kV system operating voltage levels. In the case of 4-kV nameplate motors, the limitation would be 110% of 4 kV, or 4.4 kV, which would be the maximum continuous voltage rating of the motors. Figure 2.8 shows a RAT-fed 4-kV bus from Figure 1.38 (in Chapter 1) with the 230-kV system at its highest voltage of 241.5 kV and a lowest allowable winding ratio of 54.866. Tap B would provide a ratio of 54.826, as indicated in Figure 2.7, and would be the optimum choice. With regard to 460-V nameplate motors, the limitation is 110% of 460, or 506 volts.

Volts/Hz withstand curves for generators, and transformers are readily available from a number of manufacturers. The curves for transformers and generators will generally have the same slope, but generators start with a lower continuous rating of 105% and transformers at 110% (which is the same as motors). Because motor curves are not readily available, the bullets presented here were extrapolated from one of the more conservative volts/Hz curves for transformers and should reasonably represent motor withstand times as well:

- 134% = 0.1 minutes
- 131% = 0.2 minutes

Electrical Studies   61

- 126% = 2.0 minutes
- 120% = 5.0 minutes
- 116% = 20 minutes
- 114% = 50 minutes
- 111% = 100 minutes

Motor, generator, and transformer voltage levels are limited by the level of magnetic flux that their core iron laminations can carry without overheating, primarily from the increase in eddy current losses. The term *volts/Hz* is used to account for operations at reduced frequencies that lower the impedance and increase the excitation of the core iron. At 60 Hz, the percent volts/Hz corresponds directly to the overvoltage magnitude; that is, 115% overvoltage = 115% volts/Hz.

Figure 2.9 illustrates the procedure for calculating the impact of operating a 4-kV nameplate motor at a reduced frequency of 58 Hz. For reference,

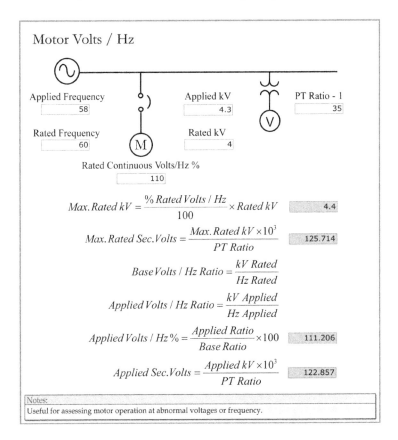

FIGURE 2.9
Motor volts/Hz withstands.

the maximum bus and associated potential transformer secondary voltages at 60 Hz are determined to be 4.4 kV and 125.7 volts, respectively. A base volts/Hz ratio is calculated by dividing the rated or nameplate kV by the rated frequency. An applied volts/Hz ratio is then determined by dividing the applied kV by the applied frequency. The applied volts/Hz percentage can be calculated by dividing the applied ratio by the base ratio and multiplying by 100. In this example, the motor's maximum continuous volts/Hz rating of 110% is exceeded with a calculated percentage of 111%, even though the 60-Hz maximum rated voltage of 4.4 kV is greater than the applied voltage of 4.3 kV. This is because the applied frequency of 58 Hz reduces the impedance of the stator windings, which increases the ampere turns and the resulting magnetic flux in the stator core iron laminations.

2.3 Conductor Parameters

The next step in performing electrical studies usually involves determining the ampere ratings and impedance data of the associated electrical conductors. In general, smaller conductors are mostly resistive in nature. As the size of the conductor [i.e., the geometrical mean radius (GMR)] increases, the inductive reactance reduces slightly and the conductor resistance reduces more dramatically. The inductive reactance is reduced because of the effect of annular current flow (i.e., eddy currents) in the extra conducting material near the skin effect region or boundary, which produces a counterflux, resulting in a lowered self-flux opposition voltage (as illustrated in Figure 2.10). Somewhere around medium size, the conductor resistance will equal the inductive reactance, and larger conductors will be mostly inductive in nature.

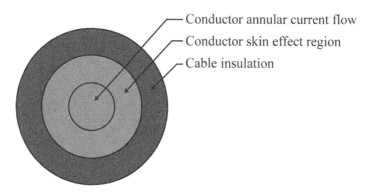

FIGURE 2.10
Conductor radius and inductance.

Electrical Studies 63

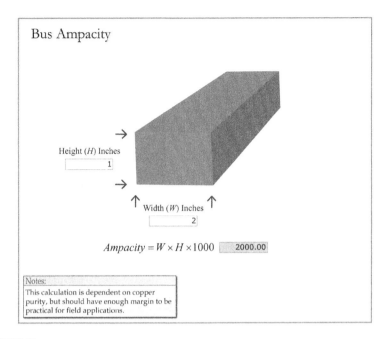

FIGURE 2.11
Copper bus ampacity.

2.3.1 Buses

Generally, copper and aluminum bus ampacities are a function of cooling, the cross-sectional areas, and the purity of the metals used to fabricate the bus bars. The calculation in Figure 2.11 is sometimes utilized in industrial plants for the construction of copper bus bar jumpers and allows a 30°C increase over ambient temperatures with the calculated current flowing. This calculation is conservative and has more than an ample margin because the standards for buses allow a much larger temperature rise.

The manufacturer of the busway normally provides the ampacity and impedance parameters for a specific design. Figure 2.12 shows the resistance and inductive reactance values for a 500-foot run of 2000-amp, low-voltage copper busway (480-V applications) manufactured by the Square D company. The resistance value assumes a conductor temperature of 80°C.

2.3.2 Insulated Cable

In general, cable ampere ratings are a function of the amount of copper, the temperature rating of the insulation, and the design of the raceway system. Although specific ampacity values can be obtained from manufacturers, tables in the National Electrical Code (NEC) are commonly used to determine the ampacity for a particular design. Cable impedance data can be

64 *Electrical Calculations and Guidelines*

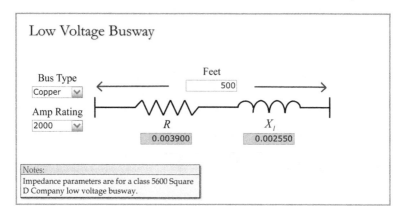

FIGURE 2.12
Low-voltage busway.

furnished by the manufacturer or taken directly from tables provided in IEEE Standard 141.

Figures 2.13 and 2.14 present the ampere rating and impedance parameters for a 500-foot run of 1000-kcm, copper-insulated, 90°C, low-voltage cable (480-V applications) installed in magnetic and nonmagnetic conduits, respectively. By comparing the resistance values, it can be seen that the resistance is higher with magnetic conduit due to eddy current and hysteresis watt losses in the rigid conduit. Because most of the heating takes place in the conduit, the cable ampere rating is not affected. Care must be taken to include all three phases in the same metallic conduit; otherwise, the conduit eddy current heating will be excessive without the subtractive flux from the other phases. The reactance values are also a little higher with magnetic conduit, mostly because the magnetic flux spreads out as it is attracted to the

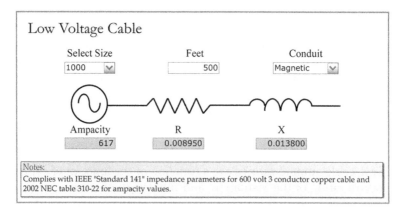

FIGURE 2.13
Low-voltage cable in a magnetic conduit.

Electrical Studies

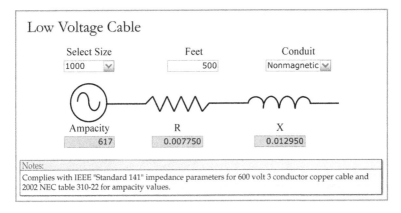

FIGURE 2.14
Low-voltage cable in a nonmagnetic conduit.

conduit, which results in a weaker subtractive effect from the other phases and increased self-flux.

Three conductors in one conduit are assumed for determining the ampacity, and a conductor temperature of 75°C is assumed for the conductor resistance values. More than three phase conductors in one conduit may require a reduced ampere rating because of the heat transfer from the other conductors; this is covered in the NEC. In somewhat balanced systems without significant harmonic content, the neutral conductor does not need to be considered in the derating. The ampacity values comply with 2002 NEC Table 310-22, and the impedance parameters are in agreement with IEEE Standard 141. The cable reactance values are higher than the busway inductive reactance shown in Figure 2.12, even though the busway spacing between phases is greater, because of the higher ampacities and geometries of the bus bars.

Figure 2.15 presents the ampacity and impedance parameters for a 500-foot run of 5-kV, 1000-kcm, copper-insulated, 90°C, medium-voltage cable (4-kV applications) installed in a magnetic conduit. Three conductors in one conduit are assumed for determining the ampacity, and a conductor temperature of 75°C is assumed for the conductor resistance value. The ampacity value complies with NEC 2002 Table 310-75, and the impedance parameters are in agreement with IEEE Standard 141. The resistance is higher due to the conduit eddy current and hysteresis watt losses.

Figure 2.16 provides the ampacity and impedance parameters for a 500-foot run of 15-kV, 1000-kcm, copper-insulated, 90°C, medium-voltage cable installed in an underground, nonmagnetic conduit. Three conductors in one conduit and a soil heat transfer capability or thermal resistivity (RHO) of 90 are assumed for determining the ampacity, and a conductor temperature of 75°C is assumed for the conductor resistance value. The ampacity value complies with NEC 2002 Table 310-79, and the impedance parameters are in agreement with IEEE Standard 141. The inductance is higher than the 5-kV cable as a result of the

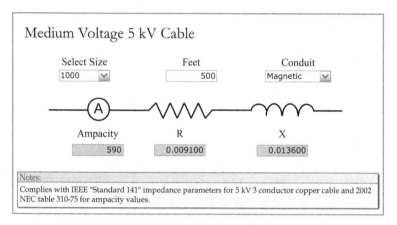

FIGURE 2.15
Medium-voltage 5-kV cable in a magnetic conduit.

increased spacing caused by the thicker 15-kV insulation, which overrides the magnetic conduit effect in the 4-kV cable. The ampacity is higher because the ground or earth acts as a heat sink, draining heat from the cable.

2.3.3 Overhead Aluminum Conductor Steel-Reinforced (ACSR) Cable

The ampacity and impedance values for overhead aluminum conductor steel-reinforced (ACSR) cables are provided in the *Westinghouse Electrical Transmission and Distribution Reference Book* and other sources. Phase conductor spacing affects the inductive reactance values. The magnetic self-flux that links with the conductor to cause the opposition voltage or inductive effect increases when the spacing is greater because the subtractive magnetic

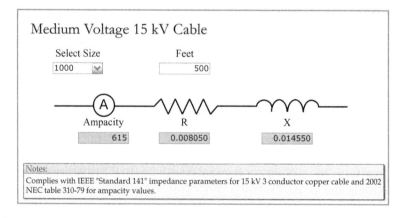

FIGURE 2.16
Medium-voltage 15-kV cable in a nonmagnetic conduit in earth.

fluxes from the other phases have a weaker effect. If all three conductors from the different phases could occupy exactly the same position or location, the inductive reactance would be roughly zero because the vector sum of all three magnetic fluxes would be approximately zero. This, of course, is not possible because the phases must be insulated from each other to prevent short circuit currents from flowing. The subtractive flux phenomenon, where the overlapping flux gradient from other phases subtracts from the self-flux originating from each conductor, is illustrated in Figure 2.17. Consequently, to determine the impedance values more accurately, the spacing between phases must be known. With overhead conductors, this spacing is usually not symmetrical, and the unbalanced spacing first must be converted to an equivalent uniform spacing or geometrical mean distance (GMD) before the symmetrical spacing parameters provided in reference tables can be applied. Figure 2.18 illustrates an equivalent uniform spacing of 3.1 feet for three-phase conductors that are separated by 2, 3, and 5 feet, respectively. This assumes that the utility periodically transposes or swaps the position of the phase conductors to balance out the three impedances; otherwise, the short circuit currents will not be equal and the three phase voltages supplied to customers will not be balanced due to nonuniform voltage drops in the line conductors. Three-phase motors are very sensitive to unbalance and need to be derated for voltage differences as small as 1%.

Figures 2.19 and 2.20 show the impedance parameters for 18-kV (3-foot) conductor spacing and 138-kV (12-foot) conductor spacing, respectively. The actual spacing may differ significantly based on local weather and other environmental conditions, distance between spans, and concerns over galloping conductors, pole or tower support details, and margins preferred by the local utility. Although shown in miles, the line distance is equivalent to 500 feet for comparison to the other conductors already discussed. Comparing Figures 2.19 and 2.20, you will find that the resistance and ampacity values are the same, but the inductive and capacitive reactance values are higher for

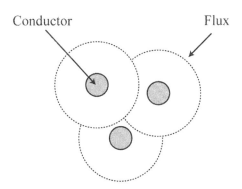

FIGURE 2.17
Spacing impact on inductance.

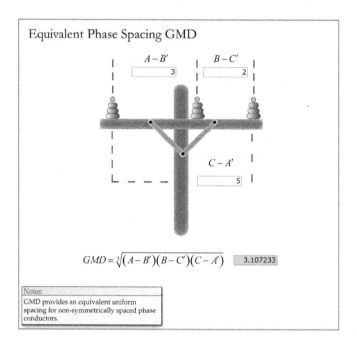

FIGURE 2.18
Equivalent symmetrical conductor spacing.

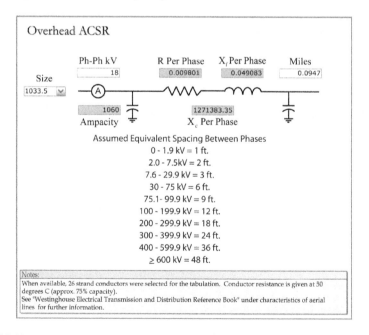

FIGURE 2.19
18-kV 1033.5 ACSR 3-foot spacing.

Electrical Studies

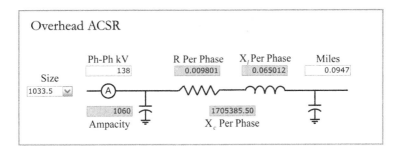

FIGURE 2.20
138-kV 1033.5 ACSR 12-foot spacing.

the 138-kV overhead cables due to the increased spacing between conductors. As mentioned before, increasing the spacing weakens the subtractive flux from the other phases, which increases the self-flux and inductance. The capacitive reactance increases because the distance between plates increases, which reduces the microfarad value.

Although logarithmic and not exactly linear, impedance parameters for a different spacing can be roughly extrapolated from two known distances, so long as the application does not require a higher level of precision (as shown in Figure 2.21). The midpoint distances should be kept as small as possible.

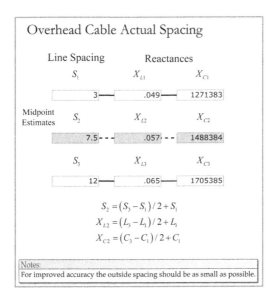

FIGURE 2.21
Roughly extrapolated conductor spacing impedance changes.

2.4 Study Accuracy

At this point in this chapter, the author has addressed the preliminary data gathering, conversions, transformer tap optimization, and conductor parameters required for studies. Now, the discussion can proceed to the actual procedures for voltage, power transfer, and short circuit calculations. Calculations require only a precision that is practical for the particular application. Most calculations do not consider the exciting current associated with transformers or stray capacitance. Some calculations consider only the reactive components, not the resistive ones. Many procedures assume nominal magnitudes for voltages and other parameters, not actual values that may be in play prior to or during electrical system faults and disturbances. However, experience has shown that the calculation methods presented in this chapter and throughout this book have enough precision to be practical for the intended applications.

2.5 Voltage Studies

2.5.1 Bus Voltage Drop

Figure 2.22 provides a procedure for calculating bus voltage drops for balanced loads. First, the circuit MVA is converted to amps using the standard three-phase power formula. Next, the impedance parameters (R and X) for the total circuit are calculated, and the portions of R and X that are associated only with the load are determined. Finally, the sum of the resistive and reactive voltage drops at the load is calculated and then multiplied by the square root of 3 (1.732) for conversion from a phase-neutral to a phase–phase voltage magnitude. In this example, because of the source impedance, the voltage at the bus is reduced from 4.3 kV to 3.8 kV. The power factor or angle will affect the results of the calculation.

2.5.2 Line Voltage Drop

Figure 2.23 provides a similar procedure for calculating line voltage drops for balanced loads. First, the total circuit R and X are determined. Then the load R and X are derived. Finally, the sum of the resistive and reactive voltage drops at the load is calculated and then multiplied by the square root of 3 for conversion from a phase-neutral to a phase–phase voltage magnitude. As you can see, the voltage at the load is reduced from 4.3 kV to 3.95 kV due to the source and cable impedance.

Electrical Studies

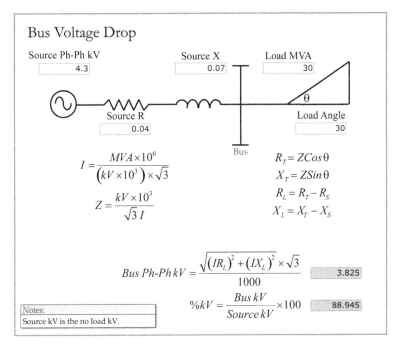

FIGURE 2.22
Bus voltage drop.

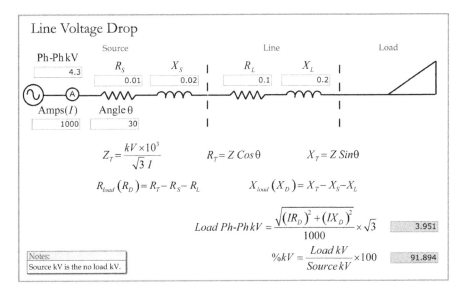

FIGURE 2.23
Line voltage drop.

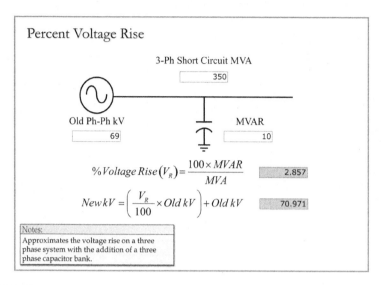

FIGURE 2.24
Capacitive voltage rise.

2.5.3 Capacitive Voltage Rise

Figure 2.24 covers how to approximate the effect that three-phase balanced capacitance has on voltage levels. If the total circuit capacitance in megavars (MVAR) multiplied by 100 is divided by the short circuit duty in MVA, the result will yield the percent voltage rise of the circuit. The voltage increase could be caused by installed capacitor banks or by the capacitance of long transmission lines that are lightly loaded. In this example, the addition of a 10-MVAR bank will increase the voltage from 69 to almost 71 kV.

2.5.4 Collapsing Delta

Figure 2.25 discloses how to calculate the unfaulted phase–phase voltages if the faulted phase–phase voltage is known. This is useful for unbalanced voltage studies and for testing some phase–phase impedance relay elements. In this case, if the C-B voltage is reduced to 50 kV, the unfaulted phase–phase voltages will be reduced to 201 kV.

2.6 Power Transfer Calculations

Figure 2.26 deals with the basic power transfer equations for three-phase watts and vars caused by voltage and angular differences (power angle) or

Electrical Studies

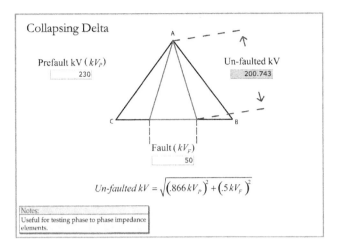

FIGURE 2.25
Collapsing delta.

the delta angle between two different sources in an electrical system, and the limiting effects of the total circuit inductive reactance ohms. The term *angle* in this case refers to angular differences between the voltages on bus 1 and the voltages on bus 2. If phase–phase voltages are used in the equations, the calculations yield three-phase watt and var magnitudes that will flow from

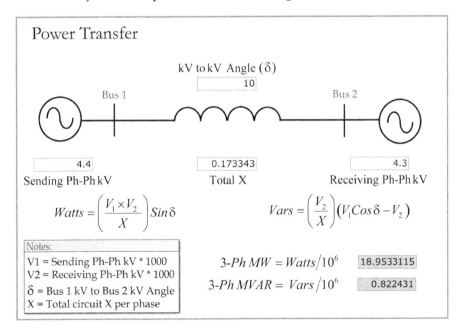

FIGURE 2.26
Power transfer equations.

one point to the other. In general, watts will flow from the leading to the lagging angle, and vars will flow from the higher to the lower voltage system. In this example, a 10-degree angular delta between bus 1 and bus 2 causes 18.9 megawatts (MW) and 0.8 MVAR to flow from bus 1 to bus 2 when the buses are operated in parallel.

2.7 Two-Generator System

Figure 2.27 illustrates a two-generator system with identical turbines and generators. The right triangle in the first quadrant represents an aggregate system load that is resistive and inductive and, with respect to the generators, can be modeled as a series circuit. The current is the reference, and the voltage phasor falls into alignment with the hypotenuse (VA) of the right triangle. If equal steam flows through each turbine and the field or excitation currents are the same, each generator would provide half the watt and half the var requirements of the system load. If the generator governors are on manual and steam flow is increased for machine A, it would feed a greater portion of the watt requirement, and the system frequency would increase if machine B did not back down.

The increased steam flow in machine A would cause its power angle to increase (in the leading direction), which would put it in advance of machine B, allowing it to feed a greater portion of the load. If the generator automatic voltage regulators (AVRs) are also on manual and excitation or field current is increased for machine A, it would feed a greater portion of the var requirement and the system voltage would increase if machine B did not back down on excitation. This is referred to as *boost vars*, and if the power factor was measured for generator A, it would show that the current was lagging the voltage by a greater amount because the load is inductive in nature.

If the excitation for A is increased enough, the total system var requirement would be fed by A and generator B would be at the unity power factor. If the excitation for A is increased beyond that point, B would become a var load for A and the current in B would lead the voltage, resulting in *buck vars*.

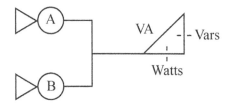

FIGURE 2.27
Two-generator system.

Electrical Studies

In this case, the right triangle for B would be inverted and shown in the fourth quadrant.

2.8 Ohmic Short Circuit Calculations

Short circuit calculations that do not involve transformation can be performed easily using ohmic values.

2.8.1 No Transformer

Figure 2.28 covers how three-phase short circuit current at the end of a line impedance can be determined easily if the short circuit duty at the source bus is known. The bus short circuit current is converted to R and X values that can be added to line R and X, and then the total current at the end of the line can be determined by using the three-phase Ohm's Law. The phase-to-phase short circuit current is normally equal to 86.6% of the three-phase value because the ratio of three-phase ohms to phase–phase ohms equals 1.732/2 (0.866), and the three-phase ohms do not need to consider the return

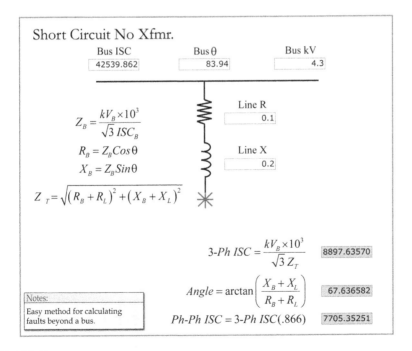

FIGURE 2.28
Short circuit: no transformer.

path impedance. The ohmic impedances for three-phase, phase–phase, and phase-ground can be calculated with the following expressions:

- Phase-neutral ohms = EN/IL
- Phase–phase ohms = EL/2IL
- Three-phase ohms = EL/$\sqrt{3}$IL
  - EN = Phase-neutral voltage
  - EL = Phase–phase voltage
  - IL = Line or phase current

2.8.2 Parallel Sources

When two sources are operated in parallel, the short circuit contribution from each source needs to be added together to determine the total short circuit current. If the angles are the same, the new total current will simply be the algebraic sum of the two currents. However, if there is an angular difference, a conductance, susceptance, and admittance method for solving parallel impedances can be applied to determine the total magnitude of the combined currents. Figure 2.29 presents a method of resolving the total short circuit current from two sources that have an angular difference. Figure 2.30 shows the reverse process of subtracting or removing a source with an angular difference from the total value. With regard to generating stations, these calculations are particularly useful for determining high-voltage switchyard short circuit currents when generating units are on-line and off-line. They are also useful for determining the combined short circuit currents of the generator and generator step-up transformer (GSUT) at the UAT primary terminals.

2.9 The Per-Unit System

Whenever transformers are involved in a study, by convention, a percentage or per-unit approach for resolving problems is used because of the different voltages involved. Because ohmic values depend on which side of the transformer (primary or secondary) you are looking at, transformer impedances are almost always expressed as a percentage. The per-unit value is simply the percentage numeral divided by 100; for example, a 7% impedance transformer would have a per-unit impedance of 0.07. Because a transformer's primary MVA is equal to its secondary MVA (neglecting excitation current), a percentage or per-unit system can be applied so long as the base values chosen for MVA, voltages, impedances, and currents are consistent. For convenience, the base values are normally selected to be in agreement with the apparatus nameplate values or with other data or studies that were

Electrical Studies

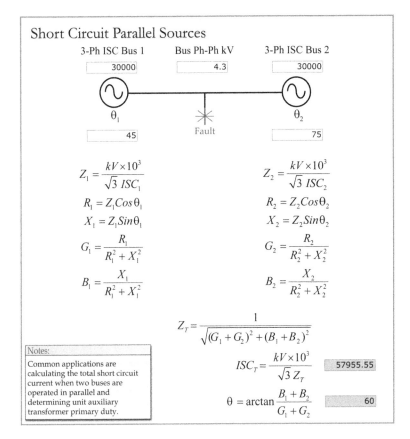

FIGURE 2.29
Short circuit: parallel sources.

previously performed. During the course of a study, it is often necessary to adjust values to a different base to be consistent with other data or to convert to other forms that are more convenient.

2.9.1 Basic Formulas

Although it is not always obvious, the basic formulas presented in Chapter 1 still apply when using the per-unit system. One complication is that a 1.0 per-unit value could represent full load amps or some other parameter, and because 1s can easily drop out of formulas, the expressions may no longer be presented in a manner that is familiar to the reader. For example:

- Per-unit (pu) amps = pu volts/pu Z
- Amps = pu amps × base amps; or shortcut method:
  - Amps = base amps/pu Z

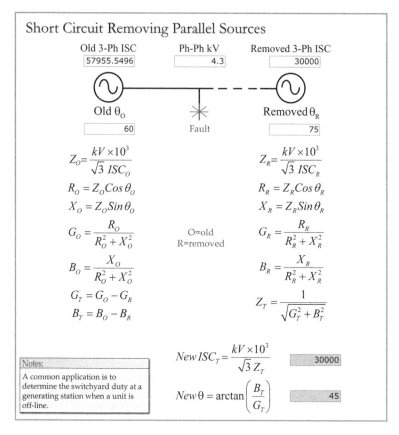

FIGURE 2.30
Short circuit: removing parallel sources.

2.9.2 Corrected Voltage Base

The percent or per-unit impedance of transformers often needs to be adjusted for applied winding taps and actual applied voltages that are different from the nameplate base values. The nameplate will show the %Z at a specified MVA and winding tap. Normally, it would be specified at the oil and air (OA) or lower MVA rating and for the mid-tap winding voltage. If the applied voltage is different or the selected tap is different, the %Z self-cooled rating is usually adjusted or corrected to reflect the difference. Figure 2.31 is used to adjust or correct a transformer's base per-unit impedance to a different winding tap. Figure 2.32 is the same calculation, used again to convert the transformer's tap-corrected base per-unit impedance to the actual applied or system voltage of 14 kV. Figure 2.33 combines both calculations into a single expression that corrects for both the winding tap and applied voltage differences.

Electrical Studies

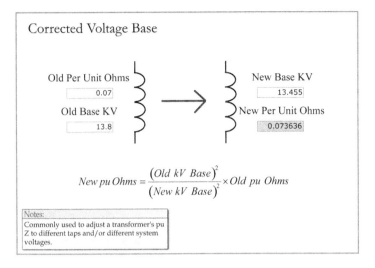

FIGURE 2.31
Transformer winding tap correction.

2.9.3 Converting Per-Unit Z to Amps

A transformer's infinite bus (unlimited supply), three-phase, short circuit current can be calculated by dividing its rated base three-phase current by its per-unit impedance, as shown in Figure 2.34. Using the primary kV base current provides the primary current for a secondary fault, and using the

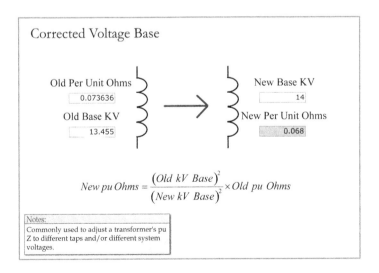

FIGURE 2.32
Transformer applied voltage correction.

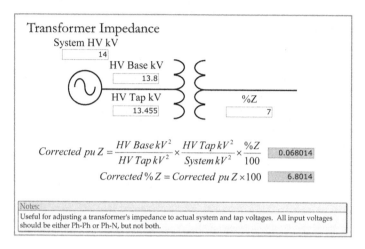

FIGURE 2.33
Transformer tap and applied voltage correction.

secondary kV base current provides the secondary current for the same fault. Figure 2.33 represents the transformer shown in Figure 2.34; the expression yields the infinite-source, three-phase, secondary short circuit current on the primary or 14-kV side and is probably within 90% or more of the actual value for faults near the secondary bushings or terminals. However, the standard practice is to complete a more refined calculation that includes the source and line impedance to the short circuit location.

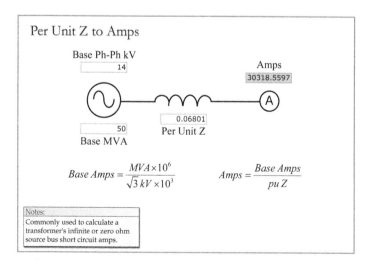

FIGURE 2.34
Per-unit Z to amps.

Electrical Studies

2.9.4 Converting Amps to Per-Unit R and X

Figure 2.35 illustrates how to calculate the per-unit R and X if the short circuit angle is known, and Figure 2.36 shows how to determine the balanced three-phase short circuit current if the per-unit R and X are given.

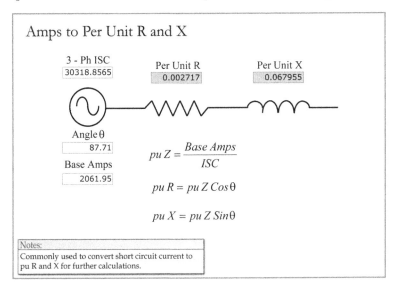

FIGURE 2.35
Amps to per-unit R and X.

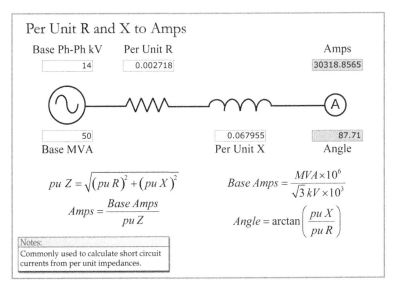

FIGURE 2.36
Per-unit R and X to amps.

2.9.5 New MVA Base

Occasionally, MVA base values need to be adjusted to accommodate different apparatus or data from another study. Figure 2.37 shows how to adjust per-unit ohms from a 50- to a 100-MVA base. As you can see, doubling the base doubles the per-unit impedance. The impedance does not need to be corrected for single-phase transformers that are used in three-phase banks. Although the combined MVA is higher, the base current does not change, and "three-phase ohms" equal "phase-neutral ohms." A 7% single-phase, 10-MVA transformer also has a 7% impedance when used in a three-phase 30-MVA bank.

2.9.6 Converting Per-Unit to Ohms

At some point in your study, per-unit ohms may need to be converted to actual ohms for protective relay or other applications. This can be accomplished using the formula presented in Figure 2.38. The inverse calculation is shown in Figure 2.39.

2.9.7 Converting Amps to Per-Unit Z

Converting amps to per-unit Z is useful for some applications, as shown in Figure 2.40; for example, dividing a motor's full load amps (FLA) by the locked rotor amps (LRA) provides the per-unit impedance of the motor, which is useful for associated studies.

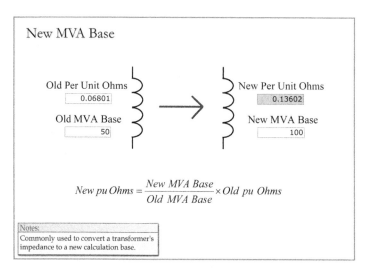

FIGURE 2.37
New MVA base.

Electrical Studies

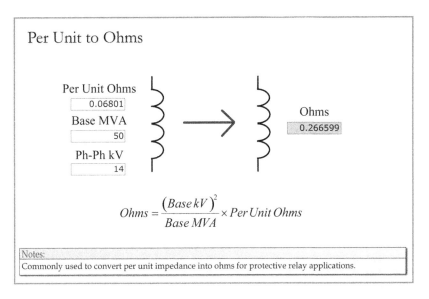

FIGURE 2.38
Per unit to ohms.

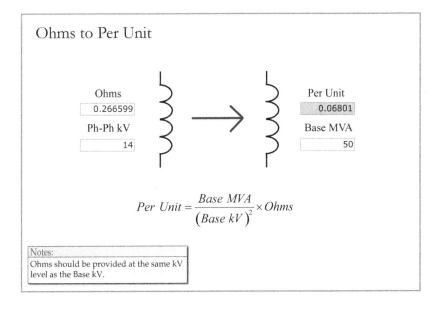

FIGURE 2.39
Ohms to per-unit.

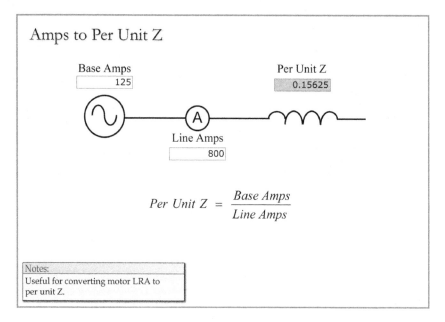

FIGURE 2.40
Amps to per-unit Z.

2.10 Per-Unit Short Circuit Calculations

2.10.1 Transformer Short Circuits

Figures 2.41, through 2.43 illustrate procedures for calculating short circuit currents for the three most common transformer configurations. By comparing the three procedures, it can be seen that they are the same for determining secondary three-phase and phase–phase short circuit currents, and the results are the same. In addition, the primary currents for secondary three-phase faults are the same in all cases. However, the wye–delta and delta–wye winding configurations cause the current to increase in one of the primary phases for secondary phase–phase faults. The delta–wye grounded transformer can also deliver ground fault current.

2.10.2 Transformer Three-Phase and Phase-to-Phase Fault Procedures

Let's start with the delta–delta transformer in Figure 2.41 because the short circuit currents are easier to calculate. First, an MVA base is selected; using the transformer MVA as the base avoids additional conversions. Next, the source short circuit current is converted to R and X, and then to per-unit values. The transformer percent impedance is also converted to per-unit R and X values using the X/R ratio, and finally the line R and X is converted to

Electrical Studies

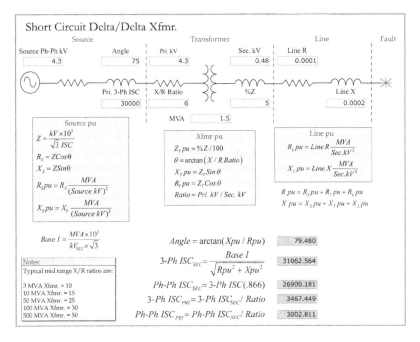

FIGURE 2.41
Delta–delta transformer short circuit.

per-unit R and X values. The three-phase secondary fault current is found by dividing the secondary base current by the total per-unit impedance, and the angle is the arc-tangent of the per-unit X/R ratio.

Dividing the secondary short circuit current by the phase–phase ratio will yield the primary current. The secondary phase–phase fault current magnitude is 86.6% of the three-phase value, and in this configuration, the secondary current divided by the transformer ratio yields the primary current. Figure 2.41 also provides typical XR ratios that can be utilized if specific transformer X/R ratios are not available. As you can see, the three-phase secondary side short circuit current is 31,063 amps.

For the wye–delta and the delta–wye transformers in Figures 2.42 and 2.43, the calculating procedures are identical to Figure 2.41, with the exception of the primary currents for phase-to-phase secondary faults. In both cases, one of the primary phases sees it the same as a three-phase 100% fault, even though it is really an 86.6% phase-to-phase fault. The delta–wye grounded configuration can also deliver ground fault current. As shown, the three-phase secondary short circuit currents in Figures 2.42 and 2.43 agree with Figure 2.41 and are also 31,063 amps.

Now let's look at the phase-to-phase current flow for the delta–wye transformer in Figure 2.43, because it is simpler. Figure 2.44 illustrates the current flow for an A-B fault. As you can see, B phase has a greater primary

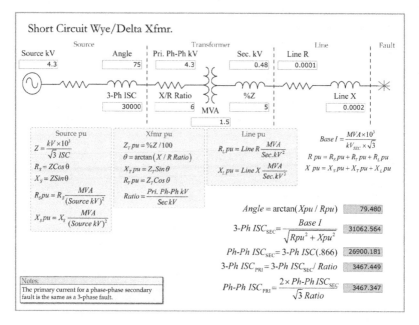

FIGURE 2.42
Wye-delta transformer short circuit.

current due to the contribution from the other two phases. Figure 2.45 shows the phasor relationships.

Figure 2.46 illustrates the current flow for the wye–delta transformer in Figure 2.42 for an A-B secondary fault. The current flow in this case is much more complicated than the delta–wye transformer configuration. In addition to the A-B winding, the B-C winding in series with the C-A winding is in parallel with the fault, and it also contributes current to the A-B fault, which increases the primary current in A phase.

Figure 2.47 illustrates the phasor relationships. The B-C and C-A vectors are reversed because current flows in each secondary winding are in on polarity instead of out on polarity. Two A-B vectors are shown, one for each winding configuration that feeds the fault.

2.10.3 Three-Winding Transformer Short Circuits

Some three-phase transformers are equipped with one primary winding and two secondary windings (three winding transformers) that feed different buses. With three-winding transformers, there is an additional parallel impedance path that can increase the secondary short circuit current magnitude due to the magnetic mutual coupling between the secondary windings. Some protection engineering groups will ignore the parallel path since the short circuit current may also be reduced by load on the unfaulted secondary

Electrical Studies

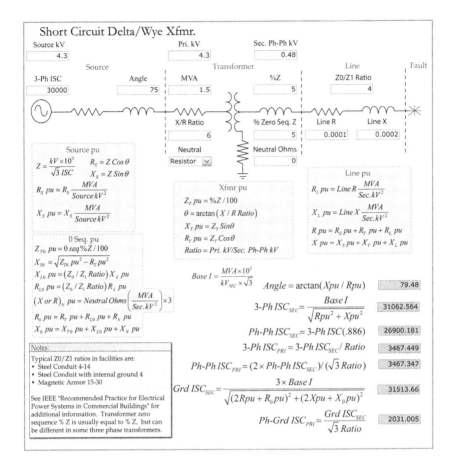

FIGURE 2.43
Delta–wye transformer short circuit.

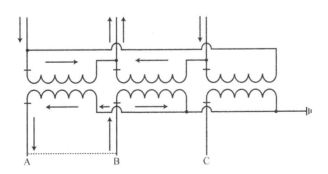

FIGURE 2.44
Delta–wye transformer phase-to-phase faults.

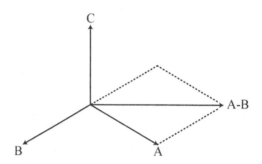

FIGURE 2.45
Delta–wye transformer phase-to-phase vectors.

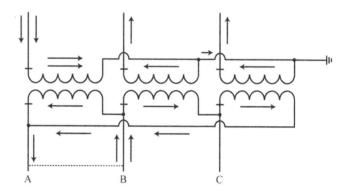

FIGURE 2.46
Wye–delta transformer phase-to-phase faults.

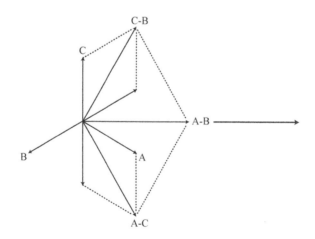

FIGURE 2.47
Wye–delta transformer phase-to-phase vectors.

Electrical Studies

winding at the time of the event. If that winding is loaded, it can cause additional voltage drops in the primary winding, which can have a reducing effect on the calculated short circuit current. From a practical point of view, the author is in agreement with ignoring the parallel secondary paths when calculating short circuit secondary currents in three-winding transformers for relay coordination purposes and is not aware of any problems created by that practice.

2.10.4 Transformer Ohmic Short Circuit Calculations

As a matter of technical interest, transformer short circuit currents can be determined by using ohmic values instead of the per-unit system. Figure 2.48 presents an ohmic approach for resolving delta-delta short circuits to the per-unit method previously presented in Figure 2.41. Supposedly, per-unit procedures are more efficient; consequently, they are the favored method for presenting power system analysis courses at universities and for practicing electrical power engineers in the industry.

By analyzing and comparing the procedures, you will find that the answers are the same, but the ohmic method has one fewer calculation step. In both

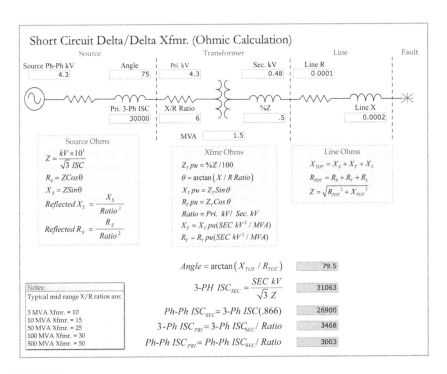

FIGURE 2.48
Comparing per-unit to ohmic short circuit procedures.

examples, the source is given as a three-phase short circuit current with an associated angle, the transformer nameplate percent impedance does not need to be corrected for different taps or applied voltages, and the ohmic value of the cable impedance to the short circuit location is determined from look-up tables.

Resolving three-phase short circuit currents and angle using a per-unit procedure involves the following:

- Source three-phase ISC to per-unit (6 steps)
- Transformer %Z to per-unit (5 steps)
- Total per-unit ohms (5 steps)

Total steps = 16

Resolving three-phase short circuit currents and angle using ohms involves the following:

- Source three-phase ISC to ohms (5 steps)
- Transformer %Z to ohms (7 steps)
- Total ohms (3 steps)

Total steps = 15

As previously discussed, the ohmic approach has one fewer step. In both cases, the transformer phase–phase ratio needs to be calculated to determine the primary current values. With ohmic calculations, the source and cable ohms do not require conversion to per-unit values. Since the transformer impedance is already expressed as a percent, the per-unit method requires fewer conversion steps in that case, but the base current needs to be calculated to determine the short circuit current.

The ohmic approach may facilitate the transfer of knowledge from more introductory courses for students entering this field of study, as well as being a proof for the per-unit system. For students who are well versed in single-phase series and parallel impedance calculations, the per-unit system may be awkward and does not encourage the normal logical process. Since base quantities are represented as 1.0 per-unit and because 1s can readily drop out of formulas, the expressions may no longer represent Ohm's Law or other terms that are familiar to students.

For example, current is normally determined by dividing the voltage by the impedance. However, with the per-unit system, the short circuit current is calculated by dividing the base current by the impedance. While the per-unit system may have an advantage with network calculations, almost everyone in the industry has been using computers for network calculations since the 1970s, and before that with alternating current (AC) and direct current

Electrical Studies

(DC) network models. However, nonnetwork calculations are often performed with handheld calculators for less complex series or radial systems, and the ohmic approach appears to be simpler in that case.

2.10.5 Sequence Impedances

Calculating ground faults and analyzing unbalanced conditions for three-phase systems can be quite complex. In 1918, Dr. Charles L. Fortescue presented a paper at an American Institute of Electrical Engineers (AIEE) convention in Atlantic City, New Jersey, on a method for analyzing three-phase systems using positive-, negative-, and zero-sequence phasors (symmetrical components). The positive sequence (1) represents normal balanced conditions and rotation; the negative sequence (2) represents unbalanced currents in the armature or stators of rotating electrical apparatus that produce a reverse rotation current in the rotors or fields of generators and motors; and the zero sequence (0) represents ground fault currents without rotation. Consequently, many in the industry will refer to balanced events as *positive phase sequence,* phase-to-phase events as *negative phase sequence,* and ground faults as *zero phase sequence.*

Although the foregoing is really an oversimplification of symmetrical components, ground fault currents through delta–wye transformers can be readily calculated with a procedure that utilizes positive-, negative-, and zero-sequence impedances. Figure 2.49 illustrates the sequence impedances that will be utilized (though for simplification, it does not show the resistance). Figure 2.50, for illustrative purposes, is mathematically the same, but it applies the ground return path impedance to each sequence leg, negating the need to multiply the ground return path impedance by 3.

2.10.6 Transformer Ground Fault Procedures

Referring to Figures 2.43 and 2.49, basically the base current is multiplied by 3 and then divided by the sum of the positive-, negative-, and

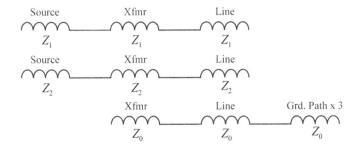

FIGURE 2.49
Normal delta–wye ground fault sequence impedance.

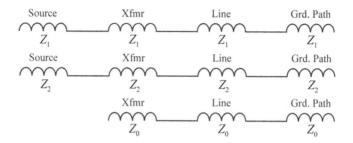

FIGURE 2.50
Equivalent delta–wye ground fault sequence impedance.

zero-sequence per-unit impedances. The positive-sequence per-unit impedances are derived from the source impedance, percent impedance of the transformer, and the line ohmic values to the point of the fault. The positive- and negative-sequence magnitudes are considered equal, and doubling the positive-sequence impedances will account for both in the equation. The zero-sequence impedance includes the transformer's zero-sequence %Z (which usually is considered the same as the %Z of the transformer, but it can be lower in three-phase core form transformers), line and ground return path impedances and any limiting impedance in the neutral. Because the base current is multiplied by 3, any impedance introduced in the transformer neutral and in the ground return path has to be multiplied by 3 to balance out the equation.

As you can see in Figure 2.43, the ground fault magnitude exceeds the three-phase fault value because the zero-sequence impedance does not include the source impedance. However, the zero-sequence impedance increases faster than the positive-sequence impedance as the distance from the transformer to the ground fault location increases; therefore, the three-phase short circuit current often exceeds the phase-ground fault current at the secondary side circuit breaker, depending on the distance and associated line impedance.

Transformers and generators are generally not designed to handle large, close-in-ground fault currents that are above the three-phase value; consequently, some transformers are equipped with neutral reactors to limit the close-in-ground fault current, and generators are normally high-impedance grounded to limit the ground current. Figure 2.43 also provides typical Z_0/Z_1 ratios for commercial and industrial facilities. The positive sequence impedances for cables and busways can be multiplied by a Z_0/Z_1 ratio to estimate the impedance for the line plus the ground return path (which does not include transformer neutral-ground impedance). Because the overall Z_0/Z_1 ratio can range from as little as 1 to as much as 50, estimating the ratio precisely is impractical because of the many variables in ground current return paths. Because of this, worst-case conditions are often assumed when making ground fault calculations for short circuit duty considerations. The suggested Z_0/Z_1 ratio of 4 assumes that the ground return path impedance

Electrical Studies

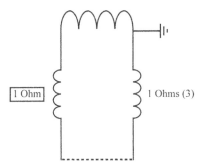

FIGURE 2.51
Z_0/Z_1 ratio.

equals the line impedance to the point of the fault and is the more conservative calculation.

Figure 2.51 shows a representation of the line impedance as 1 ohm and the ground return path as 1 ohm × 3. The overall zero-sequence impedance for the line plus the return path equals 3 + 1, or 4. Accordingly, in this case, the Z_0/Z_1 ratio is 4. For Z_0/Z_1 ratios greater than 4, much of the circuit impedance is in the line itself due to the increased spacing between the line and the return path. In analyzing the Z_0/Z_1 ratios presented in the notes sections of Figures 2.43 and 2.52, a Z_0/Z_1 ratio of 4, the lowest possible value, is assigned when a ground conductor is provided in the same conduit as the line conductors, reducing the spacing between the line and ground return path and the associated self-flux and inductive reactance of each conductor. One could then assume that for Z_0/Z_1 ratios higher than 4 that involve ground and neglecting arc resistance, most of the increase in impedance would be associated with increased self-flux in the line conductor and would be inductive in nature. Figure 2.52 provides a method of estimating the increase in line impedance if the Z_0/Z_1 ratio is known. In the case of a magnetic armor conduit, the line impedance for ground faults would be greater by at least a factor of 3.75.

2.10.7 Demystifying Ground Fault Calculations

There can be a number of myths or misconceptions associated with the ground fault theory covered in the prior section. Let's start with the following questions associated with the per-unit symmetrical component sequence impedance method for calculating ground faults:

- Is the calculation method accurate?
- Are ground faults complicated?
- Is there really a 3I0?

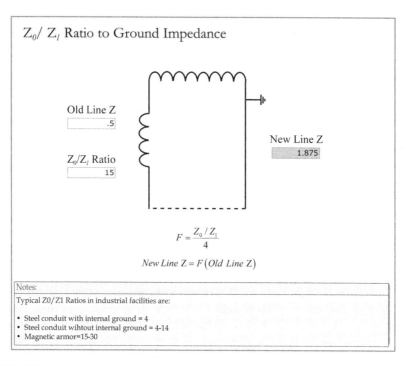

FIGURE 2.52
Z_0/Z_1 ratio to line impedance increase.

- Is there really a 3V0?
- Why is the ground fault current higher than the three-phase short circuit current at the transformer?
- Why do ground fault currents decrease faster than three-phase short circuits?
- Do the ground fault sequence impedances really represent the circuit?

As a matter of interest, delta-wye ground fault current can also be determined without resorting to sequence impedances by applying more conventional methods without major difficulty for applications that do not involve parallel sources on the wye side. It is really a relatively simple circuit, with current flows in only one of the three single-phase transformers that make up a three-phase bank. It can also be debated that this method is simpler, has fewer steps, and relies on conventional theory instead of a sequence impedance procedure that is contrary to actual basic theory.

The sequence impedance method (which provides an equal result) advocates 3V0 and 3I0, which can be calculated but do not really exist, and suggests that the ground fault current is higher at the transformer because the source impedance is not considered in the zero-sequence path (which is not the real reason). Further, the zero sequence impedance increases rapidly because it

Electrical Studies

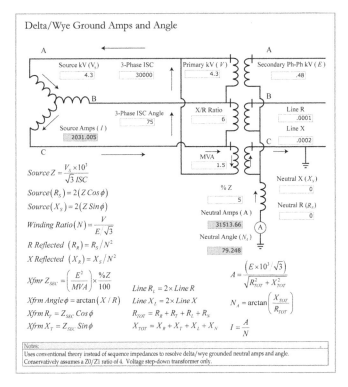

FIGURE 2.53
Step-down transformer ohmic ground fault calculation.

is multiplied by 3 (which is also not the real reason). Figure 2.53 illustrates the same delta–wye grounded transformer as Figure 2.43, but with an identical C-phase secondary ground fault; however, in this case, the actual current paths and ohmic impedances will be utilized.

As you can see by the current flow in Figure 2.53, the ground fault looks like a phase–phase fault to the source. Consequently, the phase source impedance is multiplied by 2 to consider the impedance from two different phases. Twice the source impedance is then reflected to the secondary or 480-V side (this calculation applies only to a step-down transformer since the reflected ohms would be multiplied by the square of the turn's ratio in a step-up transformer). The transformer ohms are also determined on the 480-V side, and the ground return path impedance is assumed to equal the line impedance (Z_0/Z_1 ratio = 4). Therefore, doubling the line impedance accounts for the return path impedance as well. Finally, the total neutral or ground current can now be determined by dividing the phase-neutral voltage by the total actual impedance (including the neutral-ground impedance, if applicable).

If you compare the ground fault current magnitudes in Figure 2.53 to those in Figure 2.43, you will find that they are identical. The ground fault current exceeds the three-phase value near the transformer because the ratio is higher (i.e., higher secondary current) because phase–neutral instead of phase–phase voltages are utilized in the calculation. The ground fault impedance increases faster than the three-phase impedance because the three-phase calculation does not need to consider the return path impedance. The ground fault angle can now be determined easily by the arc-tangent of the X/R ratio.

In conclusion, the sequence impedance method for determining ground fault current has a practical precision. However, for single-source delta-wye transformer calculations, it unnecessarily complicates the theory and has several extra calculation steps. It certainly does not represent actual currents, voltages, and return-path impedances and does not provide the real reasons for the various outcomes. Obtaining the angle is also not obvious. More important, it also does not facilitate the transfer and knowledge from more introductory courses in AC theory.

2.10.8 Generator Three-Phase Short Circuits

Figure 2.54 presents the three-phase short circuit levels for generators. Basically, the highest level of current occurs during the first few cycles and is calculated by dividing the base current by the *direct axis saturated subtransient*

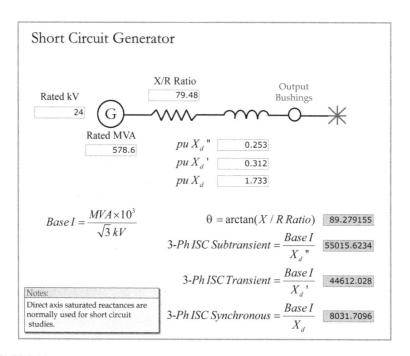

FIGURE 2.54
Generator short circuit.

per-unit reactance or impedance. This is the reactance that is typically applied in short circuit studies and represents the magnetic flux magnitude prior to the fault. The direct axis saturated subtransient reactance, **Xd"**, is used when generators are at the rated voltage (slightly saturated) prior to the fault and the pole faces are more directly aligned with load currents. A short circuit angle of 90° lagging is often assumed for large generators, but if the X/R ratio is known, a more refined angle can be determined. The intermediate level of fault current is calculated by dividing the base current by the *direct axis saturated transient per-unit reactance*, **Xd'**. This level of fault current typically lasts for tenths of seconds. The final level is determined by dividing the base current by the *direct axis synchronous per-unit reactance*, **Xd**.

Please note that this level of current is normally well below the full load ampere rating of the machine; consequently, conventional overcurrent relays (without voltage restraint or control) cannot isolate generators from short circuits if they are set with a time delay to coordinate with downstream devices. The foregoing current decrements are basically caused by the inability of excitation systems to overcome the internal voltage drops and subtractive magnetic flux (armature reaction) that develop during large inductive reactive current flow in the stator windings. The synchronous reactance is also used to determine voltage drops in operating synchronous generators, and both the armature reaction and the inductive reactance are included in the **Xd** value. As with transformers, the initial phase–phase fault current for practical purposes is normally 86.6% of the three-phase subtransient value. Technically, double the negative phase sequence reactance should be used for the phase–phase calculation, but any differences are usually negligible.

2.10.9 Generator De-Excitation

The fault current decrement is determined by the excitation system and the trapped magnetic flux in the core iron and the subtransient, transient, and synchronous reactances for low-impedance faults. If the fault impedances are higher, the decrement will take longer. This is a particular problem for short circuits on the secondary side of UATs on units that are not equipped with generator bus breakers. The UAT capacity (5%–8%) is so small and high-impedance, in relationship to the generator, that the generator continues to feed after tripping fault current, during coast down, driven by the trapped magnetic flux in the air gap. Two pole generators generally decay faster than four pole machines because of the greater inertia associated with slower machines. There are a couple of commonly used methods to reduce excitation current and the trapped magnetic flux in both machines.

Figure 2.55 illustrates a direct field breaker; when the unit is tripped off-line, a resistor, often equal to the DC field ohms (impedance matching for maximum power transfer), is inserted across the field as the main contacts are in the process of opening to discharge the field current and trapped

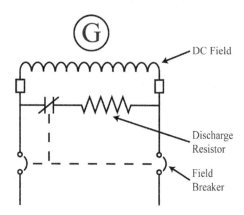

FIGURE 2.55
Direct DC field breaker.

magnetic flux more rapidly. As generator designs grew larger during the 1970s, the DC excitation current became higher and direct field breakers became more expensive. Somewhere around 300 MVA and greater, it became more economical to interrupt the excitation power supply indirectly instead of the field current directly.

Improved designs had de-excitation circuitry that reversed polarity on the field for faster fault current decrements. Although de-excitation circuitry was generally an option, it is often not used and difficult to apply on cross-compound units (high pressure [HP] and low pressure [LP] turbo-generators operated in parallel). Both the direct field breaker and the de-excitation circuitry reduce the UAT secondary fault current to approximately zero in around 4 seconds depending on design details.

Because transformers are generally built to withstand the electromechanical forces from fault currents for only 2.0 seconds, it is not unusual to fail auxiliary transformers due to external fault conditions that trip almost instantaneously from transformer differential protection. During the 1980s, a large coal plant in Nevada that was not equipped with direct field breakers or de-excitation circuitry developed a short circuit in the busway between the UAT and secondary circuit breaker. Although the UAT differential relays actuated and tripped the unit in around six cycles, the cross-compound generators continued to feed fault current to the short-circuited secondary bus for 42 seconds during coast down. Obviously, the transformer required replacement after the event, even though the fault was external and unit tripping was almost instantaneous.

2.10.10 Motor Contribution

Motors can contribute current to short circuits for a few cycles because of the rotating inertia and trapped magnetic flux in the core iron. Normally, motor

Electrical Studies

contribution does not need to be considered when coordinating overcurrent relays because of the rapid decay. However, it may need to be considered when analyzing instantaneous protection elements. It definitely needs to be included when calculating the interrupting and momentary short circuit current capabilities of switchgear circuit breakers. This will be discussed in more detail in the material on circuit breaker duty calculations in Chapter 5.

Figure 2.56 shows a procedure for calculating the symmetrical portion of motor current contribution for circuit breaker interrupting and momentary ratings and also can be used for other studies concerned with motor symmetrical current contribution. In the example, only one motor is considered in the calculation, and the old motor values represent that motor in advance of entering new data for the next motor. The motor per-unit Z is determined by dividing the motor full load amps by the maximum locked rotor amps.

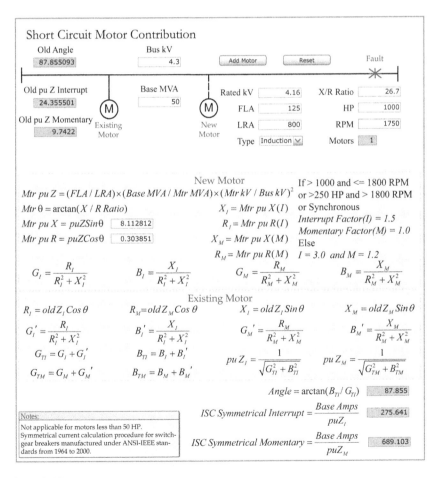

FIGURE 2.56
Motor symmetrical short circuit contributions.

This value is then converted to a common MVA base, which should be equal to or greater in magnitude than the largest motor on the bus.

Because the motor nameplate voltage is lower than the source bus voltage, the motor per-unit impedance also needs to be corrected to the bus voltage magnitude. Depending on the type of motor (synchronous or induction), horsepower, and revolutions per minute (RPM), different multiplication factors are applied to increase the motor per-unit impedances for interrupting and momentary symmetrical contributions to account for current decrements during protective relay and circuit breaker operating times or cycles and specific motor design details.

The motor per-unit impedance is then added to other motors previously calculated using the conductance, susceptance, and admittance method of deriving parallel impedances. The total symmetrical interrupting and momentary current contributions for all motors on the bus can then be determined by dividing the base amps by the total interrupting and momentary per-unit impedances, respectively. Cable impedance is usually considered negligible for medium-voltage motors but is normally considered for low-voltage motors. For low-voltage motors below 50 HP, usually the contribution is assumed to be 4 times the full load amps, or 0.25 per unit.

Bibliography

Baker, T., *EE Helper Power Engineering Software Program* (Laguna Niguel, CA: Sumatron, Inc.), 2002.

Electric Power Research Institute, *Power Plant Electrical Reference Series*, Vols. 1–16 (Palo Alto, CA: Electric Power Research Institute), 1987.

General Electric, *ST Generator Seminar*, Vols. 1 and 2 (Schenectady, NY: General Electric), 1984.

Institute of Electrical and Electronics Engineers (IEEE), *Recommended Practice for Electric Power Systems in Commercial Buildings* (Piscataway, NJ: IEEE), 1974.

National Fire Protection Association, *National Electrical Code* (Quincy, MA: NFPA), 2002.

Stevenson, W.D., Jr., *Elements of Power System Analysis* (New York: McGraw-Hill), 1975.

Westinghouse Electric Corporation, *Applied Protective Relaying* (Newark, NJ: Silent Sentinels), 1976.

Westinghouse Electric Corporation, *Electrical Transmission and Distribution Reference Book* (East Pittsburgh, PA: Westinghouse Electric Corporation), 1964.

3

Auxiliary System Protection

In general, protection engineering requires a high degree of technical expertise, as well as experience with the nuances of the operation of generating stations. Practicing protection engineers also assume significant responsibilities. They are concerned about personnel safety, and try to protect the apparatus on one hand, and not cause nuisance tripping or unnecessary outages on the other. In essence, they are performing a balancing act, doing their best to leave themselves in a defendable position. They also need to make sure that their settings coordinate with other relays or protective functions. The suggested settings in this chapter, as well as the next on generator protection, have withstood the test of time and have proven to be very effective at coordinating with other tripping functions, protecting electrical apparatuses, and mitigating nuisance tripping.

The author has completed approximately 35,000 megawatts (MW) of protective relay reviews for fossil and nuclear plants that ranged between 25 and 1250 megavolt-amps (MVA) and were placed on-line or into service between 1952 and 2013. The reviews took place between 1981 and 2015. Nondefendable protection settings or questionable relay practices are not uncommon and usually involve one or more of the following elements:

- Calculation errors
- Illogical permissive or supervisory contacts
- Improper applications
- Lack of coordination with other functions
- Possibilities for nuisance tripping
- Settings that do not actually protect the associated apparatus
- Important functions not applied

With the advent of the newer digital relay technology, oversights also may be more commonplace because of the increase in flexibility and functions, and associated complexities.

To facilitate the reviews, a software program that automates the tedious calculations associated with protective relay settings was developed by the author. Although it automates the calculations for both the main unit generator and the plant auxiliary power protective relay systems, the greatest advantage occurs with generator settings. It can take 40 hours to use a handheld calculator and graph paper to determine the various settings for generator multifunction digital relays, excluding the manufacturer's programmable logic controller (PLC)-type mapping, assuming that the required input data is readily available. With the software, the settings can now be derived in 30 minutes.

This chapter and the following one on generator protection will identify the questionable practices and oversights found for each protection element or function (except for calculation errors) and provide a rough estimate of the percentage of plants in the survey group with each type of anomaly. This chapter covers the protective relaying of low- and medium-voltage systems that provide the necessary auxiliary power for generating stations and industrial plants. Usually, protection engineers limit their scope to switchgear circuit breakers equipped with adjustable protective devices. This would normally include lighting transformer low-side breakers; 480-volt buses that directly feed 50–299 horsepower (HP) motors, and motor control centers (MCCs), but not their positions; 4-kilovolt (kV) buses that feed 300–5000 HP motors and transformer loads; and 6.9- and 13.8-kV buses that feed motors larger than 5000 HP.

With low-voltage 480-volt systems, there is no absolute guarantee that the protection can operate to clear a fault. In practical terms, auxiliary system arc voltages can get as high as 300 volts, which may be above the phase-neutral voltage value for low-voltage systems and consequently high enough to extinguish the arc. There have been cases where switchgear source air circuit breaker main contact overheating problems caused enough metal splatter and ionization to breach the insulation dielectric between phases. The resulting short circuit bypassed the current flow through the contacts, reducing the metal splatter and ionization, which allowed the fault to extinguish before the protective relays could time out. In one case that was captured by a digital fault recorder, the cycle kept repeating and an operator inspecting the switchgear room noticed a pungent ozone smell, heard the arcing, and requested the control room to trip the breaker. This phenomenon does not appear to happen with medium- and high-voltage systems since it is much more difficult to extinguish arcs.

The author is aware of a couple of cases where a 3-foot air gap on a 230-kV disconnect carried the output of four generating station units. This took place at a coastal plant, and salt air had corroded the contact mating surfaces on one phase, causing it to overheat and fuse open, and consume conductor material as it continued to carry the output of the units. Because the phase-ground and phase-phase clearance distances were more than ample, the event did not cause a short circuit. Operations became aware of it when the switchyard was illuminated by the arc at night. Because prior switching for maintenance reasons had left the four units radially or series-connected to the electrical system through a single set of disconnects, eliminating any alternate paths, the contact surfaces were forced into a failure mode.

3.1 Switchgear Overcurrent Coordination

Protective relay selectivity, or the tripping of the fewest components to isolate a short circuit condition, is achieved by having the lowest minimum trip (MT) or actuation value and the shortest time delay on the most downstream fault-clearing device. As you move upstream toward the electrical source, each fault-clearing device should have progressively higher MT values and longer fault-clearing times to ensure coordination at all possible currents. The maximum symmetrical bus, three-phase short circuit currents (excluding motor contribution) are normally used to compare tripping times for switchgear applications because other contributors have a short life (i.e., a few cycles). Faults near the bus will have the same short circuit current levels. In industrial plants, 0.3 seconds is generally considered the minimum amount of time delay between upstream and downstream devices and may be a little marginal. Experience has shown that 0.25 seconds sometimes does not coordinate, and many in the industry prefer to use 0.4 seconds between devices for a more conservative approach, in order to ensure coordination. Curves with like inverse slope characteristics will have increased time margins at currents below the maximum. The best practice is to apply the same curve type (inverse, very inverse, or extremely inverse) for the bus tie, transformer low-side breaker, and high-side or primary overcurrent protection. Where that is the case, little or no curve work will be necessary to ensure coordination. In relation to low current levels, the more inverse the curves are, the faster the relays will actuate at higher currents. If the relays or devices coordinate properly, the outage impact to the station will be minimized and there will be a good indication as to the location of the fault, allowing service to be restored quicker. The amount of time delay between positions is necessary to account for the following:

- Current transformer (CT) performance differences
- Relay calibration inaccuracies
- Electromechanical relay overtravel when the fault is interrupted
- Protective relay reset time (sequential events)
- Differences in relay curves
- Voltages that are higher than normal
- Three winding transformer impedance differences
- Circuit breaker tripping or contact part time differences
- Circuit breaker interruption differences (which current zero?)
- Higher initial currents due to asymmetry and motor contributions
- Differences in primary and secondary currents in the source transformer

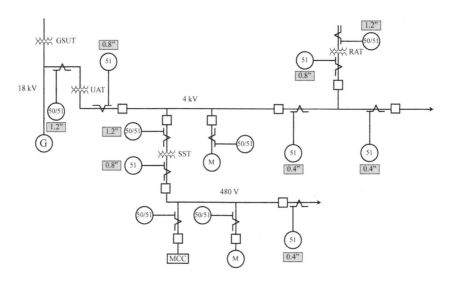

FIGURE 3.1
Typical auxiliary system overcurrent one-line.

Figure 3.1 illustrates a typical generating station auxiliary power system's one-line configuration with the associated CTs and overcurrent relays. The most downstream overcurrent relays are for bus load feeder breakers that drive motor, transformer, and MCC loads. Feeder breakers are usually equipped with instantaneous overcurrent elements (50) that will actuate in around one cycle for three-phase, phase–phase, and ground (if solidly grounded) short circuit conditions, and with breaker interruption time included, isolate faulted feeders in around six cycles, or 0.1 seconds. The fast operation clears or isolates faults before more upstream breakers can operate and have a greater impact on the auxiliary power system. Relay targets or flags will indicate which element and phase actuated to guide operations and maintenance to the problem area more quickly. The more upstream circuit breakers are not equipped with (50) instantaneous overcurrent elements and only have (51) time overcurrent functions because they need time delay to coordinate with bus load feeder breakers.

The shaded boxes in Figure 3.1 represent the amount of time delay at the maximum bus, three-phase, symmetrical short circuit current (somewhere around 30,000 amps) that is suggested for coordination purposes with switchgear applications, for primary side source transformers, associated secondary side source breakers, and bus tie breakers; and this ensures that the relays are timed at identical magnitudes and at curve points where the time margins are the lowest. This maximum symmetrical timing current value assumes nominal voltages and again does not include motor contribution or momentary asymmetrical currents due to their rapid decay during short circuit conditions (i.e., few cycles).

Auxiliary System Protection

In more recent years, the author has been finding plants where the protection was coordinated with asymmetrical short circuit currents from arc-flash studies, which also include the fault current contribution from motors. This practice significantly increases the arc-flash hazard for station personnel since the timing current is higher and will take much longer to trip from the lower or more symmetrical currents. As shown in Figure 3.1, the bus tie breaker has a time delay of 0.4 seconds in order to coordinate with bus feeder load breakers. The transformer lower-voltage or secondary side-source circuit breakers for the unit auxiliary, reserve auxiliary, and station service transformers (UAT, RAT, and SST, respectively) are upstream from their associated tie breakers and have time delays of 0.8 seconds for three-phase bus faults, and the transformer primary or high-side overcurrent elements have time delays of 1.2 seconds or longer. Delta–wye or wye–delta transformer configurations may require 1.3 seconds or more to coordinate with the low-side relays for phase-phase faults, as discussed in Chapter 2. The transformer high-side time overcurrent protection is absolutely necessary to provide breaker failure or stuck breaker protection in case the low-side circuit breakers fail to clear a bus short circuit condition.

The medium-voltage protection setting suggestions outlined here and presented in Figure 3.1, also apply to low-voltage (480-volt) switchgear. Prior to the 1970s, the rack-in-out low-voltage breakers were equipped with series overcurrent protection that relied on oil or air-type timing orifices, dashpots, and diaphragms for time delay. This type of protection proved to be very unreliable, and this author was able to secure management approval during the 1980s to upgrade the protection at 12 multiunit generating stations during unit overhauls. By that time, discrete solid-state relays were available in small, relatively low cost packages that included all three phases. Because of the extended range of the relays, high-ratio CTs (1200/5) could be used to avoid saturation problems with lower-voltage class, less expensive CTs. This approach offered the following advantages over upgrading the breaker protection with direct replacement, solid-state series overcurrent devices:

- Protection engineers prefer the more conventional type of inverse curves over the often-confusing series trip-type curves for coordination and apparatus protection.
- Operations liked the fact that the protection was no longer part of the circuit breakers and the breakers were now interchangeable, without regard to the protection settings.
- Removing the series overcurrent devices created more room and improved access for electricians that were performing maintenance or overhauling the breakers.
- Relay technicians no longer required a special, cumbersome, high-current test set in order to test the protection for proper operation.

- The material cost for the discrete relays and associated CTs was less than the manufacturer's direct solid-state series overcurrent replacements at that time.

The settings presented in the next discussion assume that discrete relays with standard CT ratios are also used for the 480-volt systems. If that is not the case, the actual CT ratios will need to be modified to be compatible with the graphics presented in the figures displayed in this chapter. For example, a 1000/1 ratio would need to be converted to 5000/5 to be in agreement with the software and figure graphics. In the original design, some engineering companies applied discrete induction-disk overcurrent relays for 480-volt bus source and tie breakers because of the difficulty in trying to coordinate station service transformer primary overcurrent relays equipped with standard inverse curves with series trip-type curves, which often cannot be coordinated at all curve points.

3.2 Overcurrent Schematic

Figure 3.2 presents a typical time overcurrent relay, alternating current (AC) schematic (a three-line simplified drawing) for source, tie, and feeder breakers. As mentioned previously, the load feeder breakers, in addition to the (51) functions, are equipped with (50) instantaneous tripping elements. In Figure 3.2, only one (51) function is provided for each phase and and the (50) elements are not applied. The phase relays are stared or tied together on the right side,

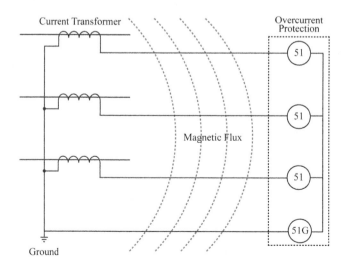

FIGURE 3.2
Typical time overcurrent schematic.

Auxiliary System Protection

and the secondary currents are routed through a residual ground relay (51G) before returning to the neutral or stared side of the CTs. The ground relay can only actuate for ground faults and cannot see phase-phase and three-phase events; consequently, it can be set to operate below load current levels if the neutral does not carry a load. Normally, only one safety ground is provided to facilitate testing of the secondary insulation. The curved dashed lines represent stray magnetic flux, which is always present with primary current flow, and links with the secondary wiring, inducing small voltages. Induced current flows into the relays are restricted by the CTs that block secondary AC current flow unless there is a corresponding current flow on the primary side. However, if the CTs are shorted and primary current is flowing, the induced voltages and associated currents could actuate the 51 G, which has greater sensitivity, if the setting thresholds or MT points are low enough.

3.3 Current Transformer (CT)

3.3.1 CT Safety Ground

Figure 3.3 shows a doughnut CT wrapped around an energized 12-kV conductor. Stray capacitance is everywhere, and there will be some level of

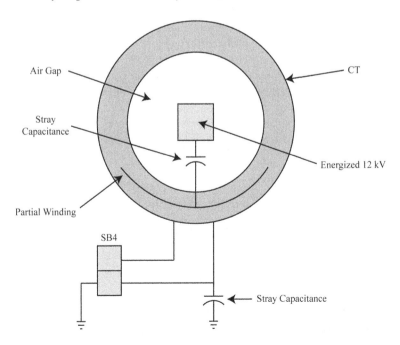

FIGURE 3.3
CT safety ground.

internal capacitance between the energized primary conductors and the secondary turns of the CTs and also stray capacitance between the external secondary wiring and ground. Consequently, there is a series voltage divider circuit between the energized primary conductor and ground. The phase-neutral voltage (6.9 kV) will have a path for current flow through the internal CT capacitance that is in series with the external wiring capacitance if only one phase is energized or during times of circuit breaker pole disagreements. If the stray capacitive farads between the 12-kV conductor and internal CT secondary wiring is higher (with a lower impedance) than the external secondary wiring stray capacitance, the majority of the voltage will appear on the higher-impedance secondary side. With three-phase systems, the current exchanges tend to be phase–phase; however, if the stray capacitance is not balanced, the resulting phase-neutral voltage differences also can cause a voltage increase on the secondary side. The instrument transformer safety ground shorts out the external wiring capacitance and prevents a voltage rise on the secondary side from capacitive coupling. Accordingly, energized CTs should be treated as hazardous even if the associated circuit breaker is open and no current is flowing. It would be particularly risky to lift the ground to perform a megger or megohm test of the external secondary wiring insulation with the CTs energized.

3.3.2 CT Open Circuit

Opening a CT circuit can be very dangerous; a high impedance on the secondary side will reflect to the primary side, causing a much-higher-than-normal voltage drop in the primary conductor at that location. This voltage drop will be stepped up by the turns-ratio of the CT on the secondary side and can create lethal voltages of several kilovolts, depending on the voltage class of the transformer and the specific amount of laminated iron in the core. At some point, the iron will saturate from the increase in excitation current, which will limit the magnitude of the voltage rise.

3.3.3 CT Reflected Ohms

In cases where parallel-source or tie circuit breakers are not sharing the current equally because of connecting bus or cable impedance differences, and where there is concern about breaker continuous current ratings, CTs can be installed with secondary resistors to balance out the current flow by reflecting the appropriate ohms to the primary side.

3.3.4 CT Burden

When protective relays or other devices are added to CT protection circuits, care must be taken to ensure that the external secondary circuit impedance or burden does not hinder the capability of the CT to accurately reproduce

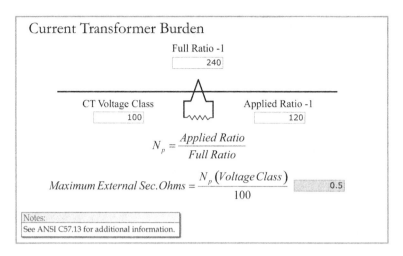

FIGURE 3.4
CT secondary burden.

a representative secondary current for proper protective relay operation during short circuit conditions. Figure 3.4 illustrates a procedure for calculating the maximum secondary external burden or ohms for relay class CTs that are built to Institute of Electrical and Electronics Engineers (IEEE) standards. As the figure shows, a 1200/5 or 240/1, 100-volt class CT with an applied ratio of 120/1 can operate properly from 0 to 0.5 ohms. If the full winding is utilized, it could operate with up to 1 ohm in the external secondary circuit. To be more specific, for the relay-class CT shown in Figure 3.4, IEEE standards indicate that the error should not exceed 10% at 20 times the rated current (100 amps secondary) with an external secondary circuit burden of 0.5 ohms or less at a 50% power factor (PF). Normally, the voltage class is provided on the CT nameplate. The voltage class can also be approximated as the point where the exciting current equals 10 amps during a CT saturation test; excitation currents above 10% or 10 amps will not allow the CT to stay within the specified accuracy. At this point, the CT-laminated iron is in partial saturation, and further increases in excitation or saturation levels may distort the sine wave, in addition to causing a reduction in output current, since the primary excitation current is not reflected to the secondary side.

3.3.5 CT Saturation

Figure 3.5 illustrates a CT saturation test with the associated volts versus amps curve (Figure 3.6). During operation, the excitation voltage is provided by a voltage drop that develops in the primary side conductor; however, as with any transformer, that voltage is also reflected to the secondary side. This test is performed during electrical commissioning to prove that

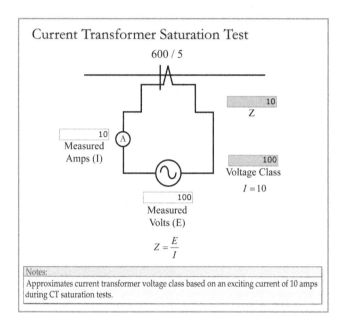

FIGURE 3.5
CT saturation test.

the CT is healthy and compares favorably with the other phases, and it provides an indication that the particular CT does not have winding or laminated iron anomalies. As shown in the figure, voltage is applied to the secondary winding and plotted against the corresponding excitation current. The 10% or 10-amp exciting current point is on the knee of the

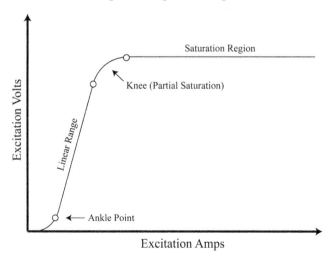

FIGURE 3.6
Typical CT saturation test curve.

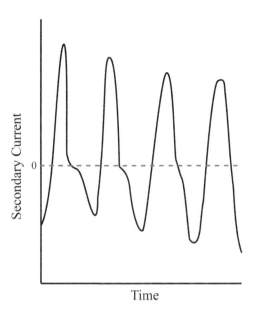

FIGURE 3.7
Saturated CT output waveform.

curve, where the iron is partially saturated. The higher the saturation, the greater is the increase in current for the same step change in voltage, due to a reduced amount of self-flux and associated impedance, as the iron reluctance advantages fade.

If the CT saturates during short circuit conditions, it may not be able to reproduce a sine wave (as illustrated in Figure 3.7), and there may not be enough energy in the output waveform (spike) to operate the associated protective relay elements. In this example, the four cycle series shows a component of direct current (DC) offset (60 Hz half wave or 120 Hz to a filter), which is common with asymmetrical currents, since the positive going zero crossing only lowers the impedance for the positive half-cycle. It will take a few cycles to decay enough for the negative going waves to normalize. As the fault current decays from the circuit X/R ratio, the positive peaks are reduced in magnitude, the width or time duration of each peak increases, and the negative current value increases.

By the fourth cycle, the current decrement allows the production of a normal 60-Hz sine wave. CT saturation possibilities from asymmetrical momentary fault currents can be mitigated by selecting higher-voltage class CTs (more iron), higher-ratio CTs (less current), or a reduction in the circuit impedance or burden (less voltage). A detailed analysis is complex because the behavior of the iron is not linear in the knee and saturation regions, and the iron is also affected by the DC component. Additionally, the behavior of the particular protective relay element to the nonsinusoidal offset waveforms may not be known. My suggestion would be not to push the limits, by providing some margin when the asymmetrical momentary or peak fault current magnitudes are above 100 amps secondary. As mentioned before, this could be accomplished by providing a total combinational or aggregate margin of 20% by increasing the CT ratio, selecting CTs from a higher-voltage class, and/or reducing the secondary circuit impedance.

3.4 Motor Overcurrent

For time overcurrent coordination purposes, it is usually easier to start with the most downstream loads and work your way upstream in case an increase in time delay is warranted for a particular situation. For industrial plants, the most downstream devices would normally be bus-load feeder breakers or contactors that drive motors, MCCs, and other transformers.

3.4.1 Stator Overcurrent Protection (51)

Simple and reliable induction-disk overcurrent relays, properly set, are very effective at providing protection over a wide range of operating conditions for standard service, noncyclic loaded, across-the-line starting induction motors, which are associated with typical generating-station switchgear applications (50-5,000 HP). As shown in Figure 3.8, the suggested MT or actuation point is 125% of the nameplate full load amps (FLA). There is no specific time associated with the MT threshold in an induction-disk relay; it is simply the amount of current where some movement of the disk is perceptible, and it could take all day to close the tripping contacts. Because motor cables are normally sized to continuously carry 125% of rated FLA, the cables should be fully protected.

Many motors are purchased with a 1.15 service factor or have been rewound with higher-temperature insulation systems that should provide an equivalent service factor of 1.15 or greater. Consequently, this setting does not provide running overload protection between 1.01 and 1.24 for some motors, and 1.16 and 1.24 for most motors. Many designs do not have the capability to be overloaded beyond their service factor rating. In other words, even if the control dampers or valves are fully open, the rated service factor FLA may not be exceeded. The service factor is not a continuous rating, and it allows a temporary overload for an undefined time period. A service factor of 1.15 permits an additional 10°C temperature rise at full load. Although the stator windings are normally the weak link thermally with respect to current flow during running, there are many additional stator failure modes, and altogether, approximately 37% of motor failure mechanisms are associated with the stator. With many years

Auxiliary System Protection

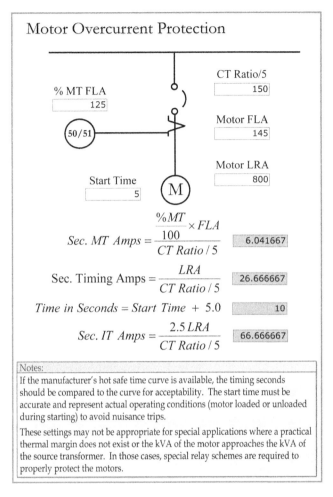

FIGURE 3.8
Motor overcurrent settings.

of electrical experience in large generating stations, this author cannot verify a single switchgear-fed, 50-HP or larger motor that failed thermally from high current flow because the control system dampers or valves were opened beyond the HP capacity of the motor during normal operating conditions. In many cases, one would think that process alarms for increased flows or pressure would alert the operations of the particular control system anomaly before motor overload damage could occur.

Accurate statistics on the specific causes of stator winding failure are not readily available. In addition to high-temperature conditions that can chemically degrade stator insulation systems from high current flow and the associated amps squared R (I^2R) watt losses, there are many other stator failure mechanisms, including the following:

- Shorted turns
- Switching transients
- Ground transients
- Overvoltage
- Bus transfer forces
- Thermal expansion and contraction
- Electromechanical forces
- Partial discharges
- End turn vibration
- Blocked cooling passages
- Hot spot locations
- Shorted iron laminations
- Moisture and contamination intrusion
- Dirty filters or other obstructions that prevent proper air flows

The manufacturer's thermal withstand margins are often based more on experience than on refined calculations. Normally, insulation systems do not start to degrade chemically until a temperature threshold is reached that is reasonably close to the rating of the insulation system. In general, the life expectancy of insulation systems can be very difficult to assess since various manufacturers may use chemical formulations that are not identical for a particular class or type of insulation.

3.4.2 Rotor Overcurrent Protection (51)

The rotor is normally the weak link during starting, but the stator can be the weak link in cases where the HP rating of the motor is relatively small for the particular bus voltage, permitting the use of smaller-than-normal stator conductors. From a protection perspective, it does not matter which one is the most limiting since the settings should be based on the motor thermal limitations in either case. The timing is set for either 5 seconds longer than a normal start time at rated voltage maximum locked rotor amps (LRA) or under the motor hot safe time curves (i.e., the motor initially at the rated operating temperature) if the acceleration and hot safe time curves are available. Many motors will stay close to their LRA current until they reach roughly 90% speed before the starting current rapidly drops off. Obviously, setting the relay curves halfway between the acceleration and hot safe time curves splits

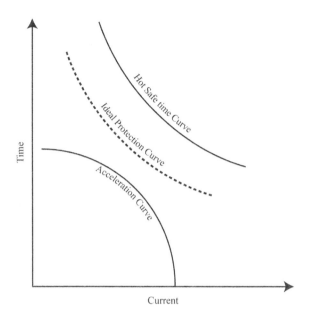

FIGURE 3.9
Motor acceleration, hot safe time, and protection curves.

the margins for preventing nuisance tripping and protecting the apparatus, as shown in Figure 3.9. It is not unusual for motors greater than 5000 HP to lack sufficient thermal margin to fit the relay curves between the acceleration and safe time curves. For larger motors that lack ample margins, there are many schemes for providing adequate protection. The simplest involves using a zero-speed switch to trip the circuit open if the motor fails to rotate in a specific amount of time.

The hot safe-time and acceleration curves were not automatically provided by manufacturers in the past unless specified in the motor purchase order; consequently, many existing plants do not have these curves. The suggested settings should prevent motor squirrel-cage rotor damage for locked rotor conditions, including stalls, and control malfunctions that do not properly unload the motor during starting. IEEE recommends 2–5 seconds beyond a normal start time at LRA, depending on the relative acceleration durations. However, in the absence of specific hot safe time curves, because of nuisance tripping concerns, this author is reluctant to use a 2-second margin based on past experience with the variability of stopwatch times, system voltages, relay curves, and control system unloading. Care must be taken that the normal start time represents actual operating conditions to prevent nuisance tripping, that is, the motor being loaded or unloaded during starting. Additionally, the design margins for rotors may be dictated more by experience than refined calculations.

3.4.3 Short Circuit Protection (50)

Short circuit coordination with the bus source breaker protection is achieved with the instantaneous overcurrent setting. A setting of 250% of the rated voltage maximum LRA is shown in Figure 3.8. The 250% figure is the maximum IEEE recommendation for providing a margin above asymmetrical LRA values that are generally assumed to be 160% higher, in order to prevent nuisance tripping during starting conditions. Depending on the actual circuit X/R ratio, the asymmetrical current could be as high as 1.73 or less than 1.6. The worst case for asymmetrical currents is when the sine wave is at zero crossing at the time the breaker contacts close, as illustrated in Figure 3.10. This reduces the developing current rate of change and also eliminates the opposing voltage from the prior sine wave polarity, further reducing the impedance and momentarily causing a higher-than-normal current to flow in the phase that is close to zero crossing. Conversely, energizations at the sine wave peak will have the highest rate of current change and the associated impedance from the increase in self-flux.

The same phenomenon happens when short circuits occur when a particular phase sine wave is close to zero crossing. With motors, as with transformers, there may be additional effects from residual magnetism in the core iron laminations, depending on the relative polarities. Because of the high setting, the instantaneous element should actuate only from high-magnitude short circuit currents, not from asymmetrical starting currents or unusual operating conditions. Except in somewhat rare cases, there is no reason to set the instantaneous trip (IT) point lower because more-than-adequate margins between the 250% setting and the available fault current levels are normally provided. If the instantaneous element fails to actuate, the (51) time overcurrent function will time out to trip the breaker, with a longer time delay, which may or may not coordinate with the source bus tie breaker.

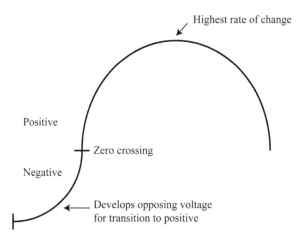

FIGURE 3.10
Asymmetrical currents.

3.4.4 Digital Motor Protection

The newer digital relays have some advantages over the older, induction-disk relays. By National Electric Manufacturers Association (NEMA) standards, motors are designed for two starts in succession, with the rotor coasting to rest between starts when initially at ambient temperature. The coasting to rest requirement is provided to give higher-inertia motors more time to cool because the stored heat in the rotor from a longer acceleration period will be greater; conversely, the coastdown period will also be longer. The motor is usually near the rated operating temperature (after the first start), and only one more start is permitted. Many plants have an operating procedure that requires a 1-hour cooling period before permitting a third start; the motor can be at rest or running during the cooling interval.

Digital relays provide various degrees of thermal modeling of the rotors, the stators, or both. This thermal modeling may prevent motor abuse from too many starts in succession or insufficient cooling periods, and it also may protect larger motors that lack sufficient thermal margins for standard overcurrent protection. However, in the case of induction-disk relays, the number of starts or the cooling period could easily be programmed into the plant's digital control system. The thermal modeling is complex, has long time constants that are difficult to test, and may not accurately represent the motor. Improved thermal modeling can be achieved by bringing stator resistance temperature detector (RTD) inputs into some relay models, but they may not accurately reflect hot spot temperatures and may not be able to respond fast enough to protect motors from severe rotor events (that is, stalled or locked rotor conditions).

As mentioned before, the manufacturer's thermal design withstand margins are usually based more on experience than on refined temperature modeling calculations. This author's preference would be to utilize the overcurrent functions for tripping, as presented earlier in this section, and the thermal functions for alarm-only operation and for preventing third starts until the motor has cooled to the rated operating temperature; and allow tripping in the case of larger motors (normally above 5000 HP) that lack the necessary thermal margins between the acceleration curves and hot safe time curves for adequate overcurrent protection.

3.4.5 Motor Overcurrent Oversights

The settings for roughly 75% of the switchgear-fed motors reviewed during the author's surveys of 35,000 MW of large generation, were found to be questionable, as follows:

- MT thresholds were commonly above 125% of FLA and as high as 250%.
- Timing at LRA likely exceeded the hot safe time curve limits.

- Instantaneous elements were set too low and did not provide sufficient margins to ensure mitigation of nuisance tripping during motor asymmetrical starting current conditions.

3.5 Motor Control Center (MCC) Source Overcurrent (50/51)

MCC source feeders can be somewhat complex and a little tricky to set without increasing the time delay on upstream source breakers for the bus. If any of the station MCCs are located adjacent to the switchgear, the short circuit at the MCC bus would be approximately the same as the main 480-volt bus. This means that the feeder breaker instantaneous element would need to be disabled and a time delay of 0.4 seconds at the source bus three-phase short circuit current (ISC) levels would be required to coordinate with the MCC load positions. An additional time delay of 0.4 seconds then would be required for all upstream bus source breakers for overcurrent coordination reasons, which would significantly increase personnel arc-flash safety concerns, as well as possibilities for greater apparatus damage and fires. The increase in upstream time delay could be mitigated by installing phase inductive reactors at the MCC to limit the short circuit current at the bus and is highly recommended because of personnel safety concerns. With low-voltage systems (480 volts), the fault current drops off fairly fast as the distance and cable impedance from the switchgear to the MCC increases. Assuming three-phase short circuit magnitudes of 30,000 amps for both 4-kV and 480-volt main unit buses, and a 4/0 cable run of 200 feet each, the 4-kV current would be 27,918 (89%) and the 480-volt three-phase short circuit would be 13,010 amps (43%) at the end of the cable.

If the switchgear bus three-phase short circuit current (ISC) is at least twice the magnitude of the short circuit current at the MCC bus, the instantaneous element can be applied. A fault current ratio of 2 or greater (main bus ISC/MCC bus ISC) provides enough margin (no iron involved) to prevent a 200% MCC feeder instantaneous element from operating for asymmetrical faults at the MCC bus. If the ratio is less than 2, the instantaneous element may need to be disabled to mitigate the possibility of nuisance tripping and time delay may need to be added to the switchgear bus source breaker relays to ensure coordination.

Figure 3.11 covers the MCC source breaker overcurrent settings. The suggested MT is 110% of the lowest ampacity rating of the MCC feeder to prevent future loads from affecting the setting, and also because there is no obvious rationale to set the MT thresholds at or below the continuous current ratings. The suggested time delay of 0.4 seconds at the MCC bus maximum three-phase symmetrical short circuit current should coordinate with downstream MCC positions, as they are normally equipped with some form of IT overcurrent protection. In this example, the asymmetrical short circuit value

Auxiliary System Protection

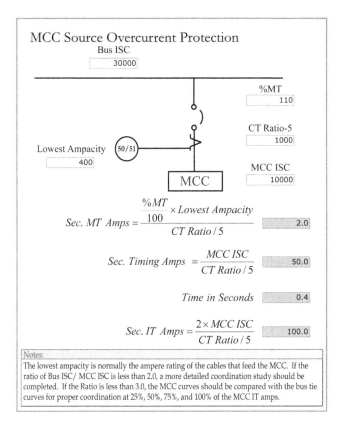

FIGURE 3.11
MCC source feeder overcurrent.

at the MCC bus would be around 1.6 times 10,000 or 16,000 amps (although the 200% setting in Figure 3.11 could be reduced to 175%, or 17,500 amps) and provide adequate margins above the likely asymmetrical magnitudes. In either case, it would not have the sensitivity to detect cable faults for the full distance of the run; and this author tends to err in the direction of preventing nuisance tripping. The main purpose or advantage of the instantaneous setting is to facilitate coordination with source bus tie relay settings. The bus tie breaker (Figure 3.15) will be set to trip in 0.4 seconds at 30,000 amps, with a MT threshold of 2200 amps, in comparison to the MCC source feeder breaker, which has a MT of 440 amps and is timed to trip in 0.4 seconds at 10,000 amps. Consequently, the curves for the MCC feeder breaker should actuate before the bus tie breaker can time out to trip.

If the instantaneous elements cannot be applied, the MCC source feeder breaker relay curves should be compared with the switchgear bus tie breaker at 25%, 50%, 75%, and 100% of the three-phase main bus fault current to ensure coordination at possible current levels. Almost all bus source and tie overcurrent relays use inverse curves. The greater the inverse, the faster the curves will trip at higher short circuit currents. Shown here is a comparison of a very inverse to an extremely inverse curve with identical tap (MT) and time dial setting:

- **Very Inverse Tap 5, TD 4**
  - 1 second @ 25 amps
  - 0.58 seconds @ 50 amps
- **Extremely Inverse Tap 5, TD 4**
  - 1 second @ 25 amps
  - 0.33 seconds @ 50 amps

As you can see, both curves will trip at the same time, of 1.0 seconds at 25 amps or a midlevel current. However, the extremely inverse curve will trip much faster at the higher current level of 50 amps (0.33 versus 0.58 seconds).

Figure 3.12 presents an extremely inverse curve for the source bus tie breaker to illustrate the coordination problem. In analyzing the bus tie relay curve for the applied setting, the multiples of tap would be 75/5.5 (or 13.6) and the time dial would need to be around 7.0 in order to trip in about 0.4 seconds for a bus fault of 30,000 amps. Provided next are the approximate

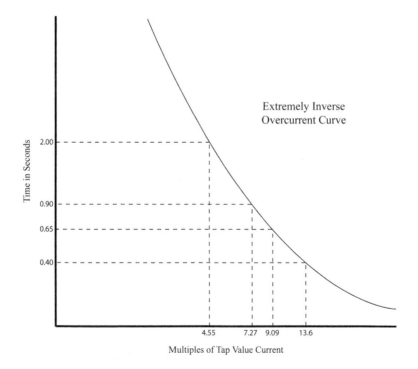

FIGURE 3.12
Source bus tie breaker, extremely inverse curve.

source bus tie overcurrent tripping times for faults on the MCC bus or source feeder cables to the MCC that are not cleared by an instantaneous protection element:

- 10,000 amps = tap multiplier of 4.55 = 2 seconds
- 16,000 amps = tap multiplier of 7.27 = 0.9 seconds
- 20,000 amps = tap multiplier of 9.09 = 0.65 seconds

Obviously, the calibrated 20,000-amp fault (in the cable to the MCC) may not have enough margin to guarantee operation of the instantaneous element if it does not pick up precisely at its MT threshold. Faults closer to the source bus should not be a problem because the IT element would have additional pickup margins to ensure coordination with the bus tie breaker. However, a symmetrical current cable fault of 20,000 amps would operate the bus tie breaker in 0.65 seconds. The MCC source feeder is timing at 0.4 seconds at 10,000 amps and will likely operate faster with 20,000 amps (depending on the actual curve). Although the MCC setting most likely will coordinate, the operating time for the MCC feeder at 20,000 amps should be verified to assess coordination margins.

As you can see in the foregoing examples, protection timing is very important. Accordingly, when testing protective relays, the timing should be adjusted as close as can practically be achieved to the critical timing current presented in the original relay settings specified by a protection engineer, not by testing another point on the curve or extrapolating timing from curves and time dial settings. Unfortunately, not using the critical timing current specified in the relay settings is a relatively common poor practice in the industry. It is also not that uncommon to find that stations do not have the original protection engineering setting documentation, and determine the settings from as-found testing, which is also a nondefendable practice.

3.5.1 MCC Source Feeder Protection Oversights

MCC feeder overcurrent relays can be difficult to set properly, and about 75% of the as-found settings during the author's reviews were not coordinated properly, did not adequately protect the equipment from overload conditions, or both. If the instantaneous elements cannot be applied because the MCC is located close to the source bus, personnel arc-flash hazards and potential damage to electrical apparatus will increase significantly.

3.6 Bus Tie Overcurrent

The next upstream device, depending on the bus configuration, would normally be the bus tie breaker. This breaker is used to feed the bus from a second source (reserve or common bus) if the normal source is not available (because of a unit trip or it is in startup or is out for maintenance). Only feeder breakers or the most downstream device should be equipped with instantaneous overcurrent elements. Source and tie breakers need to coordinate with downstream devices; consequently, instantaneous overcurrent elements cannot be used for those positions. Bus tie overcurrent relays can be a little tricky to set because the tie breaker needs to handle the starting current for all connected motors on the bus during automatic bus transfer operations, and also bus parallel currents when the voltage or power angles between the two buses are different. Depending on the transformer impedances, 10° voltage phase angle differences between the UAT and RAT can cause more than rated current to flow across the tie breaker. The reserve or common bus can also be fed through the unit bus, which makes it downstream of the UAT low-side breaker protection.

3.6.1 Roughly Estimating Bus Transfer Motor Currents

As mentioned previously, many motors stay approximately at locked rotor current levels until reaching roughly 90% speed. Conversely, motors that are deenergized can be in starting current very quickly when reenergized during bus transfers. Figure 3.13 illustrates a procedure for roughly estimating combined starting current magnitudes from motor loads during bus transfer conditions. For illustration purposes, four identical motors are shown in the procedure. The combined locked rotor current is roughly estimated at 85% because the bus voltage will be somewhat depressed from the high motor starting load. The start time is also roughly estimated at 85% because the motors are still rotating and not stopped when the transfer occurs. As illustrated in Figure 3.13, the combined current stays relatively high, at 2720 amps for 4.0 seconds, before it reduces to a full-load ampere level. The higher current level and time can then be compared to the bus tie relay curves to see if there is a possibility of nuisance tripping. Obviously, having a digital record of actual currents during bus transfer conditions would be preferred over estimating the current levels.

3.6.2 Delta Bus Transfer Currents

Figure 3.14 calculates the amount of current that will flow when two systems are operated in parallel with a delta voltage angle of 10°. A similar calculation was presented in Chapter 2, with some minor differences. For convenience, the circuit X_L has been replaced with short circuit values for each bus and then summed; the total circuit X_L is determined with a three-phase version of Ohm's Law. The procedure also uses standard power formulas to determine megavolt-amps and the magnitude of current flows between the two systems. In the case of a generating station, unit output power can get to the transmission or subtransmission systems through the generator step-up

Auxiliary System Protection

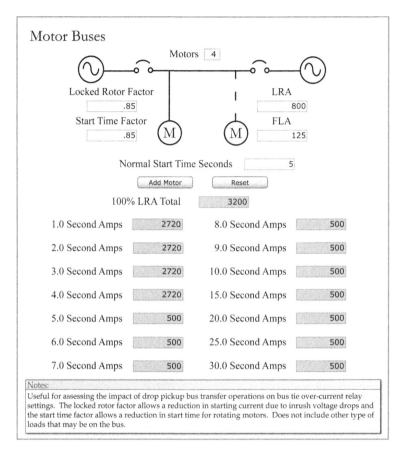

FIGURE 3.13
Motor loads during transfer.

transformer (GSUT), and also through the UAT and RAT, which are connected in series during parallel operations.

3.6.3 Bus Tie Overcurrent (51)

Figure 3.15 covers a procedure for calculating settings for bus tie overcurrent relays. Medium-voltage (4-kV) load feeder breakers are usually rated for 1200 amps continuous, and source and tie breakers at 2000 or 3000 amps. In this example, the bus tie breaker is rated for 2000 amps, and the suggested value for the secondary MT point is 110% of the lowest ampacity, or 2200 amps. The lowest ampacity could be the transformer, switchgear bus, circuit breaker, or cables/busway between the transformer and low-side breaker or beyond the bus tie breaker. The timing is set for 0.4 seconds at the maximum three-phase symmetrical short circuit current. This setting

124   *Electrical Calculations and Guidelines*

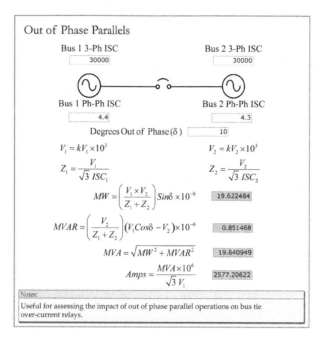

FIGURE 3.14
Bus parallel currents.

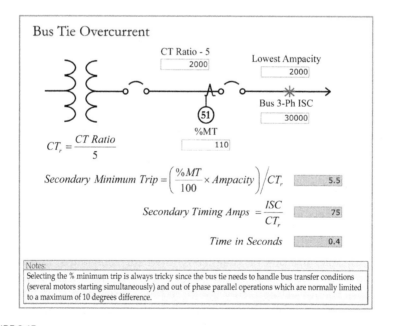

FIGURE 3.15
Bus tie overcurrent settings.

assumes that the timing does not need to be increased to accommodate a feeder load that does not utilize or have lower set instantaneous overcurrent elements.

The 2200-amp MT value is below the bus transfer motor currents of 2720 amps in Figure 3.13, and it is also below the bus parallel currents of 2577 amps calculated in Figure 3.14. The motor bus transfer current decrement can be compared to the relay curves to see if it will be a problem. In this case, the delta angle parallel currents will be a problem unless automatic circuitry or operator action limits the duration of the parallel. In general, parallel operation should be limited to a few seconds, and bus parallels with delta angles greater than 10° should not be permitted. A safe angle range can be calculated using the procedure presented in Figure 3.14 and supervised administratively with operator synchroscope angle estimates or by sync check relays. If the delta angles are too large, the unit output megawatts will need to be reduced to bring the angles into a more favorable range. The delta angles are normally greater for designs that generate into bulk power (230 kV), but utilize subtransmission (69 kV) to feed the RAT. During temporary parallel operation of the unit with the reserve 4-kV buses, the combined short circuit currents will usually exceed the current interruption capability of feeder breakers. Most plants assume a risk that a severe fault will not occur during the short duration of the parallels. If it did occur, the source and tie relays would time out to clear the fault, but load feeder breakers could be severely damaged from the combined short circuit currents.

3.6.4 Bus Tie Protection Oversights

Approximately 40% of the surveyed bus tie breaker overcurrent relays were not properly coordinated with upstream and downstream protective devices. Additionally, the settings may not have considered bus parallels where high currents can flow due to power angle differences and bus transfers where several motors are in the starting current simultaneously.

3.7 Transformer Secondary Side Overcurrent (51)

Setting transformer secondary side overcurrent relays is relatively easy. As Figure 3.16 illustrates, the suggested MT value is 125% of the lowest ampacity, or 15% higher than the bus tie breaker. The timing is set for 0.8 seconds or 0.4 seconds longer than the bus tie overcurrent relays at the same maximum three-phase symmetrical short circuit current. If the relay curves are the same and have identical inverse characteristics, comparing the curves to ensure coordination at all points will not be necessary.

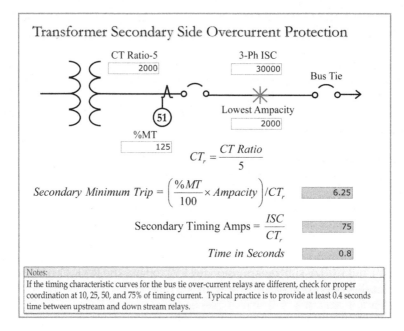

FIGURE 3.16
Transformer secondary-side overcurrent settings.

3.7.1 Transformer Secondary Side Protection Oversights

Roughly 40% of the as-found settings, during this author's reviews of approximately 35,000 MW of generation, were not coordinated properly with downstream protective relays.

3.8 Transformer Primary Side Overcurrent (50/51)

In addition to protecting the transformer, the primary side overcurrent (51) relays provide backup protection for transformer differential relays, if equipped, and breaker failure or stuck breaker protection for the secondary side breaker if it fails to properly isolate or clear bus short circuit conditions. In the case of fossil and nuclear generation auxiliary transformers, other than the primary or high-side overcurrent, there is usually nothing in the upstream generator protection package that would have the sensitivity to detect short circuits that are downstream of the transformer differential zone of protection (that is, medium-voltage bus faults). Because the fault could be in the transformer or upstream of the secondary circuit breaker, fault isolation requires complete transformer deenergization. Setting transformer

Auxiliary System Protection

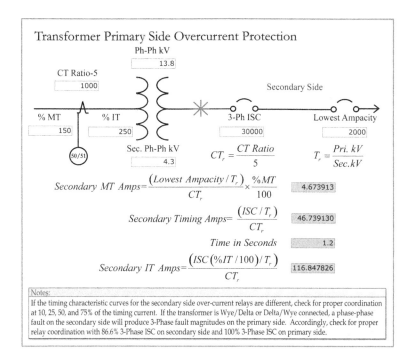

FIGURE 3.17
Transformer primary-side overcurrent settings.

primary-side overcurrent relays is also relatively easy. Figure 3.17 is applicable for RATs, UATs, and SSTs. The suggested MT value is 150% of the lowest ampacity or the oil and air (OA) or self-cooled rating, or 25% higher than the secondary side overcurrent relay MT thresholds for two winding transformers. With three winding transformers, the combined OA MVA rating of all three windings would normally be used on the primary side.

Figure 3.18 presents a calculation to determine the overload capability of power transformers built to IEEE standards. The calculation does not provide accurate results if the overload currents are close to FLA and assumes an I^2T number of 1250, with the lower transformer MVA or OA rating used to determine 1.0 per-unit FLA. The results indicate that a power transformer can withstand 150% load for $1250/(1.5^2$, or 556 seconds (9.26 minutes).

Back to Figure 3.17, the timing is set for 1.2 seconds, or 0.4 seconds longer than the secondary-side overcurrent relays, at the maximum three-phase symmetrical secondary bus short circuit current divided by the transformer turns ratio. Generally, due to electromechanical magnetic force limitations, transformers are built to withstand external close-in short circuit conditions for 2 seconds. Accordingly, the primary overcurrent relays should deenergize the transformer in less than 2 seconds at maximum secondary-side symmetrical fault current levels. If the relay curves have the same inverse characteristics as

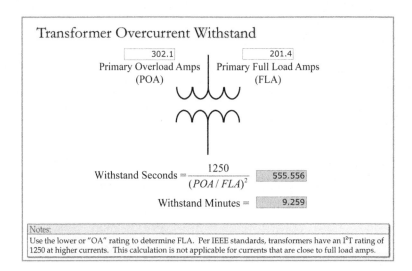

FIGURE 3.18
Transformer overcurrent withstand.

the secondary-side overcurrent relays, comparing the curves to ensure coordination at all points will not be necessary. However, if the transformer configuration is delta–wye or wye–delta, a phase–phase fault on the secondary side will cause increased currents to flow in one phase on the primary side, as explained in the preceding chapter. If this is the case, there must be an adequate timing margin between the primary and secondary relay curves, with 86.6% of the maximum three-phase fault current on the secondary side and 100% of the maximum three-phase secondary fault current on the primary side, and the time delay may need to be increased to 1.3 seconds or more.

Instantaneous overcurrent (50) elements can be applied for transformer high-side protection. As shown in Figure 3.17; the suggested setting is 250% of the maximum three-phase symmetrical short circuit current for a secondary-side fault. Although not absolutely guaranteed, experience has shown that a 250% setting will not cause nuisance tripping during transformer magnetizing inrush conditions and is well above the possible secondary asymmetrical short circuit current magnitudes. The instantaneous element will not see short circuit events on the secondary side, which is desirable for coordination reasons.

3.8.1 Transformer Primary Side Protection Oversights

The medium-voltage reserve and unit auxiliary transformers in the surveyed units were all protected by differential relays, but about 20% of the units were not equipped with primary overcurrent protection. Primary overcurrent elements are essential because they also provide breaker failure protection. If the low-side breaker fails to interrupt a bus fault that is outside the differential zone, only the primary overcurrent relays would normally have the sensitivity to detect and isolate the short circuit condition. If equipped, the phase overcurrent elements for about 60% of the surveyed units had MT thresholds that were well above 150% of FLA; essentially, overload protection was not provided.

3.9 Residual Ground Protection (51G)

Many modern generating stations and large industrial facilities connect resistors from unit and reserve auxiliary transformer wye-neutral bushings to ground to limit the magnitude of available medium-voltage ground fault current. This has a number of advantages, including a reduction in hazards to employees and the mitigation of motor iron damage. Other advantages will be addressed later in this chapter, in Section 3.10. The amount of ground fault current reduction is limited by the transformer secondary bus breaker CT ratios and the sensitivity or range of the protective relays applied for that purpose. A typical limited ground fault current value for a large generating station auxiliary system is around 1000 amps. Because the 100% ground fault current is in the load range, the magnitude can be determined by simply dividing the phase-neutral voltage by the neutral resistance. The residual ground relays are connected between the CT neutral point and the star point for the phase relays, as shown in Figure 3.2. Only ground current and not phase–phase or three-phase current will be detected by these relays.

Figure 3.19 illustrates a complete residual ground scheme, with the exception that only one feeder breaker is shown. Relay settings for feeders with different CT ratios will need to be calculated separately. Instantaneous elements should not be applied because of concern over inadvertent tripping from CT saturation caused by asymmetrical currents during feeder motor starting and from transformer magnetizing inrush conditions. Error currents from saturated CTs will likely affect only one phase and, consequently, will look like ground fault current to the residual ground relay. A time delay of 0.3 seconds on downstream feeder breaker ground protection is suggested to allow asymmetrical currents to decay to more reasonable levels. Because the relays do not see load current, the MT values can be set low.

In this case, 2.5% of the maximum ground fault current has been selected for the feeder breaker MT with a time delay of 0.3 seconds at the 100% ground fault current. The next upstream device, the bus tie position, has a suggested MT of 5% with a timing of 0.7 seconds at the maximum ground current. The source breaker has a suggested MT value of 7.5% and a time delay of 1.1 seconds at maximum ground current. The final or most upstream ground relay is connected to a CT on the source transformer neutral and has a suggested MT of 10% and a time delay of 1.5 seconds at the maximum ground current

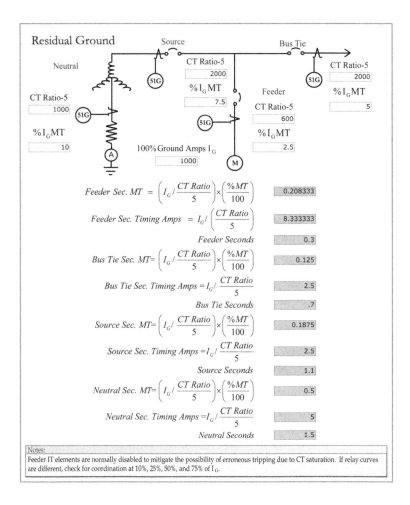

FIGURE 3.19
Residual ground overcurrent settings.

and requires complete deenergization of the transformer. The neutral relay is necessary to detect ground faults upstream of the transformer secondary side breaker. Again, the suggested settings depend on the range or sensitivity of the associated relays, and an iterative approach may be required before the final settings can be determined. The transformer differential relays most likely do not have the sensitivity to detect ground faults on the secondary side.

3.9.1 Residual Ground Protection Oversights

About 75% of the limited residual ground schemes that the author has reviewed had protection oversights: the instantaneous elements should be

Auxiliary System Protection 131

disabled to mitigate nuisance tripping possibilities, as discussed previously. It is also not unusual that these schemes are not coordinated properly.

3.10 High-Impedance Grounding

High-impedance (resistance) ground schemes are typically applied in generating stations, refineries, combustible areas, and industrial facilities where phase-neutral voltages are not utilized to feed loads. Three types of ground detection schemes are normally employed: residual ground (already discussed), neutral grounding transformers for wye-fed systems, and grounded wye–broken delta grounding transformers for delta-fed systems. Residual ground schemes are commonly used for medium-voltage applications (4 kV and higher), neutral high-impedance grounding is applied for almost all medium-size to large generators to detect stator ground faults, and also on older, medium-voltage auxiliary systems; and wye-grounded–broken delta schemes are used primarily on low-voltage (480-volt) delta systems, and to detect grounds between generator breakers and their associated step-up transformers.

This author is an advocate of high-impedance grounding because of the significant number of advantages that it provides over solidly grounded systems, especially in the area of personal safety. High-impedance grounding provides the following benefits over solidly grounded systems:

- Significantly reduces personnel electrical hazards
- Mitigates the likelihood of fires and explosions
- Reduces exposure to induced voltages
- Reduces exposure to step and touch potentials
- Prevents the flow of high-magnitude ground fault currents from the accidental or inadvertent dropping of tools or material on energized phase conductors
- Reduces unplanned electrical shutdowns
- Mitigates motor iron damage
- Reduces short circuit duty near the source
- Is inexpensive to install or retrofit

As discussed in Chapter 2, unlimited ground fault currents are higher than three-phase short circuit amperes near the source transformer. Limiting the ground fault magnitude can lower the short circuit duty at the secondary source breaker, prevent motor iron damage, and reduce the likelihood of fires and explosions. Accordingly, hazards to plant operators, electricians, and technicians who work around energized medium-(4 kV) and low-voltage (480 volt) systems are significantly reduced with high impedance grounding. 480-volt systems are particularly problematic because the bus short circuit levels are often higher than the medium-voltage buses, and conductors with exposed terminals permeate the entire plant in MCCs, transformer cooling system cabinets, welding outlets, switchboard panels, and numerous other termination enclosures. In addition to burns and explosive forces, arc-flash energy can scorch lung tissue, causing death by suffocation.

With the exception of generator protection, the ground detector schemes only alarm, and consequently, they do not result in the immediate shutdown of plant equipment or the facility. However, the voltage developed on the unfaulted phases to ground increase by a factor of 1.732 (the square root of 3) for a 100% ground, which normally exceeds the continuous voltage rating of the cables and insulation systems associated with the particular auxiliary power system. Cable insulation ratings are often phase–phase and may not denote the phase-ground capability; that is, a 600-volt three-conductor cable may be continuously rated for only 600/1.732 (or 346 volts) to ground, and 5-kV cable is usually rated at 2.9 kV to ground. Accordingly, facilities applying alarm-only, high-impedance grounding schemes should have procedures in place to quickly isolate grounded feeders or buses before they develop into damaging high-current double line to ground, phase–phase, and three-phase short circuits from the increase in voltage on the unfaulted phases. Experience has shown that the temporary increase in phase-ground voltage is not a problem for healthy cables and motors. However, there are reports of surge capacitor failures from the increase in phase-ground voltages. The surge capacitors represent a capacitive load, and the increase in phase to ground voltage on the unfaulted phases thermally overloads the wye-grounded capacitors from the higher current flow.

3.10.1 Induced Voltages

Another advantage of high-impedance grounding is the limiting of hazardous induced voltages into structures and raceway conductors by unlimited ground faults. Figure 3.20 presents an induced voltage calculation that discloses that an unlimited ground or double line to ground fault in a cable that parallels an adjacent deenergized circuit in a duct bank or cable tray for as little as 300 feet can induce over 1 kV into the deenergized circuit from the magnetic flux linkage. As the figure shows, the amount of induced voltage is affected by the current magnitude, the separation distance between the circuits, and the location of the ground current return path. Increasing the current, parallel distance, return path separation, and reducing the distance between circuits will all independently increase the magnitude of the induced voltage.

Figure 3.21 illustrates an electrician disconnecting the cables on a deenergized motor. Prior to the Occupational Safety and Health Act (OSHA) of

Auxiliary System Protection

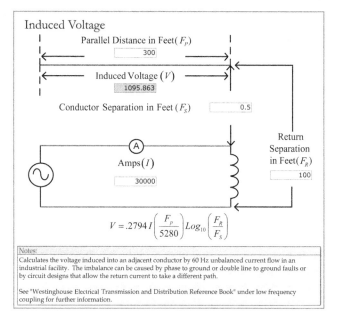

FIGURE 3.20
Induced voltages.

1970, it was standard practice to rack out switchgear circuit breakers as part of an electrical clearance or tag-out procedure, and then disconnect motor splices or terminations for maintenance purposes with the bare hands. Following OSHA, many plants began purchasing switchgear grounding breakers that could be racked into cubicles to effectively ground the motor and feeder cables at the switchgear end. This procedure, which is still commonplace in the industry, has a number of disadvantages. First, breakers are individually adjusted or custom fitted to mate with their specific rack-in-out

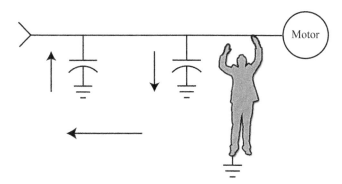

FIGURE 3.21
Induced voltage hazard.

mechanisms and associated primary disconnects, and racking general-purpose (noncustom-fitted) grounding breakers may damage or wear out electrical current carrying disconnects and racking mechanisms.

Second, the grounding breaker makes it more dangerous for workers because of its remote location. Safety grounds are applied to protect workers from induced voltages and inadvertent energizations. In the case of induced voltages, the grounds need to be adjacent or close to the workers, and grounds at the remote switchgear end increase the induced voltage hazard. For modeling purposes, one-half of the distributed cable capacitance is shown at each end of the motor cables in Figure 3.21. A 1-kV induced voltage in the motor cables will push current through both the worker and distributed capacitance at the motor location, and then through the distributed capacitance at the switchgear end to complete the circuit. Grounding the switchgear end bypasses the current limiting capacitive reactance at that location, causing additional current to flow through the worker. Obviously, in the case of an inadvertent energization, the grounding breaker cannot be in place if it occurs. The safer alternative would be to rack out the breaker, treat the circuit as energized, and then disconnect the motor while wearing protective safety gloves that are rated for the particular voltage. Personal grounds can then be applied at the motor end of the cables once they are disconnected from the motor.

Although high-impedance grounding eliminates harmful induced voltages from ground faults, it does nothing to eliminate the possibility of higher induced voltage magnitudes from a double line to ground faults. If the ground is not isolated before it develops into a double line to ground faults, the induced voltages could be much more serious or hazardous. For example, A-phase could fail to ground initially, and then C-phase from the higher voltage, on a different feeder on the other side of the facility. This would significantly increase the return path separation and the calculated induced voltage magnitude.

3.10.2 Transient Voltage Mitigation

Figure 3.22 illustrates the before-and-after conditions of grounding a floating (nongrounded), three-phase power source. When a ground occurs, the voltage on the unfaulted phases to ground increase by the square root of 3 (i.e., the phase-phase value). If the ground is spitting or intermittently arcing, the insulation system capacitance can cause voltage doubling as the configuration repeatedly and rapidly goes from the ungrounded to the grounded mode. It is generally thought that in industrial facilities, the transient voltages developed during arcing ground conditions can get as high as 6 per unit and are typically in the 1–20-kilohertz (kHz) range. This level of transient voltage can fail the insulation on unfaulted phases, causing high-damaging double-line to ground short circuit currents to flow.

When one phase of a floating system is grounded, it unbalances the insulation capacitance and a capacitive or charging current flow can be measured

Auxiliary System Protection

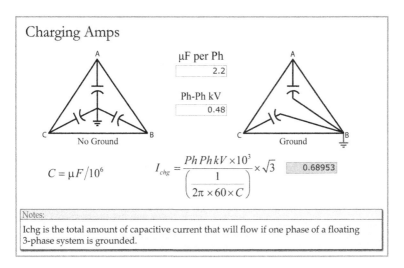

FIGURE 3.22
Capacitive charging amps.

to ground. As can be seen in Figure 3.22, the ground shorts out the capacitance for one phase, and the remaining two will allow ground current to flow. The current for each can be determined by dividing the phase–phase voltage by the associated X_C and then multiplying by the square root of 3 to determine the total capacitive current flow, since they are 60° apart. The general rule of thumb is to have the ground detector scheme provide a resistive current that equals the capacitive or charging amps that flow when one phase is 100% grounded. This will reduce the maximum voltage transient from 6.0 to around 2.4 per unit volts, and is equivalent to installing resistors in parallel with the insulation capacitance to keep the charge levels and resulting voltage increasing levels lower.

Although any amount of resistive current provided by the ground detector will reduce the transient voltage from maximum, resistive currents above the system charging ampere level will not appreciably reduce the transient magnitude further. Normally, the weak link for 60-hertz events is the surface area of insulators and bushings. However, higher frequency transients lower the impedance of the air gap capacitive reactance, and double line to ground faults that breach the air gap may have been initiated by arcing spitting grounds. Before the charging amps can be calculated, a microfarad capacitance value per phase needs to be determined. The amount of capacitance in the total system can be approximated from tables in various publications, from manufacturer data, or by measuring the three-phase capacitance to ground with a capacitance meter during an outage with all associated loads, transformer windings, and buses included in the measurement. Low-voltage capacitance meters will generally yield the same capacitance values

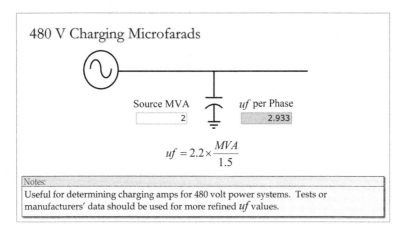

FIGURE 3.23
Typical 480 volt system capacitive microfarads.

as the higher-voltage insulation PF test sets. The result can be divided by 3 to roughly approximate the per-phase capacitance.

A per-phase microfarad level of 2.2 and a charging current of 0.7 amps are typical for a 1500-kVA, 480-volt, three-phase transformer with connected loads, as indicated in Figures 3.22 and 3.23. This was determined by performing an applied ground test on a 480-volt floating (ungrounded) system and measuring the capacitive current flow. Figure 3.23 roughly extrapolates the charging microfarads per phase for a 2000-kVA, 480-volt, three-phase transformer with connected loads at 2.9 microfarads per phase; the charging current would be around 0.9 amps.

A 4-kV, three-phase 10 MVA transformer with connected loads will generally have around 0.86 microfarads of capacitance per phase or 2.4 amps of charging current, as measured by a low-voltage capacitance meter. The 4-kV system has a lower capacitance than the 480-volt system, even though its MVA is over six times greater. The increased thickness of the 4-kV insulation reduces the microfarad value. Figure 3.24 roughly extrapolates the per-phase capacitance for a 15 MVA 4-kV system as 1.29 microfarads; the charging current would be approximately 3.6 amps.

The microfarads for both two- and four-pole HP and LP 20-kV, 280 MVA cross-compound generator systems were each measured at approximately 0.25 microfarads per phase with a 10-kV PF insulation test set. Again, the microfarad value is lower because of the increased thickness of the insulation system, even though it is approximately 187 times higher in MVA than the 480-volt system. Figure 3.25 extrapolates the per-phase capacitance of a 400-MVA generator system at about 0.357 microfarads. The charging current would be about 3.3 amps for the 280 MVA generators (including the isolated phase bus and connected transformer primary windings) and 4.7 amps for the 400-MVA generator. The estimates are coarser for bulk power generators

Auxiliary System Protection

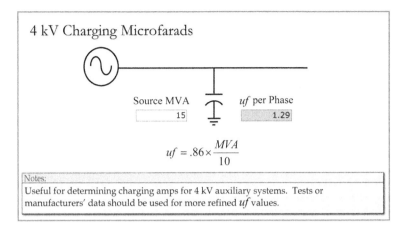

FIGURE 3.24
Typical 4-kV system capacitive microfarads.

because their rated voltages typically range from 12 to 26 kV. However, the insulation thickness differences are somewhat offset because the charging current magnitudes increase as the voltage goes up.

As discussed earlier, ideally, the ground detector would be sized to provide a resistive current that equals the capacitive current that flows when one phase is 100% grounded. Accordingly, the total ground current would be the vector sum of the capacitive insulation system current and the resistive current from the ground detection scheme (90° apart), or one of the currents multiplied by 1.414 if they are equal to each other.

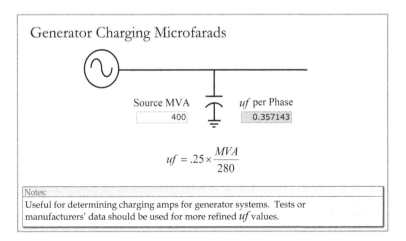

FIGURE 3.25
Typical generator system capacitive microfarads.

3.10.3 Primary to Secondary Capacitive Coupling

As mentioned before, stray capacitance is everywhere. Figure 3.26 illustrates the stray capacitance between the primary and secondary windings of a power transformer and the stray capacitance between the secondary-side insulation system and ground. The amount of inner-winding capacitance can be reduced by including a conductive shield between the primary and secondary windings (grounded at one location only), which is often provided for GSUTs. In the case of delta-connected secondary windings that are floating or ungrounded, if only one phase is energized, a voltage divider circuit is formed between the stray internal transformer capacitance and the external insulation system secondary capacitance to ground. If the external stray capacitance is lower, resulting in a higher impedance, because the secondary-side circuit breaker is open, high voltages from the primary side can couple to the secondary side through this capacitance.

There are cases where potential transformers installed on the secondary side failed from the high-capacitive coupled phase-neutral voltage during single-phase switching. A properly designed ground detector will appear as a resistance in parallel with the secondary stray capacitance, reducing the secondary-side impedance and forcing most of the primary-side phase-neutral voltage to be dropped across the transformer internal capacitance. This phenomenon tends to be more associated with single-phase switching, circuit breaker pole disagreement, or blown fuses. When all three phases are energized, the capacitive current flows primarily between phases and high-magnitude, capacitive-coupled 60-hertz voltages do not tend to occur.

3.10.4 Neutral Grounding (59G)

In Figure 3.27, the procedure for sizing the various components for a transformer neutral, high-impedance, alarm-only ground detection scheme is illustrated. First, the capacitive-charging amps are multiplied by the grounding transformer ratio to determine the amount of resistive current needed

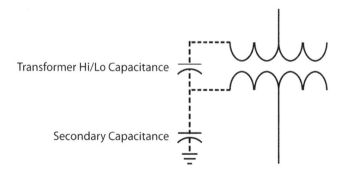

FIGURE 3.26
Primary to secondary capacitive coupling.

Auxiliary System Protection

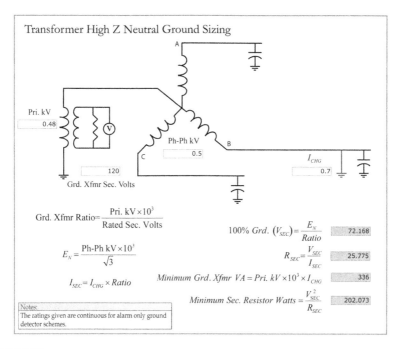

FIGURE 3.27
Neutral grounding transformer sizing.

on the secondary side. Next, the phase-neutral voltage is derived and then divided by the grounding transformer ratio to determine the secondary voltage for a 100% ground condition. The resistor ohmic value can now be obtained by dividing the 100% ground secondary voltage by the desired current. The minimum resistor watt rating can be calculated by dividing the 100% ground secondary voltage squared by the resistor's ohmic value (E^2/R). And finally, the minimum volt-amp (VA) of the grounding transformer can be obtained by multiplying the primary charging amps (actually the resistor current) by the primary voltage rating of the transformer. The primary voltage rating of the grounding transformer should be equal to or greater than the system phase-neutral voltage. A secondary voltmeter should be installed for monitoring purposes and to ascertain ground severity, and it will read zero if no ground is present. A voltage relay should also be installed in parallel with the secondary resistor to alarm for ground conditions. The normal practice is to set the minimum pickup point at 10% of the 100% phase-neutral voltage on the secondary side and alarm in 1 second. In general, resistors run hot, and high-temperature insulation wiring may be required for the resistor terminations.

Figure 3.28 shows the same procedure for sizing the grounding components for a large generator, except that the components do not need to carry the current continuously because the generator stator ground-tripping relay

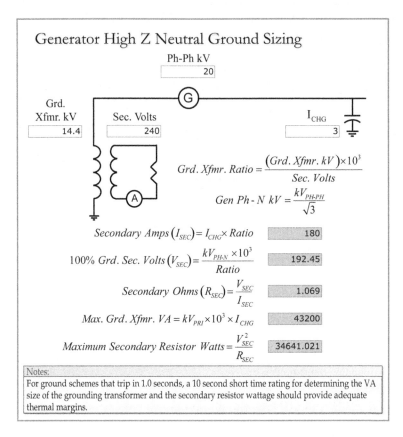

FIGURE 3.28
Generator neutral grounding transformer sizing.

will deenergize the system in a short time. At minimum, a thermal margin of 10 should be applied. That is, the apparatus should be capable of carrying 100% ground fault current for a period that is 10 times longer in duration than the 100% ground time delay of the generator stator ground-tripping relay.

An I^2T calculation can be used to determine the short time current-carrying capabilities of the grounding scheme components (resistor and grounding transformer), as shown in Figure 3.29. Assuming that the manufacturer specified a short time thermal rating of 100 amps for 40 seconds, a new rating could be calculated for the actual ground fault current level of 180 amps. First, square the 100 amps (10,000) and then multiply it by 40 to derive an I^2T value of 400,000. Then, apply the expression shown in Figure 3.29 to convert the 100-amp short time thermal rating to an equivalent 180-amp short time rating. In this case, the apparatus should be able to carry 180 amps for 12 seconds without loss of life.

Figure 3.30 covers a calculation procedure for determining the ground detector primary current, fault ohms, and percent ground for different

Auxiliary System Protection

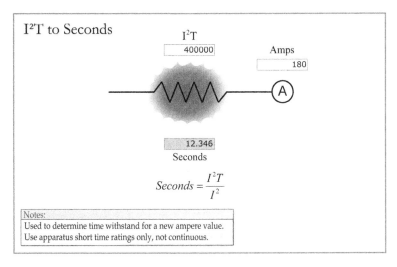

FIGURE 3.29
I²T calculation.

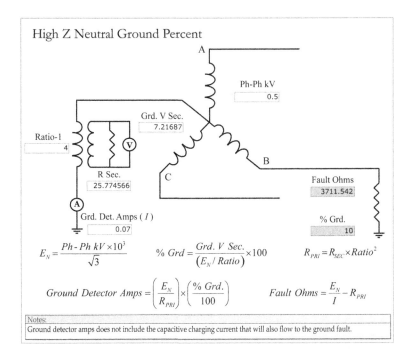

FIGURE 3.30
Neutral ground detector fault ohms.

values of secondary voltages in order to assess the severity of the fault. The percent ground is determined by dividing the actual secondary voltage by the 100% ground secondary voltage and multiplying by 100. The primary 100% ground detector current will be reduced by the same percentage, and then the number of ground fault ohms necessary to limit the ground current to that value can be determined. The fault ohms can be calculated by subtracting the reflected secondary to primary ohms from the total ohms required to limit the ground detector primary current to the determined value; the remaining amount would represent the fault ohms.

3.10.5 Grounded Wye–Broken Delta Grounding (59G)

Because delta transformer windings are not equipped with neutral points, a different approach is needed for ground detection. The most common method is to install a grounded wye–broken delta ground detector scheme to derive a neutral point, as shown in the lower ground detector transformers in Figure 3.31. Without the presence of a system ground, voltage will not be developed on the secondary side. If you trace the direction of phase–phase current flow in the primary windings during normal operation, it can be seen that the two developed secondary voltages oppose each other and cancel each other out. When a 100% system ground fault occurs, it is equivalent to shorting out one of three transformers. This, in effect, eliminates the opposing voltage for two of the three phase–phase combinations and allows a voltage to develop on the secondary side, as shown in the phasors of Figure 3.32. Because of the connections, the two secondary phase–phase voltages

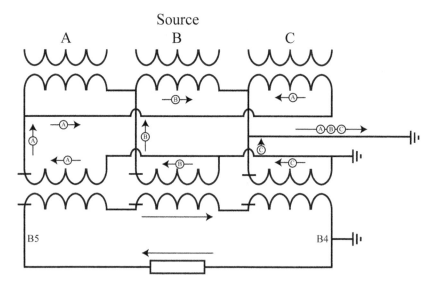

FIGURE 3.31
Grounded wye–broken delta ground detector current paths.

Auxiliary System Protection

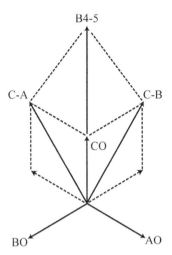

FIGURE 3.32
Grounded wye–broken delta ground detector phasors.

end up being 60° apart and, consequently, will develop a total secondary voltage that is greater by the square root of 3, or 1.732 times the magnitude of one of the phase-phase voltages. As you can see in the figure, the developed secondary voltage ends up being in phase with the faulted phase-neutral voltage on the primary side. A voltage relay should be installed in parallel with the secondary resistor to alarm for ground conditions. A voltmeter should also be provided to determine the severity of the ground. The normal practice is to set the alarm relay minimum pickup point at 10% of the 100% system ground secondary voltage, with a time delay of 1 second.

Because the three transformers are connected in series on the secondary side, current will flow equally through all three secondary windings as presented in Figure 3.31. Therefore, each winding or phase will contribute equal primary current to the ground fault location. Accordingly, in the calculation in Figure 3.33, the desired total primary charging amps are divided by 3 and then multiplied by the winding ratio to determine the desired secondary current. Then, the maximum ohmic value for the secondary resistor can be determined by dividing the 100% ground fault secondary voltage by the desired secondary current. Again, the 100% ground fault voltage is the phase-to-phase system voltage divided by the transformer ratio multiplied by 1.732, or 216.5 volts, as indicated in the figure. The minimum VA for each transformer can be determined by dividing the primary charging current by 3 and multiplying the result by the rated primary voltage of each transformer. Squaring the 100% ground secondary voltage and dividing it by the secondary resistor's ohmic value (E^2/R) will develop the minimum wattage for the resistor.

Figure 3.34 provides a calculation procedure for determining the amount of ground detector current, fault ohms, and percent ground for different

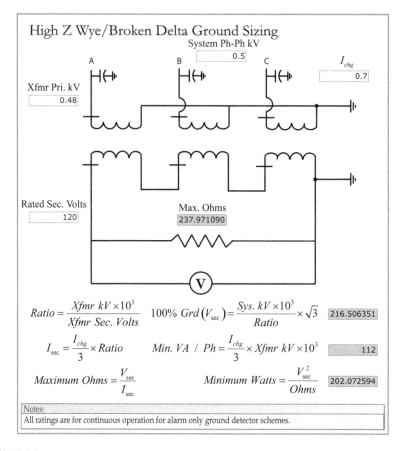

FIGURE 3.33
Grounded wye–broken delta ground detector sizing.

values of secondary voltages to assess the severity of the fault. If you compare the 10% ground fault ohms to Figure 3.30 for the neutral grounding scheme, you will find that they are the same. The percent ground is determined by dividing the actual secondary voltage by the 100% ground secondary voltage and multiplying by 100. The primary ground detector current will be reduced by the same percentage. The amount of fault ohms necessary to limit the ground to that amount of current can then be determined by dividing the phase-neutral voltage by the ground detector current and then subtracting the reflected ground bank primary ohms. In this case, the single transformer-reflected secondary to primary ohms needs to be divided by 9 to account for the impact of all three windings.

Wye/delta transformer windings should not be energized with the delta open or broken. A transformer with a broken or open delta can become electrically unstable (depending on circuit impedance parameters), causing very high currents to flow. This condition is known as *neutral instability,* or

Auxiliary System Protection

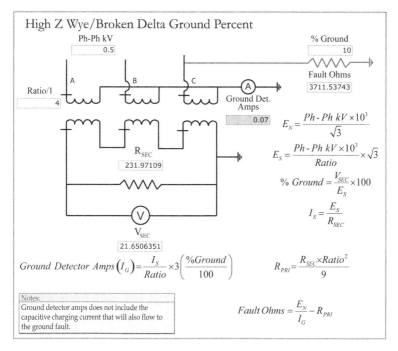

FIGURE 3.34
Grounded wye–broken delta ground detector fault ohms.

three-phase ferroresonance. The individual transformers can go in and out of saturation in a three-phase fashion, causing very high current flows into the bank (roughly 600% of FLA). Installing the secondary resistor helps to close the delta or stabilize the connection. With the secondary resistor, instability is usually not a problem, but as a precaution, primary fuses should always be installed with this type of connection. Another problem is that blown fuses can cause voltages to develop on the secondary side, which could actuate the voltage relay. For this reason, tripping with this type of scheme is generally not recommended. Also, blown fuses can be hard to detect because the winding configuration also tries to reproduce the missing voltage; its ability to do so depends on the impedance parameters of the circuit. The most effective way to detect blown primary fuses quickly is to install fuses equipped with plungers that will actuate microswitch alarms when they blow. Fuses with the foregoing features are commercially available for 480-volt applications.

However, if fuse blowing turns out to be a problem, indicating possible instability, a resistor can be installed in neutral to prevent the transformers from saturating. Figure 3.35 shows a procedure for calculating the impact of installing a neutral resistor. In this example, a 50-ohm resistor is connected between the ground detector neutral point and ground, and the secondary resistor is reduced from 232 to 200 ohms to counter the effect of the neutral resistor and still maintain approximately 0.7 amps of resistive current

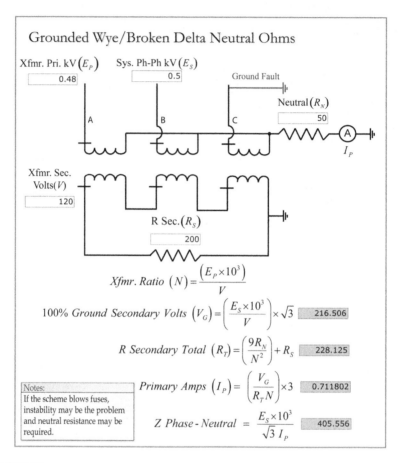

FIGURE 3.35
Grounded wye–broken delta neutral ohms.

in the neutral for a 100% ground fault. The selected neutral ohms need to be multiplied by 9 to account for all three transformers and then reflected to the secondary side. Then, the adjusted ohms can be added to the secondary resistor ohms for a total ohmic value. The total primary amps can now be determined by dividing the 100% ground secondary voltage by the total secondary resistance (including the reflected neutral ohms) multiplied by the transformer turns ratio, and then multiplying the result by 3 to account for the contribution from each phase.

3.10.6 High-Impedance Ground Detection Oversights

For about 20% of the author's surveyed units, the low-voltage 480-V systems were found solidly grounded. As mentioned before, with solidly grounded systems, the low-voltage bus short circuit magnitudes are usually higher than three-phase faults and normally also higher than three-phase, medium-voltage bus short circuits. A typical value is 30,000 amps, and some were found to be as high as 48,000 amps. Since the 480-volt systems permeate the entire plant in welding outlets, transformer cooling system cabinets, and MCCs, with numerous exposed terminal blocks in many other scattered enclosures and cabinets, there is a serious concern about employee safety. Faults normally start as an arcing failure to ground from moisture or other contamination and the arc by-products and ionization from the high current levels can quickly breach the air-gap insulation between phases, causing a three-phase short circuit. The resulting arc flash blast can scorch the lung tissues of people in the area, causing death by suffocation and other major injuries from the explosive forces and high temperatures. There are many other issues involving combustibles (coal handling), step and touch potentials, hazardous induced voltages, and motor iron damage.

The vast majority of existing generating stations are equipped with high-impedance, alarm-only ground detection circuits for their 480-volt systems. However, about 75% of these schemes have design oversights involving alarm thresholds, monitoring instrumentation, and transient voltage mitigation. The major advantage of high-impedance schemes is that the 100% ground fault current is only around 1.0 amps, which does not provide enough energy to breach the insulation between phases. If an employee or contractor accidently drops a tool or material onto an exposed screw or conductor, causing a ground fault, it will be a harmless incident. It also does not cause the loss of the particular load, which may be critical to the unit.

3.11 Transformer High-Speed, Sudden-Pressure Protection (63)

In addition to the overcurrent protection presented earlier, oil-filled transformers that are 5 MVA and larger are usually equipped with high-speed, sudden-pressure relays and differential protection. Two types of sudden-pressure relays are commercially available: a gas type for nitrogen-blanketed transformers and an oil or liquid type for conservator tank designs that is usually installed in the piping that connects the higher mounted conservator tank with the main transformer tank. Both devices will detect internal faults very quickly and electrically isolate the transformer to limit internal damage and tank ruptures in order to mitigate explosions and fire hazards to personnel and adjacent equipment. The intent of both transformer designs is to mitigate moisture intrusion by keeping either the nitrogen gas or the oil pressure higher than the outside air pressure. Sudden-pressure relays (63) for nitrogen pressure designs are usually mounted on the top center of the tank to reduce seismic sensitivity during earthquakes, although some manufacturers provide seismic baffles that permit mounting the sudden-pressure relay on the top side wall of the tank in the nitrogen environment. Nitrogen-type relays will compare the pressure in a reference chamber that is fed through a small orifice to the transformer tank pressure. For some designs, routine testing involves removing a pipe plug to actuate the relay and then sealing off the plug opening with the thumb and recording the contact reset time to ensure that the timing orifice is clear, and the relay operates correctly. A ladder is normally required for access, and if the transformer has exposed bushings, a formal electrical clearance is also required. Piping can be installed to bring the plug to ground level, reducing test time, clearance or tag out procedures, and personnel fall exposure.

3.12 Transformer Current Differential Protection (87)

A basic differential scheme for one phase only is illustrated in Figure 3.36. The dashed lines in the primary circuit represent the electrical apparatus being protected, which could be a transformer, motor, generator, bus, or feeder. Basically, the current going in needs to equal the current going out, or else a fault is indicated. The arrows in the CT secondary circuit denote the relative direction of the current flow during normal load flows or through fault conditions. The currents will circulate between the two CTs, and no current will flow in the operating coil so long as the secondary currents are balanced. If an internal fault occurs inside the CT boundaries (zone of protection), the load-side secondary current will either reverse in direction (multiple sources) or cease to exist if there are no other sources. In either case, current will now be forced to flow through the operating coil because the load-side CT will either produce a current in the opposite direction or become a high impedance in the absence of a corresponding amount of primary current. If the operating coil current is high enough, the relay will actuate to isolate the faulted area.

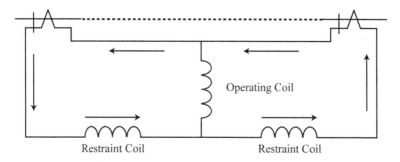

FIGURE 3.36
Current differential protection.

Auxiliary System Protection

Current differential protection offers a couple of advantages: high-speed tripping (generally around one cycle, plus breaker opening and interrupting time). Because they do not need to coordinate with downstream devices, they can be set well below the load current values. The biggest area of concern is false operations from CT error currents. The error currents can be quite significant during through fault (high-current) conditions that may saturate the CTs unequally and also have a DC component from the asymmetrical current flow. To counter this problem, the relays are designed with slope percentages that typically range from 10% to 25% for generators, motors, buses, and feeders, and 20%–60% for transformers. Slope is the ratio of one or two restraint coil currents to the operating coil current, and the actual calculation details vary among manufacturers. Some designs offer variable slope, with the slope percentage increasing as the current levels increase. The advantage of slope is that it takes more current to actuate when the currents are higher, thereby providing a greater margin for error currents.

Transformer current differential protection must accommodate magnetizing inrush current conditions during energization; which can get quite high, depending on the position of the sine wave, the circuit resistive ohms or X/R ratio, the short circuit duty of the source, the MVA of the bank, and the amount and polarity of the residual magnetism remaining in the core iron. The residual magnetism effect can be reduced with better magnetic circuit designs, and improved iron lamination material properties or alloy details that were developed during the last 50 years. Excluding residual magnetism, as discussed in the motor section, a maximum inrush current occurs (Figure 3.10) when the sine wave is at zero crossing because of the lack of development of an opposing voltage (self-flux) from the prior position of the sine wave, and the slower rate of current change, associated with zero-crossing sine waves. RMS peaks as high as 20 times FLA are possible, but not likely, with more modern designs.

As illustrated in Figure 3.37, much of the inrush current decays in the first few cycles, but the remaining level may take several seconds to fully

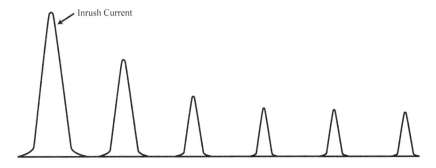

FIGURE 3.37
Transformer magnetizing inrush decrement.

dissipate. Initially, it looks like 60-Hz half-wave rectification (120 Hz to a filter) until the circuit L/R decrement allows the negative or opposite-going sine wave to produce current flow. The actual calculations for inrush magnitudes are quite complex and are not linear. Since the inrush current is only on the primary side and not balanced out from corresponding secondary side current flows, it looks like a fault to a normal differential protection scheme. For this reason, transformer differential relays are normally desensitized to 120-Hz differential currents to prevent false operation from magnetizing inrush conditions.

Figure 3.38 shows a typical nondigital or electromechanical transformer differential protection scheme that applies wye-connected CTs on the primary side and delta-connected CTs on the secondary side. Because a delta-wye transformation causes a 30° phase shift between the primary- and secondary-side phase-neutral voltages, the CT secondary currents will not

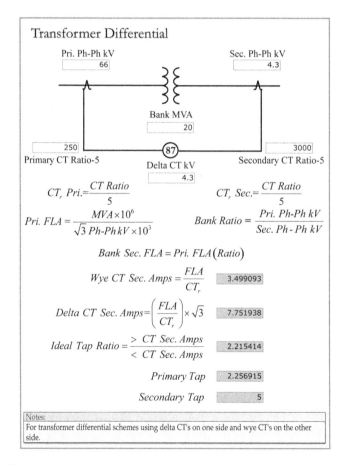

FIGURE 3.38
Transformer differential relay taps.

be in phase. In this case, applying delta CTs on the wye or secondary side will bring the secondary currents back into phase because the delta connection will shift the secondary currents by 30° (leading or lagging, depending on how the CTs are connected). Transformer differential relays are equipped with ratio tap blocks to compensate for the different current magnitudes on each side of the transformer. Current differences between the primary and secondary sides of the transformer that cannot be balanced out by the CT ratios and relay taps will flow through the operating coils of the relay. If this current is high enough (indicating an internal fault or incorrect taps), the relay will operate to electrically isolate the transformer.

To select the taps, one must first determine the CT secondary FLA on each side of the transformer. The current inside the delta CTs should be multiplied by the square root of 3 to account for the current contributions from the other CTs. Then, an ideal ratio can be determined by dividing the larger CT secondary current by the smaller CT secondary current. Finally, the closest ratio can be selected from the available relay taps. In this case, a 5.0 tap on the secondary side and a 2.3 tap on the primary side will closely match the ideal ratio. This is not intended to be a full explanation of the application of transformer differential relaying, which can be quite complex when considering unbalanced faults, but only as an example on how to determine the closest tap ratio for the relays. The newer digital relays can accept wye–wye CTs and account for differences through software programming.

3.13 Bus Transfer Schemes (27R)

Figure 3.39 shows a one-line drawing of a typical generating station. When automatic systems trip the unit off-line because of process problems, CB1 trips

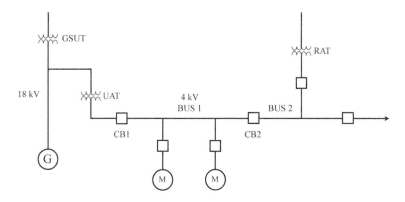

FIGURE 3.39
Typical generating station one-line.

open and CB2 will close to repower the 4-kV bus 1 after all the required supervisory permissive or logic functions are met. This section will describe the reasons for each logic or contact permissive in automatic bus transfer schemes.

Residual voltage relays (27R) are normally applied to supervise or block automatic bus transfers until the voltages are at a safe level. These relays prevent motor damage during automatic bus transfer operations from excessive voltages that can overflux motor core iron, but more important, they can cause high stator currents, resulting in increased electromechanical forces on the windings. When a bus with motor loads is deenergized, induction and synchronous motors behave like generators, producing residual voltages during coastdown for a short time as a result of the trapped magnetic flux in their core iron at the time of deenergization. If bus tie breaker CB2 closes before the residual voltage has a chance to decay to a safe level, and if the vector sum of the motor residual voltage and the new source voltage exceeds 1.33 per unit (or 133%), the motors can be damaged or the remaining life reduced. Because the duration is short (usually around 1 second or less) and generally beneath the core iron overexcitation time withstands, the main concern is that the electromechanical forces could be damaging to motors at reduced speeds when in a starting current mode of operation.

Figure 3.40 presents a suggested residual voltage relay set point calculation that will prevent damage to motors during automatic bus transfer drop and pickup operations. Basically, the relay is set for 0.33 per unit of the motor nameplate or rated voltage. When the set point value is reached,

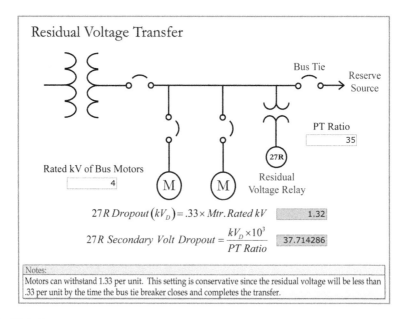

FIGURE 3.40
Residual voltage relay settings.

Auxiliary System Protection

the automatic circuitry will dispatch a close signal to the bus tie breaker, which will prevent the vector sum of residual and new source voltages from exceeding 1.33 per unit. This setting has an extra margin because the residual voltage will be below the set point (0.33 per unit) by the time the bus tie breaker CB2 contacts actually close.

Residual voltage relays should be designed in a way that the frequency decrement will not affect the set point or actuation of the relay. This can be accomplished with electromechanical coils by feeding them through bridge rectifiers.

Many older generating stations have a number of disadvantages in the design of their bus transfer schemes, in addition to not monitoring residual voltage levels. Among the many bus transfer design considerations in fossil plants is a particular concern about maintaining the fans after unit tripping to ensure that trapped combustible gases are quickly blown out of the boiler to mitigate the possibility of explosion. Because of this concern, there is a design motivation to transfer as quickly as possible. A suggested DC elementary design for bus transfer schemes is illustrated in Figure 3.41, which resolves the disadvantages of the older schemes. The author designed the scheme for a major utility during the 1980s, after a severe bus fault and resulting fire significantly damaged about a dozen breakers and cubicles in a common 4-kV switchgear room for two units. An investigation of the older design that was also applied at 11 other multi-unit fossil plants disclosed the following issues:

- Failure to coordinate with bus overcurrent protection
- Failure to protect motors during transfer conditions
- Failure to prevent unintended transfers after (86) lockout relay resetting

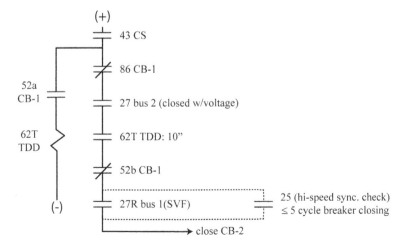

FIGURE 3.41
Bus transfer elementary.

The older bus transfer scheme failed to coordinate with overcurrent relays that should have blocked or prevented a transfer into the faulted bus. The older scheme tripped the auxiliary transformer breaker CB1 on undervoltage (faults depress the voltage) with a short time delay, and then closed the bus tie breaker CB2, energizing the fault a second time from bus 2 or the reserve before the overcurrent relays could time out and block the transfer. Second energizations cause significantly more damage and increased possibilities for fires because the electrical apparatus has stored heat from the first event, and then gets hit again with the short circuit current heating starting at a higher temperature.

In reference to Figure 3.41, the existing schemes were already equipped with 43 transfer switches to allow operations to take the automatic bus transfer in and out of service, 86 lockout relay contacts to block transfers into faulted buses, and 27 undervoltage relay contacts on bus 2 to ensure that the reserve or new bus voltages were acceptable. Three new contacts were added to the existing schemes: the 62 TDD (one shot timer that is armed when the auxiliary transformer breaker CB1 is closed), and allows 10 seconds for transfers to occur before removing the automatic bus transfer from service to prevent undesired transfers when plant operators reset 86 lockout relays; the 52b (auxiliary transformer breaker CB1 open), which allows fast transfers to coordinate with overcurrent relays (if fault relays initiate the opening of breaker CB1, bus transfers will be automatically blocked); and the 27R residual voltage relay, which dispatches a close signal to bus tie breaker CB2 if the residual voltage is in a safe range. The 25 high-speed sync-check relay is optional when transfers faster than 1 second are desired and if tie breaker CB2 closes in 5 cycles or less (some breakers can take 20 cycles or more to close). Special sync-check relays are available to measure the angular differences between the buses before the event and, if the range is acceptable, the 25 relay assumes that the angles cannot get too far out before the fast breaker contacts close, which prevents the vector sum from exceeding 1.33 (or 133%). The sync check permissive still needs to be backed up by a residual voltage relay in case the angles are not acceptable.

3.13.1 Bus Transfer Scheme Oversights

Roughly 90% of the plants surveyed were found to have questionable bus transfer schemes. With conventional fossil plants, there is a design motivation to transfer as fast as possible to maintain the fans and blow the combustible gases out of the boiler to mitigate any possibilities for boiler explosions. However, the schemes often do not coordinate with bus fault overcurrent relays, and if the transfer scheme deenergizes the fault before the overcurrent relays can time out to block transfers into faulted buses and then reenergizes it from the reserve or other source, the damage will be much more severe, often resulting in fire. Since the heat has not dissipated from the first event, the second event starts at a much higher temperature and the fault damage will be much greater; in the worst case conditions, the entire switchgear lineup may need replacement if the fire is not controlled. The second issue is motor damage; if the vector sum of the residual coastdown voltage from the bus motors and the new source voltage exceeds 133%, the motors can be damaged or the remaining life of the motors may be reduced from the excessive electromechanical forces.

Bibliography

Baker, T., *EE Helper Power Engineering Software Program* (Laguna Niguel, CA: Sumatron, Inc.), 2002.

Blackburn, L. and Domin, T., *Protective Relay Principles and Applications* (Boca Raton, FL: CRC Press), 2014.

Higgins, T., Station auxiliary transfer schemes, *Georgia Tech Relay Conference. Formally called Georgia Institute of Technology.* Paper, 1979.

Reimert, D., *Protective Relaying for Power Generation Systems* (Boca Raton, FL: CRC Press), 2006.

Westinghouse Electric Corporation, *Applied Protective Relaying* (Newark, NJ: Silent Sentinels), 1976.

Westinghouse Electric Corporation, *Electrical Transmission and Distribution Reference Book* (East Pittsburg, PA: Westinghouse Electric Corporation), 1964.

4

Generator Protection

Considering the large capital investment and the numerous possibilities for electrical system disturbances, control and equipment failures, and maintenance and operating errors, which can force generators to operate outside their original design specifications, it makes good business sense to apply protective relays that can prevent generator damage from irregular operating conditions. This chapter will cover the gathering of the information needed to calculate generator protective relay settings, the setting calculations for the various protective functions, typical generator/turbine withstand times for abnormal operating conditions, and the math associated with various types of impedance elements. As in the prior chapter on auxiliary system protection, comments on the percentage of protection oversights found during the author's reviews of approximately 35,000 megawatts (MW) of large generation will be provided under each protective relay function or standard device number.

4.1 Generator Relay Data

A good starting point is to compile the data needed for setting the various generator protective functions. Figures 4.1 through 4.3 provide examples of organizing and massaging or converting the information to more useful forms for generators, step-up transformers, and associated electrical systems. The data shown are for a 578.6-megavolt-amp (MVA), 24-kilovolt (kV) generator that is connected to a 765-kV bulk power transmission system. Organizing the data in this manner will save significant time in developing and documenting the basis for the relay settings.

The top half of Figure 4.1 shows the generator data needed for the relay settings, including generator nameplate data; the direct axis-saturated subtransient, transient, and synchronous per-unit reactances; and the various primary- and secondary-instrument transformer ratings. With the exception of converting primary ohms to secondary or relay ohms and vice versa (as shown in Figures 4.4 and 4.5), the MVA and per-unit conversions were covered in Chapter 2 and will not be discussed further here. The bottom half of Figure 4.1 (outputs) provides nominal currents, instrument transformer ratios, and various reactance ohms (not per unit) that will be used to develop

158   Electrical Calculations and Guidelines

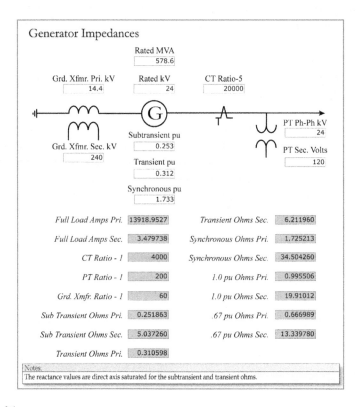

FIGURE 4.1
Generator data.

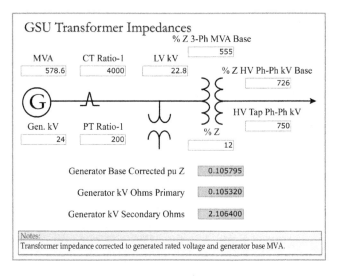

FIGURE 4.2
Transformer data.

Generator Protection

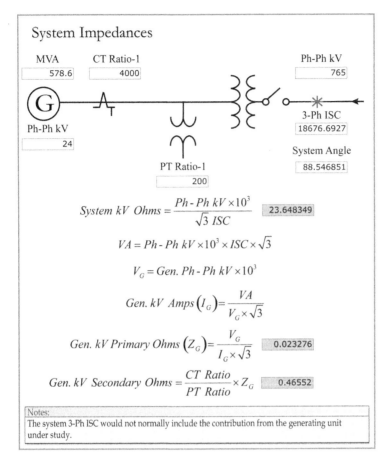

FIGURE 4.3
System data.

the settings for the backup impedance, loss-of-excitation, and out-of-step relay functions.

Figure 4.2 presents the step-up transformer secondary or relay ohms for the backup impedance and out-of-step relay functions. The transformer per-unit base impedance of 555 MVA needs to be converted to the generator base of 578.6 MVA and corrected for voltage differences on the generator side of the transformer and tap differences on the high-voltage side of the transformer. With the exception of secondary or relay ohms, the per-unit MVA conversions and corrections were already discussed in Chapter 2.

Figure 4.3 develops the high-voltage switchyard secondary or relay ohms for the generator out-of-step function. Normally, this calculation would not include the short circuit contribution from the generator under study. First, the system kV ohms and volt-amps (VA) are determined, then the generator side amps and ohms are calculated, and finally, the secondary or relay ohms

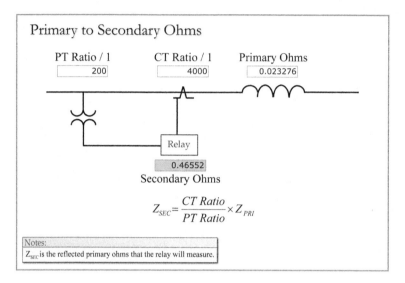

FIGURE 4.4
Primary to relay ohms.

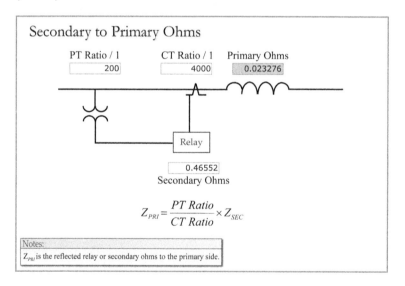

FIGURE 4.5
Relay to primary ohms.

that represent the high-voltage electrical system are developed. The system short circuit angle is not used in the calculation, but it is needed for setting the out-of-step protective relay function.

Figure 4.4 presents the expression for calculating secondary or relay ohms. The inverse calculation in Figure 4.5 is used to convert relay ohms to primary-side ohms.

4.2 High-Voltage Switchyard Configurations

Before delving into the generator protective functions, a review of the associated high-voltage switchyards might be appropriate. Figures 4.6 through 4.8 illustrate the more common switchyard configurations for handling multiple lines and generating station units. In all three configurations, the opening of line breakers increases the circuit X_L and momentarily reduces the power transfer capability on the remaining lines until the power angle changes to match the output of the turbines as the units accelerate. In the interest of simplicity, only two lines and one generating station unit are shown for each configuration, and the generator symbol includes the associated step-up transformer.

The ring bus configuration in Figure 4.6 is the simplest and least costly. It also has the lowest maintenance costs and exposure to failure because there are fewer components. As the figure shows, only three breakers are required for one generating unit and two lines. Because each element (i.e., unit or line) has two breakers, the opening of one breaker does not interrupt power flows. The generating unit position is equipped with a motor-operated disconnect

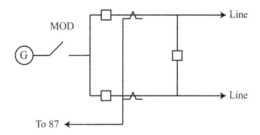

FIGURE 4.6
Ring bus configuration.

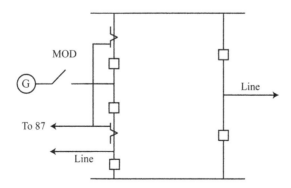

FIGURE 4.7
Breaker-and-a-half configuration.

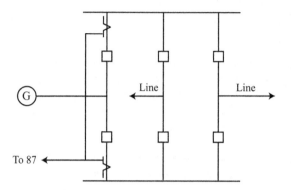

FIGURE 4.8
Double breaker double bus.

(MOD) that can be opened after the unit is taken off-line, permitting the reclosure of the unit breakers to improve the integrity of the ring bus. The 87 differential elements could be for the line from the generator step-up transformer (GSUT) to the switchyard, but they may also overlap the GSUT to provide transformer protection and, more commonly, also will overlap the generator to back up the generator protection.

Schemes that include the generator are called *overall differential* or *unit differential*. With overall or unit differentials, current transformers (CTs) are normally applied on the auxiliary transformer primary side to balance out any auxiliary load current. In all cases, the 87 differentials must remain in service when the unit is off-line (unless both unit breakers are open) because it now provides high-voltage switchyard bus differential protection for the two unit breakers and the conductors between the breakers and the open MOD. The main disadvantage of a ring bus configuration is if one breaker is out of service for maintenance, a single line fault could interrupt the power flow to the other line. Separate bus differential relays are not normally required, as the unit and line protection overlaps and protects the bus sections.

Figure 4.7 illustrates a breaker and a half configuration. Although this design requires five breakers for one generating unit and two lines, it provides additional electrical system security. The configuration has top and bottom buses that are usually protected by high-impedance bus differential schemes. High-impedance differential schemes are favored on large buses (multiple lines) to mitigate false operations from error currents associated with unequal CT loading and performance by forcing more of the false differential to flow in the CT circuit and less in the relay. The generating unit position is equipped with an MOD that can be opened after the unit is taken off-line, permitting the reclosure of the unit breakers to improve the integrity of the switchyard. As with the ring bus configuration, the 87 differential elements must remain in service when the unit is off-line and the unit breakers are closed because this configuration now provides high-voltage switchyard bus differential protection for the breakers and for the conductors between the breakers and the open MOD. A second generating unit or a third line can be accommodated with the addition of only one more breaker.

Figure 4.8 displays a double-breaker, double-bus configuration. In this example, six breakers are required for two lines and one generating unit. This provides the highest level of reliability. If a breaker is out for maintenance, a fault on a transmission line will not prevent power flow from the generating unit to the unfaulted line. Each bus section (top and bottom) would normally be protected with a high-impedance bus differential scheme. The MOD is not required because the unit breakers can remain open when the unit is off-line. The addition of another line or generating unit would require two breakers.

4.3 High-Voltage Switchyard Protection Concerns

There are a number of concerns with protection at large generating station high-voltage switchyards. The main concerns are as follows:

- The high-voltage switchyard breaker failure schemes do not always consider the impact on the bulk power electrical system (BES) if the time delays associated with fault clearing are at or more than the critical clearing times of the large interconnected generation. Approximately 90% of the author's surveyed units had breaker failure fault isolation or clearing time delays that exceeded the critical clearing time of the connected generation. Two pole machines typically have critical clearing times of 10 cycles and four pole machines around 12 cycles for three-phase faults in their local high-voltage switchyards. Longer fault clearing times could cause the unit to lose synchronization with the bulk power transmission system and cause a major electrical system upset.
- Many, if not most, of the high-voltage switchyard protective relay schemes applied in large generating station switchyards are not backed up by high speed protective relay schemes, and have been inadvertently removed from service or otherwise disabled in the past. About 90% of the surveyed units did not back up the bus, unit, or reserve transformer differentials in their high-voltage switchyards.
- Many protection functions that could mitigate disturbances to the bulk power electrical system are commonly not applied.

The worst case, close-in short circuit scenarios can cause islanding of generation and loads and widespread outages. In some cases, involving multiple lines and generating units, studies may need to be performed to see if system overreaching line relays at remote substations have the ability to isolate local generating station high-voltage switchyard faults that do not clear properly due to infeed impedance and ground current distribution.

Transmission faults unload the machine because the watt load is displaced by vars. The loss of load accelerates the machine and, at some point, as the power angle and speed differences increase, the unit becomes unstable and can no longer maintain synchronization with the electrical system, which significantly aggravates the disturbance and can cause a major electrical system upset. This is called the *critical clearing time,* and it is usually in the range of 6–20 cycles, depending on the type of fault, specific location, and severity.

The second issue is backup protection; high-voltage transmission lines almost always have backup schemes that are overreaching and do not rely on high-speed communication systems. Figure 4.9 shows a typical transmission line backup protection scheme that utilizes zone 1 and zone 2 directional distance or mho-type impedance relays. Usually, zone 1 will trip without time delay and is set at 80% of the impedance ohms of the line. Depending on the length of area transmission lines, zone 2 will be set at 120% of the line impedance or greater, with a typical time delay of 0.4 seconds, or 24 cycles, for coordination purposes. This time delay usually exceeds the critical clearing time of generating units, and nearby on-line generation will likely lose synchronization with the system before zone 2 relays can time out to clear faults beyond the lines. It is not unusual to find generating station high-voltage switchyard bus differentials, unit differentials, transformer differentials, and feeder differentials that are not backed up and rely entirely on zone 2 clearing if the single protection schemes fail to interrupt bus, transformer, or feeder faults. Additionally, depending on the number of transmission and generating unit lines in the switchyard, the zone 2 relays and directional ground overcurrent relays at the remote ends of the lines may not be able to detect faults beyond the lines due to current distribution and infeed ohms.

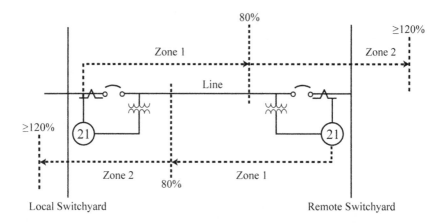

FIGURE 4.9
Transmission line backup impedance zones.

Generator Protection

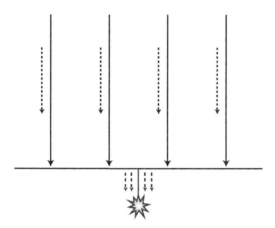

FIGURE 4.10
Fault current distribution and infeed ohms.

Figure 4.10 represents the current distribution and infeed problem. In the case of directional ground overcurrent relays, the amount of ground current in each line is reduced because the fault current is shared by all lines. If nearby generation is also feeding the ground fault, the fault current flowing in each line would be reduced further. The concerns are that the ground protection at the remote ends of the lines may not have enough sensitivity to detect local ground faults beyond the lines, and if they do, the fault clearing timing may be too long to avoid generator instability.

The final issue is infeed ohms, which is more of a problem if different sources are involved because the voltage drops are not congruent with a single source (i.e., generating units, ties to other utilities, and more remote substations). Figure 4.11 shows the calculations for infeed ohms with two sources, A and B, with A having a per-unit impedance of 0.5 ohms and an additional 0.5 ohms to the fault location, or 1 per-unit ohms total if B is not connected. First, the total circuit per-unit impedance and current is determined, then the current contribution from A only is proportioned, and finally, the per-unit impedance seen by A to the fault location is proportioned at 1.5 per-unit ohms or 150% higher when B is included.

The worst-case location for infeed at a generating station would normally be faults near the high-voltage bushings of reserve auxiliary transformers (RATs) that are fed from the same high-voltage switchyard. In that case, the zone 2 relays at the remote end of the transmission lines may not be able to detect the fault to back up the RAT transformer differential, feeder differential protection, or both. As mentioned before, the line zone 2 time delays may be too long to avoid instability problems with the generating units. Additionally, generator protection functions that can see switchyard faults (negative phase sequence and backup impedance) may have time delays that are too long to avoid transient or dynamic instability and loss of

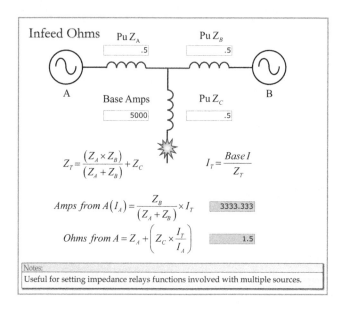

FIGURE 4.11
Infeed ohms.

synchronization with the electrical system. If the generating unit is equipped with out-of-step protection (note that many are not), the unit will be tripped off-line and isolated from the fault.

Usually, high-voltage switchyards are equipped with breaker failure schemes that will clear everything necessary to isolate faults if a circuit breaker fails to interrupt the fault current in a timely manner. Normally, two conditions are required: a specified level of fault current and an energized breaker trip coil circuit. If these conditions are met, a timer output will operate to trip the necessary adjacent breakers or units to clear the fault. It is not unusual to find breaker failure fault clearing times that are longer than the critical clearing time. The fault clearing time includes the initiating relay time, plus the breaker failure time delay, plus the longest breaker interrupting time. Consequently, close-by generation may go out of step before the breaker failure scheme can isolate or clear the fault.

To counter some of the foregoing problems, this author is an advocate of installing zone 1 impedance relays to back up high-voltage switchyard transformers and feeder differential relays. The relays measure voltage and current at the generating station switchyard and look into the GSUTs or RATs; the zone 1 relay can be set with a small time delay (three to six cycles) to ensure that the differential schemes also operate. With regard to the GSUT, it should be verified that the impedance relays cannot operate for loss-of-excitation events and cannot see unit auxiliary transformer (UAT), medium-voltage faults for coordination.

With the RAT, the relay can be set to look halfway into the transformer to ensure that it cannot overreach and see medium-voltage faults. The advantage of applying impedance elements is they are not dependent on the strength of the electrical system, which varies depending on the number of sources available at the time of the fault. Ohms are ohms, and the impedance of a transmission line or transformer does not change when generating units or lines are placed into or out of operation. Early impedance relays were not directional and consisted of a circle that was symmetrical around the point of origin. These impedance elements were supervised by directional contacts or relays. When directional electromechanical impedance relays were initially developed, the term *mho* was adopted because the torque equations included the reciprocal of ohms. The elements normally encircle the first quadrant, which represents a series current lagging circuit.

During the 1990s, when this author was overseeing or providing electrical engineering support for a large fleet of fossil generation, two serious events came to his attention within a few months of each other. The first involved finding the direct current (DC) molded case breaker open for the overall unit differential protection scheme for a 230-kV unit, and the second was associated with a protective relay technician removing the overall unit differential from service to facilitate testing the protection on a breaker-and-a-half configuration during an ongoing overhaul for a 525-kV unit. Consequently, in both cases and for the duration of both events, there were no in-service protective functions for the high-voltage lines from the plant to the switchyards or from the high-voltage disconnect to the unit breakers. Alarmed, the author rushed to install zone 1 impedance relays that would look from the high-voltage switchyard into the unit to back up the overall differential relays. An investigation disclosed that there were no high-voltage current or potential transformers available for the 55 fossil units that the utility operated, and the costs to install new high-voltage instrument transformers would be substantial.

With a generating unit, the major concern is reaching through the unit auxiliary transformer and not coordinating with 4-kV or other medium-voltage auxiliary power bus faults (which are relatively high impedance), which provides a lot of margin for the impedance relay settings. The author proposed using the existing high-voltage, delta-connected switchyard current transformers for the overall differential schemes and medium-voltage generator isolated phase bus potential transformers to drive the impedance relays. The relays could easily be installed on the unit protection panels minimizing the required wiring. Based on studies by the system protection group, it was determined to be feasible and impedance relays were installed on all the bulk power units. Conventional impedance relays could be used for the double-breaker, double-bus schemes, but overcurrent elements were also needed when the units were off-line and the isolated phase bus potential transformers were deenergized for ring bus and breaker-and-a-half configurations. While not ideal, the highly unusual schemes resolved the immediate problem and did operate properly for a number of short circuit conditions in the unit differential zone of protection.

With regard to the high-voltage switchyard bus differentials, the addition of a second set of high-impedance or current differentials would also be an expensive process. With the newer digital relays, there may be opportunities to feed the existing secondary line currents into digital differential relays. Although this scheme would not overlap and include the breakers, the breakers are normally protected by backed-up line protection.

4.4 Generator Protective Functions

This section will develop the protective function settings for the 578.6-MVA steam turbine generator and its associated step-up transformer and high-voltage electrical system, as previously shown in Figures 4.1 through 4.3. Many of the electrical protection functions in a generator package are applied to protect the prime mover or turbine, not just the generator. The suggested settings are for 50-MVA and larger, two- and four-pole cylindrical rotor generators connected to typical utility transmission or subtransmission systems. Slower speed salient-pole generators and smaller units connected to weaker systems or units with different or unusual electrical configurations may present special circumstances that may not be covered in this chapter. The following material is presented in numerical order according to the standard device numbers assigned for different relay functions.

4.4.1 Backup Impedance (21)

The 21 function is a backup impedance relay with one or two zones that look from the generator to the electrical system. The 51-V time overcurrent element performs a similar function, and it is preferable to apply the 21 backup impedance elements only, since it is more discriminating than relying on current alone. The 51-V is a voltage-controlled or restraint time overcurrent relay that permits setting the minimum trip threshold well below full load amps (FLA) to allow for generator fault current decrements, but it will not operate unless the voltage is depressed by a fault condition.

No industry consensus exists on how to set the zone 2 impedance function; various experts propose different philosophies involving the shortest or longest line. Most of the papers on the subject advocate using the zone 2 function to trip or isolate the unit for close-in, high-voltage system faults that do not clear quickly enough. Fairly long time delays are necessary to ensure coordination with downstream electrical system protective devices (i.e., breaker failure time delays, transformer bank overloads, zone 2 line protection time delays, and so forth).

Generator Protection

Because of the long time delay, this function may not be able to operate for close-in faults because the unit will likely lose synchronization with the system before the relay can time out. The out-of-step slip cycle can reset and pick up impedance elements as the impedance locus or swing travels through its cycle. Generators are thermally protected (high current flow) from high-voltage phase-phase and phase-ground faults by negative phase sequence relays, but these relays will not respond to three-phase balanced events. Based on the foregoing, it is proposed that the zone 2 element be applied to protect the generator from balanced, three-phase, prolonged system disturbances. For cross-compound units or generators operated in parallel at their medium-voltage buses, the settings may need to be modified to account for infeed impedance.

Figure 4.12 provides suggested settings for two element generator backup impedance relay schemes. Zone 1 is set to reach halfway into the GSUT

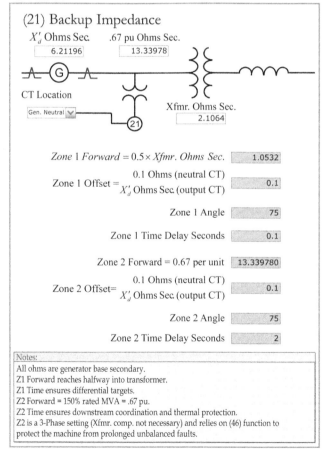

FIGURE 4.12
Generator backup impedance settings.

(1.053 ohms), with a 0.1-second time delay. The short reach eliminates concerns about coordinating with downstream devices, and the short time delay is suggested to allow time for the differential relays to operate first. Operation of a primary relay provides important information about the location of the fault and verification that the relays are performing as designed. If the CTs are on the generator output side, it is customary to offset the mho circle by the direct axis-saturated transient ohms to include the generator. If the CTs are on the neutral side, some manufacturers suggest using a 0.1-ohm offset to provide extra margins for faults near the generator output terminals in order to ensure proper operation when the relay driving voltage is very low. As in revenue metering, the location of the potential transformers (PTs) determines the precise point of measurement because the circuit voltage drops are measured accurately. However, the CT location does have a directional impact on whether the event is in a forward (toward the electrical system) or a reverse (toward the generator) direction. An angle of 75° is often used to accommodate circuit and arc resistance. The suggested reach for zone 2 is 150% of the rated generator MVA (0.67 per unit), or 13.34 ohms with a time delay of 2 seconds, to ensure downstream coordination. Because this is a three-phase setting, there is no concern about unbalanced events and compensation to account for the impact of the delta–wye transformation on unbalanced faults is not required. The suggested offset and angle are the same as the zone 1 setting.

Generators are typically designed to carry 130% of rated stator amps for 60 seconds and 180% for 20 seconds. Accordingly, the foregoing setting of 150% or greater for 2.0 seconds has more than adequate thermal and electromechanical margins. As covered in Chapter 3, transformers can handle 150% for a little over 9 minutes. A 2-second time delay is suggested because it is the short circuit electromechanical force-withstand time for transformers, and it should also coordinate with all downstream protective devices on a typical utility system. It is doubtful that turbine/generators can provide 150% of the rated megawatt load for 2 seconds, and utility load shedding programs should operate to bring the megawatt demand back into balance if the watt deficiency is severe enough. Governor droop settings for steam units are usually 5%. This is really control system terminology, but a simplistic way of looking at it is that it takes a 5% speed change to get the control valves fully opened or closed. During an earthquake islanding event in California, a digital fault recorder captured a 215-MW generator that was subjected to twice the rated load, and the voltage collapsed to zero in six cycles. The general rule of thumb is that generating units should not try to pick up more than 10% of the rated megawatts at any one time.

The foregoing discussion on megawatts is really a moot point since the angle is set at 75° (mostly vars) and the ohmic reach would be substantially lower (higher current) at nomal load angles. Angles greater than 75° would also need a slightly higher current to operate. At 90°, the reach point would be 12.89 ohms, or about 3.4% lower. However, generators can

Generator Protection 171

carry substantial vars that may thermally overload path components during major system disturbances. Since the normal utility system relies on capacitor banks and not generation to carry the system var requirements, excessive var loading for 2 seconds would be more indicative of a system fault that did not clear properly. Due to infeed concerns, it is not advised to increase the setting above the 150% MVA level or reduce the ohmic value further.

If the mho circle is not offset, the circle ohmic or impedance points can be calculated by simply taking the cosine of the angle deviation from the circle diameter or maximum reach point, as illustrated in Figure 4.13. The relay will actuate whenever the ohmic value is inside the circle.

Figure 4.14 shows an offset mho impedance circle for generator backup applications. Ohmic values on the circle can be calculated by using the trig method presented in the illustration.

The as-found Generator Backup Impedance settings from the surveys or reviews were not rational 100% of the time. The settings were intended to isolate close-in faults with a time delay of 0.8 seconds–1 second to coordinate with zone 2 time delays. The relays cannot possibly operate with that much time delay for close-in faults because the unit will go out of step, which will reset the relay during each slip cycle before it can actuate. My suggestion is to set the relay to protect the unit from prolonged three-phase system disturbance, as outlined in the foregoing paragraphs.

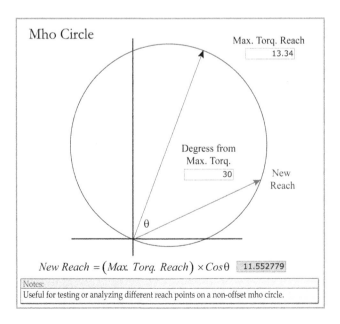

FIGURE 4.13
Mho circle reach points.

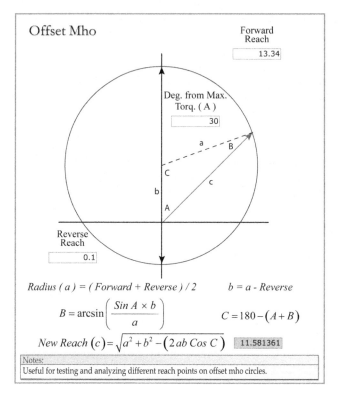

FIGURE 4.14
Offset mho circle reach points.

4.4.2 Volts/Hz (24)

This is probably the most important protective function for generators and connected transformers due to the short withstand times and the high costs associated with repairing core iron damage. The easiest way to catastrophically destroy a generator or its connected transformers is by overfluxing the iron laminations in their cores. The term *volts per hertz* (volts/Hz) is used because of the impact that underfrequency operation has on the level of flux in the core iron. In the case of transformers, a reduced frequency will lower the exciting impedance, causing higher exciting current, more ampere-turns, and increased magnetic flux in the core iron.

The increased magnetic flux will increase the saturation level of the iron and the magnitude of eddy currents in the laminations, causing higher core iron temperatures. As the laminated iron saturates, the magnetic flux will spill into structure areas that are not designed to carry the flux, and those areas will also overheat from annular current flow. In the case of generators, the core iron flux is not a direct function of frequency or speed, and the no-load DC field is safe over the entire speed range. However, a generator is like a tachometer: the slower the generator's speed, the lower the output voltage. If the automatic voltage regulator (AVR) is in service, it will try to correct for the reduced output voltage by increasing the DC field current, which can then overflux the core iron, depending on the amount of voltage regulator correction. Consequently, the expression *volts/Hz* will work equally well for generators and transformers. Generators are normally designed to carry 105% of rated voltage continuously and transformers 110% at no load and 105% at full load, but credit can be taken for the internal voltage drop in transformers. Accordingly, the generator volts/Hz relays will normally provide protection for connected transformers, depending on the load currents, voltage drops, and rated winding voltages. The continuous rated frequency voltages of 105% or 110% at 60 Hz are also the continuous volts/Hz ratings for the apparatus.

This protection is essential for generating stations because of the numerous ways that overexcitation can occur through operating mistakes or equipment failure. Although many generator voltage regulators are equipped with volts/Hz limiters, and some with volts/Hz tripping logic, the normal practice is to provide volts/Hz protection in a separate package due to concerns that there may be common modes of failure that can affect both the voltage regulator control, as well as its protective functions. The short time withstand capabilities of the apparatus for more severe events does not allow time for operator interdiction before catastrophic damage can occur.

During startup, in some designs, excitation is applied at very low speeds, which does not stress the machine, so long as the AVR is out of service and the no-load field current is not exceeded by more than 5%. As a generator is loaded, its DC field is increased to compensate for internal voltage drops and the subtractive flux (i.e., the armature reaction) from the higher current flow in the stator windings. Depending on the design details, the no-load field current is somewhere in the neighborhood of 50% of the full-load field current. Typically, a generator can withstand a full-load field with no load on the machine for only 12 seconds before damage occurs to the core iron. The typical withstand at 115% volts/Hz is around 5 minutes. The shape and slope of the withstand curves for generators and transformers are usually similar, with the main difference being the continuous rating starting points.

Figure 4.15 covers the procedure for calculating the impact of operating generators at abnormal voltages or frequencies. For reference, the maximum primary and secondary 60-cycle voltages are calculated. A base volts/Hz ratio is determined by dividing the rated kilovolts by the rated frequency. An applied volts/Hz ratio is then determined by dividing the applied kilovolts by the applied frequency. The actual or applied volts/Hz percentage can then by determined by dividing the applied ratio by the base ratio and multiplying by 100. In this example, the generator's continuous volts/Hz rating of 105% is significantly exceeded, at 50 Hz operation, with a calculated percentage of 115%, even though the applied voltage of 23 kV is well below the 60-Hz maximum rated voltage of 25.2 kV.

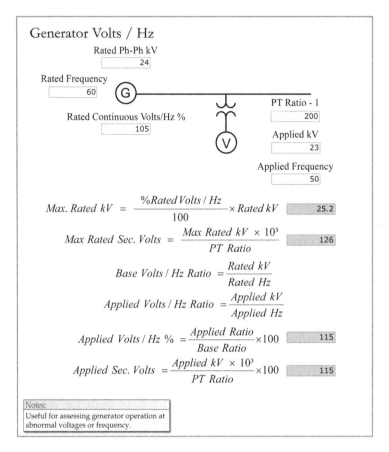

FIGURE 4.15
Generator volts/Hz.

Figure 4.16 shows the procedure for calculating the impact of operating the GSUT at the same voltage and frequency values used in the prior example for the generator. In this case, although the maximum rated 60-cycle voltages for the generator and transformer are both around 25 kV, the underfrequency event is a little more severe for the transformer with a higher volts/Hz percentage of 121%, compared to only 115% for the generator. This is because the base ratio for the transformer is lower due to the reduced kV rating of its primary winding.

Figures 4.17 and 4.18 illustrate the Westinghouse overexcitation withstand curves for generators and transformers, respectively. Although these curves tend to be a little more conservative than some manufacturers' curves, they are all in the same ballpark. Upon examining the generator curve, you will find that the generator can withstand the volts/Hz percentage of 115% (Figure 4.15) for 5 minutes. The generator curve is flat at 125% and greater with a withstand time of only 0.2 minutes or

Generator Protection

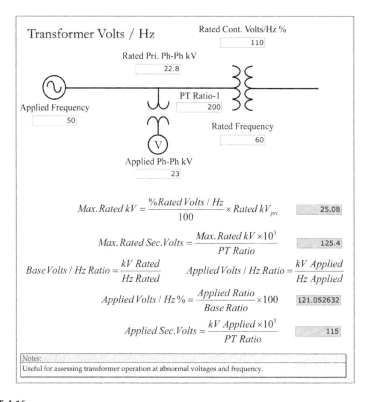

FIGURE 4.16
Transformer volts/Hz.

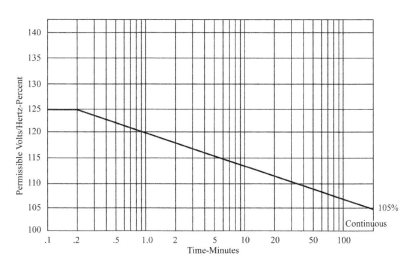

FIGURE 4.17
Westinghouse generator volts/Hz withstand curve (used with permission of Siemens Energy, Inc.).

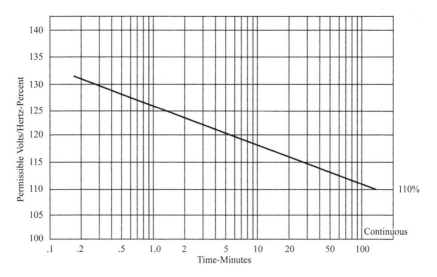

FIGURE 4.18
Westinghouse transformer volts/Hz withstand curve (used with permission of Siemens Energy, Inc.)

12 seconds. The connected transformer, however, is experiencing a volts/Hz percentage of 121%, and the associated withstand time extrapolated from the transformer curve is about 4.5 minutes. Therefore, in this case, the transformer is a little more limited.

Figure 4.19 provides suggested settings for a three-element volts/Hz protection scheme. The first element is set to alarm at 106%, or one percentage point above the continuous rating. The second element is for a curve with a minimum pickup point of 106% and a time delay of 45 seconds at 110%, and the third element has a minimum pickup point of 118% and a fixed time delay of 2 seconds. This setting complies with the recommendations of General Electric (GE) and has more than ample thermal margin to ensure protection of their machines. Consequently, in this case, it will protect both the generator and connected transformer under study. Experience has shown that the proposed settings will not cause nuisance tripping on a normal utility system; the suggested settings were applied by a large generation company that had experienced major generator core iron damage on two machines from overexcitation and did not want to push the thermal margins. Although overfluxing the core iron is considered a thermal event with time delays before damage, a large generator manufacturer did experience an almost instantaneous failure in their test pit. The iron is stacked on key-bars and an overfluxing event caused enough of a potential difference between bars to push high damaging current through the back iron area.

Figure 4.20 covers a simple yet effective single-element volts/Hz tripping scheme that utilizes an instantaneous plunger-type electromechanical voltage relay. The relay coil will mimic a volts/Hz function since the ampere

Generator Protection 177

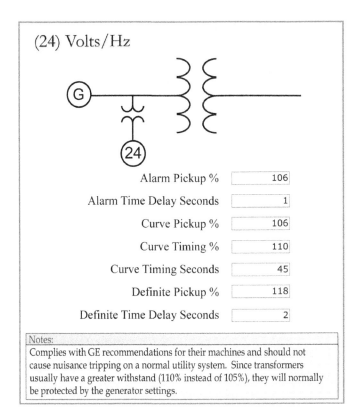

FIGURE 4.19
Volts/Hz settings.

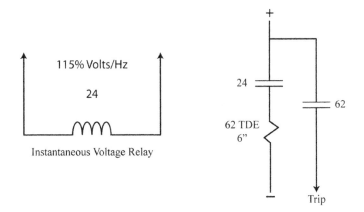

FIGURE 4.20
Simple single-element volts/Hz tripping scheme.

turns increase as the frequency is reduced from the lower impedance, which results in a lower actuation or trip point. Because only one element is provided, the suggested setting is 6 seconds at 115%. The withstand times below 115% are greater than 5 minutes and operator interdiction should be able to protect the machine. Since 6 seconds is well under the 12-second withstand time for severe events, the setting provides more than adequate margins for events 115% and higher.

Good practice is to have the AVR fed from a different set of potential transformers than the volts/Hz protection. The volts/Hz relays cannot operate if they do not receive potentials due to blown fuses or an operating failure to rack in the potential transformers or install the fuses. Depending on design details, if the AVR is on the same potential transformers as the volts/Hz relays, partial or full loss of potentials can cause the AVR to go into a full boost mode, and the volts/Hz protection may not be able to operate depending on which phases are affected by the loss of voltage and where the relays are connected. Volts/Hz is more of a problem during startup because of the underfrequency operations prior to synchronizing. Depending on the design details of the particular excitation system, volts/Hz excursions well over 140% are possible. If the excursions are high enough, auxiliary transformers are likely to saturate and trip on differential protection. There have also been cases where generating units have been tripped off-line while connected to the electrical system as a result of overexcitation levels that have exceeded 115%.

As discussed earlier, the easiest way to destroy a generator is from overexcitation; consequently, volts/Hz is probably the most important function in the generator protection package. The iron withstand time for having a full-load field on an unloaded machine is approximately 12 seconds, which is not enough time to allow operator intervention. There are numerous scenarios that can cause stator iron lamination overexcitation damage of large generators. Based on the author's reviews, about 75% of the time, this protection element was either not applied, or the time delay was too long to actually protect the machine, or the setting would prevent operations that were within the continuous rating of the generator.

4.4.3 Sync Check (25)

Another relatively easy way to damage a turbine/generator is to synchronize or parallel out of phase with the electrical system. Out-of-phase synchronizing operations can damage or reduce the remaining life of turbine/generator rotors and stationary components. Angular differences as little as 12° can instantly apply 1.5 per unit, or 150%, of full load torque to the turbine/generator shaft system. The 1.5 per-unit value was measured by a General Electric shaft torsional monitoring data acquisition system (EPRI project) at a large coal plant that was parallel to the 525-kV bulk power system with a 12° angular difference during synchronization. Plant operations acknowledged that the turbine deck really shook. Although turbines are

Generator Protection

generally built to withstand angular differences above 10°, the design margins are not normally disclosed, and most manufacturers recommend limiting out-of-phase synchronizing operations to no more than 10° maximum, including the angular advance that occurs during breaker closing. Generator sync-check relays should supervise both manual and automatic modes of operation to prevent turbine/generator damage from operator errors or from malfunctioning automatic synchronizing relays. For this reason, it is normal practice to have the sync-check relay function provided in a separate or different package than the automatic synchronizing relay (if equipped) to avoid failure modes that can affect both functions.

In the majority of designs, a clockwise rotation of the synchroscope indicates that the turbine/generator has a higher speed or frequency than the electrical system. This condition is desirable to reduce the possibility that the unit will be in a motoring mode of operation and trip on reverse power protection when the breaker is closed. The voltages should be somewhat matched during synchronizing, with a slightly higher generator voltage to ensure var flow into the system instead of the generator.

Figure 4.21 provides suggested settings for generator dynamic sync-check relays. The proposed default angles are 5° advance and 5° late.

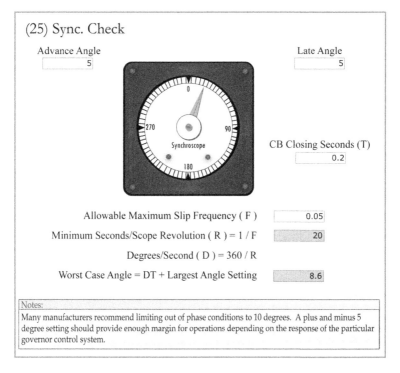

FIGURE 4.21
Synchronizing check relay settings.

The calculations consider circuit breaker closing times and maximum allowable slip rates and determines the minimum seconds per scope revolution and the worst-case angle. The minimum seconds per scope revolution is provided as a guide for operations or for setting autosynchronizing relays; the scope revolutions cannot go faster and be within the operating range of the sync-check relay. The worst-case angle assumes the breaker close signal is dispatched at the maximum late angle and that the breaker control scheme seals in. It then looks at the maximum allowable slip frequency (with a proposed setting of 0.05 Hz) and calculates the worst-case out-of-phase angle when the breaker contacts actually close. In this case, the worst-case angle of 8.6° complies with manufacturer's recommendations not to parallel if the angle exceeds 10°. Experience has shown that the proposed settings are practical and within the operating capability of most turbine-governor control systems.

Other available settings in the newer digital relays might include ratio correction factors for GSUT taps because generator potentials are normally compared to switchyard high-voltage potentials, and those ratios may not match precisely and cause unintended var flows. Although this author is less concerned about var flows from voltage differences because they do not represent real power and the shaft torques are minimal, as a general rule, plant operators should limit voltage mismatches to less than 5%.

Some of the newer digital sync-check relay functions also include slow breaker closing protection. Once the breaker control signal is dispatched, the control circuit seals in, and there is no way to abort the close operation because there is a 52a contact in series with the trip coil that prevents the coil from being energized until the breaker is actually closed. The slow breaker function could be set to operate breaker failure relaying to clear the adjacent breakers if the angular differences reach 10° indicating that the breaker is slow to close for mechanical reasons.

The maximum amount of symmetrical alternating current (AC) that flows during synchronizing at rated frequency can be approximated by the expression in Figure 4.22. Generator side voltage and ohms from Figures 4.1 and 4.2 were used in the calculation and reflected to the 765-kV side. The system's three-phase 765-kV short circuit ohms were transferred from Figure 4.3.

Figure 4.22 shows that the 765-kV current would be approximately 983 amps, and generator amps would be approximately 32,348 at 30° At 60°, 90°, and 180°, the approximated 765-kV currents for the parameters presented in the illustration would be 1897, 2682, and 3793 amps, respectively. The generator side current at 180° out would be around 124,767 amps. This does not include the DC component or peak asymmetrical current, which will also be present. Obviously, the generator and transformer windings need to be able to handle the peak electromechanical forces. The event is transitory in nature as the asymmetrical current decays and the generator pulls into step

Generator Protection 181

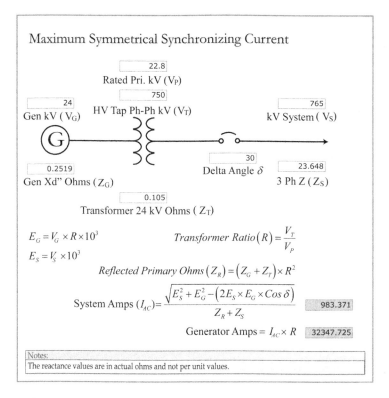

FIGURE 4.22
Maximum symmetrical synchronizing current.

with the system, and the power angle becomes congruent with the prime mover. The air gap torque is difficult to calculate and depends on electromechanical forces, circuit resistance, and the amount of power transfer from angular differences. Possible damage assessment is particularly complicated and associated with the peak torques, and the natural frequencies of the shaft and other mechanical components as the event decays. The associated apparatus may also have reduced life from other events or excursions, startup/shutdown cycles, or design or repair oversights, and major equipment damage may occur if the incident is severe enough.

With regard to the author's protection reviews or surveys, for about 50% of the units, the as-found angle windows were too large to actually protect the machine. It is true that 100% of the units did not apply slow breaker closing protection. If a unit breaker is slow to close for mechanical reasons during synchronizing, the turbogenerator system could be severely damaged as the angle keeps increasing due to the speed differences between the unit and the electrical system. Slow breaker closing during synchronizing can also cause a large upset to the high-voltage bulk power electrical system.

4.4.4 Reverse Power (32)

Reverse power protection for steam turbine generators is normally applied to protect turbine blades from overheating. The longer low-pressure blades can overheat from windage or air impingement in the absence of steam flow. Withstand times for steam turbines in a motoring mode of operation are usually in the neighborhood of 10 minutes. Motoring is not harmful to generators, so long as the proper level of excitation or DC field current is maintained. Steam turbines generally consume around 3% of their rated megawatts when motoring. This function is sometimes used on combustion turbines for flameout protection and to limit watt consumption from the electrical system, as motoring power is usually much greater for combustion turbines. With hydroturbines, there is concern about blade cavitations during conditions of low water flow.

Some designs have reverse power supervision of unit breakers to mitigate the possibility of overspeed conditions, and others use reverse power permissives in their automatic shutdown circuitry to delay opening unit breakers until the unit is in a motoring condition to mitigate overspeed possibilities. Figure 4.23 calculates the setting for a reverse power function that is applied to protect steam turbine low pressure (LP) blades from overheating. Because the actuation point needs to be well below the unit motoring requirement, a minimum pickup point of 0.5% of turbine rated megawatts (3.19 watts secondary), with a time delay of 20 seconds, is suggested. The suggested

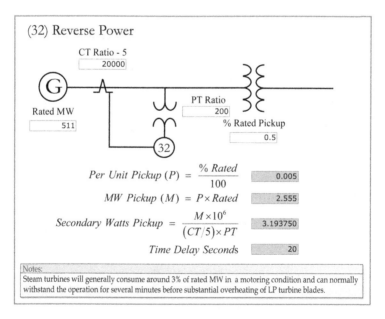

FIGURE 4.23
Reverse power relay settings.

Generator Protection 183

time delay is provided to reduce the possibility of nuisance tripping during synchronizing operations. Since the withstand times are so long, an argument can be made for even more time delay, but many in the industry will opt for a shorter time delay of 10 seconds or less. In the case of combustion turbines, the time delay may need to be reduced even further if the relay also provides flameout protection. The suggested settings should be compared to the turbine/generator manufacturer's motoring data to ensure that adequate margins are provided.

About 80% of the surveyed units were set with shorter time delays than necessary and could unnecessarily trip the unit during synchronizing operations. The longer low-pressure steam turbine blades can overheat from air impingement during motoring operations, with a typical withstand time of 10 minutes. Most of the as-found settings had time delays to trip in 10 seconds. The author's preference is to use a time delay of 20 seconds or longer to reduce nuisance tripping possibilities from momentary motoring excursions during synchronizing.

4.4.5 Disconnect (33M)

High-voltage switchyards that are in a ring bus or breaker-and-a-half configuration will have manual or motor operated disconnect switches (MOD) that can be opened to isolate generating units from the yard (see Figures 4.6 and 4.7). This allows operations to reclose the unit-connected high-voltage breakers to restore the integrity of the switchyard buses without energizing the generators. An inadvertent closure of a disconnect switch could energize an at-rest generator and cause severe damage to the machine if the duration is long enough. This can easily be avoided by installing a midposition limit switch that will trip the unit breakers open when the swing actuates the limit switch.

About 50% of the surveyed units were not equipped with limit switches for inadvertent energization protection.

4.4.6 Loss of Field (40)

Loss-of-field relays are applied primarily to protect synchronous generators from AC slip frequency currents that can circulate in the rotor during underexcited operation. Underexcitation events are not all that uncommon; consequently, this is an important function, and probably the third most significant element in a typical generator protection package. Generators can lose field or operate in an underexcited fashion for a variety of reasons, such as voltage regulator or power electronic failures, loss of excitation power supplies, opening of the laminated main leads under the retaining rings, opening of the DC field circuit breaker, arcing at the slip rings, and differences between system and generator voltages when the voltage regulator is on manual. The loss-of-field relay will not normally operate unless the generator loses synchronization with the electrical system and slips poles. However, an all-var load, if high enough, could get into the mho circle without slipping poles.

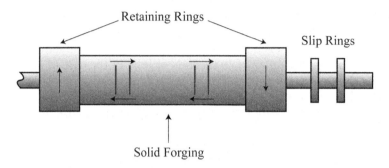

FIGURE 4.24
Cylindrical generator two-pole rotor slip frequency currents.

Figure 4.24 illustrates a two-pole cylindrical generator rotor and the associated slip frequency current flows during loss of excitation. Because a two-pole machine rotates at 3600 revolutions per minute (RPM), the centrifugal force is very high. The figure shows a pole face (opposite polarity on the other side); the field turn slots cannot be seen in this view and would be located at the top and bottom areas. The copper field turns (bars) are contained in the forging slots by aluminum or steel wedges that are in a dovetail fit. The circumferential slots in the pole face area are provided for two pole machines only, and not normally included with four pole turbine generators, in order to equalize rotor flexing. The retaining rings at each end contain the copper end turns, which are under stress from the rotational or centrifugal forces.

Synchronous generators are not designed to operate as induction machines, and they are not very good at it. Steady-state instability occurs when the generator slips poles, which happens when the power angle meets or exceeds 90°, as shown by the dashed line in the capability curves of Figure 4.25. This occurs because the magnetic coupling between the rotor and stator is reduced by the underexcited operation, causing the requisite forces to become too weak to maintain the output power, and synchronization with the electrical system is lost when the machine speeds up beyond its rated RPM. The speed difference cause induced slip frequency currents (0.1%–5%) depending on the initial load, to flow in the rotor body. The circulating currents can be damaging to the highly stressed retaining rings, circumferential slots, forging, and wedges due to overheating and arcing. The circulating current also tends to concentrate at the ends of the circumferential pole face slots in two pole machines and may overheat those areas. Although rotor-forging damage from the slip frequency currents is the primary concern, they can also cause pulsating torques that can affect the life of mechanical components. The output power on average is about 25% at full load; the var increases typically create stator currents in the range of 110%–220%, which can cause thermal damage to the stator windings after 20 or more seconds.

In addition to instability, the generator capability curve in Figure 4.25 shows the operating limitation of a typical generator at rated hydrogen

Generator Protection

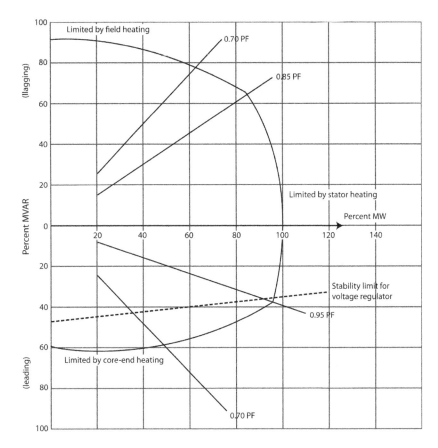

FIGURE 4.25
Generator capability curve.

pressure. The lagging area represents overexcitation, or boost vars, and the primary limitation is field winding overheating from the increase in field current I^2R losses. The leading area represents underexcitation, or buck vars, and is limited primarily from magnetic flux impingement of the core iron end packets. The stator core iron is divided into laminations in the direction of the rotor magnetic flux linkage to reduce eddy current losses by elongating the path and increasing the impedance for annular current flow.

During normal operation, the leakage air-gap flux (vector sum of the rotor and stator magnetic fluxes) strikes the end iron at right angles near the retaining rings. This perpendicular flux can cause a heating effect that is 100 times greater than the normal eddy current flow produced by a flux that flows from the pole face into the edges of the laminated stator iron. Improved designs are able to reduce the end iron perpendicular flux with higher reluctance nonmagnetic retaining rings, the addition of copper flux shields beneath the stator end windings that produce a counter flux from

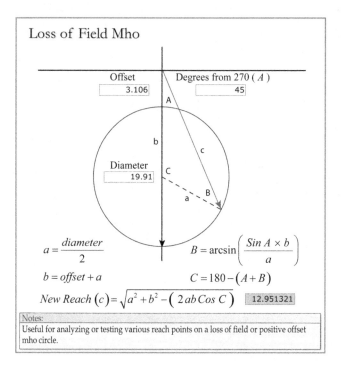

FIGURE 4.26
Loss-of-field mho circle.

the corresponding eddy or annular current flow in the shield, and increased air-gap spacing configurations. However, underexcited operation can also reduce the saturation level of the retaining rings, lowering the magnetic circuit reluctance, with a corresponding increase in the perpendicular flux.

Loss-of-field relays will operate for impedance swings that are inside their mho circles. The normal practice is to positively offset the mho circle by half the generator transient reactance ohms, as shown in Figure 4.26, to avoid stable swings that may get into this area during system disturbances and faults. Underexcitation will cause the generator to absorb lagging vars from the system (quadrant 4) that appear to be leading if the reference direction is watt flow from the generator to the system. In addition to the system feeding the var requirements of the generator, it feeds the var requirements of any connected transformers (that is, generator step-up, auxiliary, and excitation transformers). Under normal conditions, generators supply watts and lagging vars to their associated step-up transformers and operate in the first quadrant. The figure also shows the trig calculation procedures for determining the ohmic values of different reach points on the mho circle.

Figure 4.27 illustrates an impedance swing caused by a loss of field event at full load. It starts in the first quadrant in the load range and then swings into

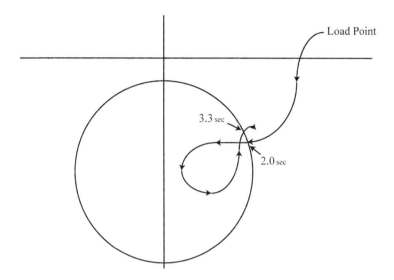

FIGURE 4.27
Loss-of-field impedance swing.

the impedance circle. It can take 2–8 seconds to enter the mho circle, where it will continue to move and eventually exit the circle 1–10 seconds later, depending on the associated parameters, and then the slip cycle repeats. The load on the unit at the time of the loss of field event affects the impedance locus of the event. The higher the load, the faster the impedance swing is. Several other factors also affect the impedance locus, including strength of the electrical system, turbine/generator inertia constants, and type of event (shorted or open field circuit).

Figure 4.28 provides the suggested settings for a two-zone loss of excitation scheme. Both zones are offset by the same amount (3.1, or half the generator transient reactance ohms). The suggested diameter for the zone 1 circle is 1 per unit, or 19.9 ohms, with a time delay of 0.17 seconds. The suggested diameter for the zone 2 circle is the generator synchronous reactance, or 34.5 ohms, with twice the time delay (0.34 seconds). The offset and mho circle diameters are quite standard across the industry. If only one zone of protection is applied, the zone 2 reach with the zone 1 time delay is recommended. However, the recommended time delays for two zones of protection differ, depending on the particular source paper, from as little as no time delay for zone 1 to as much as 1 second or longer for zone 2. For many years, the standard practice was to apply an electromechanical loss-of-field mho impedance relay with only one zone of protection, which was set according to zone 2 in Figure 4.28, but with a built-in time delay of around 0.17 seconds. This setting did an excellent job of preventing rotor damage and did not cause nuisance tripping. Because the zone 2 mho circle represents a less severe event, doubling the time delay seems reasonable. If

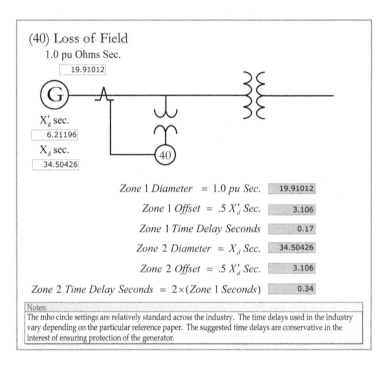

FIGURE 4.28
Loss-of-field relay settings.

the time delay is too long, the impedance swing could exit the mho circle before the function can time out.

The minimum excitation limiter (MEL) or reactive ampere limit in the AVR should be set by excitation engineers to prevent excursions into the loss-of-field mho circle. Ideally, a 20% margin is used; that is, the MEL will start its control function well before the loss-of-field relay can operate. For North American Electric Reliability Corporation (NERC) requirements, the actual MEL curve should be determined by testing the AVR. This is really a specialty area involving the particular MEL setting parameters, conversions to ohms, control response time delays, machine stability limits, slip frequencies, and core end iron overheating. Figure 4.29 illustrates the loss-of-field mho circle, generator capability curve points, and an ideal MEL control curve, all plotted on the same base using secondary or relay ohmic values. The ohmic conversion steps, as shown in the figure, have been presented previously. The figure shows four capability curve output boxes C1–C4 in ohmic values; the inputs for each output box are MW and MVAR at the maximum capability curve values. The points on the generator capability curve, C1–C4 respectively, in Figure 4.25 are:

- 1 PF (511 MW)
- 0.95 PF leading approximately (496 MW and 38 MVAR)

Generator Protection

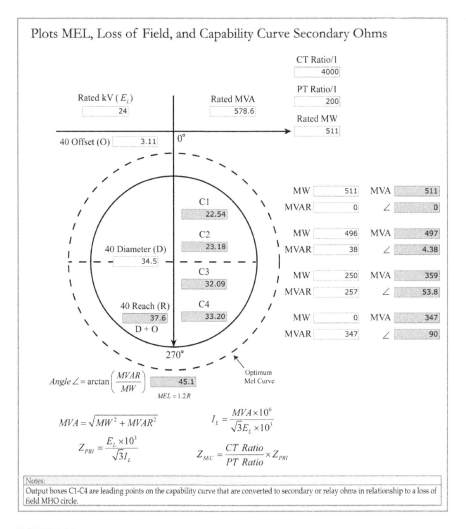

FIGURE 4.29
Generator capability curve, loss-of-field, and Mel plots.

- 0.70 PF leading approximately (250 MW and 257 MVAR)
- 0.00 PF leading approximately (347 MVAR)

The respective MVA values are:

- 511
- 497
- 359
- 347

And the respective secondary ohmic values are:

- 22.54 ohms
- 23.18 ohms
- 32.09 ohms
- 33.20 ohms

The maximum watt-only point is outside the loss-of-field mho circle. The 0.95 PF leading point is close to the AVR stability limit and is also outside the loss-of-field mho circle, but further var increases could actuate the loss-of-field relay if it loses synchronization and slips poles. The 0.7 leading power factor point is where there is a particular concern about overheating the core iron end packets, and it is also outside the loss-of-field mho circle. The maximum leading var-only area is inside the loss-of-field mho circle and would not require pole slipping to actuate the loss of field protection, assuming that the MEL control function does not sufficiently limit the leading vars into the generator.

Loss of field is probably the third most important protection element for large generators because of the relatively high rate of occurrence. For approximately 20% of the units in the author's reviews, the zone 2 time delays were too long to ensure detection. When a generator loses excitation, it speeds up to function as an induction machine. Synchronous machines are not designed to operate as induction generators, and the induced currents can damage highly stressed rotor components in cylindrical rotor machine designs. There is a slip cycle associated with induction generator operation, and some of the as-found zone 2 time delays were too long to ensure the capture of all underexcitation events. The loss-of-field swing can exit the mho circle in as little as 1 second, as presented in Figure 4.27. Consequently, in order to provide margin, the zone 2 time delay should be less than 1 second.

4.4.7 Generator Deexcitation (41)

As discussed in Chapter 2, a fault between the unit auxiliary transformer and its low side breaker will trip instantaneously (six cycles or 0.1 seconds) from the differential protection. However, for designs that do not utilize isolated phase medium-voltage bus breakers, the generators will continue to feed after tripping fault current that is driven by trapped magnetic flux in the machines during coastdown. If the unit has a direct DC field breaker, a resistor will be inserted by a third pole during tripping in order to force the trapped flux and associated fault current to zero in approximately 4 seconds. Larger machines use indirect field breakers that interrupt the AC supply to the excitation system, and in most cases, they can be equipped with deexcitation circuitry that reverses the field polarity to mimic discharge resistors that also will bring the current to zero in approximately 4 seconds. Without the deexcitation circuitry, it can take longer than 40 seconds for the coastdown fault current to decay, and the unit auxiliary transformer will likely be destroyed in the process. It also represents a significant hazard to personnel who may be in the area.

Units that equipped with excitation transformers on or near the generator output terminals also have the coastdown problem, even if they are equipped with isolated phase bus circuit breakers, since the breaker is downstream of the fault. Usually, the excitation transformers are not equipped with high side breakers or fuses and a short circuit can represent a significant hazard to personnel that are in the area because of the coastdown energy. Installing current-limiting fuses on the primary side of excitation transformers is highly recommended.

Approximately 50% of the units surveyed with indirect field breakers did not apply deexcitation circuitry to reduce the coastdown energy into faulted auxiliary and excitation transformers. In addition to severe transformer damage, it would be extremely hazardous for personnel who happened to be in the immediate area.

4.4.8 Negative Phase Sequence (46)

Although a rare event, negative phase sequence protection was applied for all generators that were 50 MW and larger, which the author has reviewed or surveyed in the past. The negative phase sequence element is used to protect the generator rotor from reverse-rotation, double-frequency currents that are induced into the rotor forging if the stator currents are not balanced (Figure 4.30). Significant unbalances can result from a circuit breaker pole that does not make or close, or a phase conductor that is open prior to energization. It also provides overcurrent heating protection for system phase–phase and phase–ground faults that do not clear properly. High-voltage system ground faults look like phase–phase faults to generators that feed delta–wye step-up transformers. However, because the negative phase sequence tripping time delays are quite long, the unit will likely lose synchronization and go out of step before the negative phase sequence relay can actuate to isolate the generator from local switchyard faults. As with loss of field, these currents

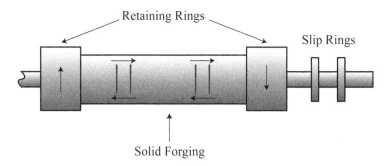

FIGURE 4.30
Negative phase sequence, 120-Hz reverse rotation-induced currents.

can cause overheating damage to the retaining rings, wedges, and other highly stressed rotor components and are potentially more damaging than loss-of-field, induced slip frequency currents because the higher frequency of 120 Hz causes the current to concentrate more on the outside surfaces (i.e., skin effect). The reverse rotation–induced rotor currents can also produce a slight negative torque, resulting in torsional oscillations.

Per ANSI/IEEE Standard C50.13, generators are rated for both continuous and short-time thermal capabilities for carrying negative phase sequence currents. The ratings vary depending on the size of the generator and whether it is directly or indirectly cooled. Directly cooled generators have gas or water passages that cool the stator conductors and special gas passages or zones that cool the rotor windings. Cylindrical rotor and salient pole generators equipped with amortisseur windings rated 960 MVA and less will normally have a continuous rating of at least 8%, as per standards. Salient pole generators with amortisseur windings and indirectly cooled cylindrical rotor machines will typically have a continuous rating of 10%. All generators rated 800 MVA and less will have a short time of 1 per-unit I^2T negative phase sequence thermal rating (or K) of at least 10. Indirectly cooled cylindrical rotor machines will have a K of 20, and salient pole generators a K of 40. In general, smaller and indirectly cooled machines will have a greater thermal capacity for negative phase sequence currents because direct cooled machines have a relatively larger capacity or megavolt-amps and the cooling system is not as effective for the rotor. Generators larger than 800 MVA will have a reduced thermal capability to carry negative phase sequence rotor currents. Although tripping from this function is somewhat rare on large utility systems, negative phase sequence protection is always provided due to the overheating limitations of generator rotors.

Figure 4.31 shows a FLA of 13,919 on A and B phases and zero current on C phase. This is really an A-B current; consequently, the A-phase angle is shown at 0° and the B-phase angle at 180°. To resolve the symmetrical component negative phase sequence I_2 current magnitude, the angle for B-phase needs to be advanced by 240° and the angle for C-phase by 120°. After the angular advancements on B-phase and C-phase, the negative phase sequence magnitude of 8036 can be determined by dividing the vector sum by 3.

As a matter of technical interest, Figure 4.32 covers a similar process for determining the positive phase sequence magnitudes. For the positive phase sequence, the B-phase angle is advanced by 120° and the C-phase angle by 240°. As shown in the figures, with phase–phase events, the positive and negative components are equal.

Also as a matter of technical interest, Figure 4.33 shows the calculation for the symmetrical component zero phase sequence component. The zero phase sequence magnitude is simply the vector sum of the actual angles (no angular advancement) divided by 3. In this case, the vector sum is zero, as A and B phases are displaced by 180°. Accordingly, a phase–phase event does not contain a zero phase sequence component. The actual ground current

Generator Protection

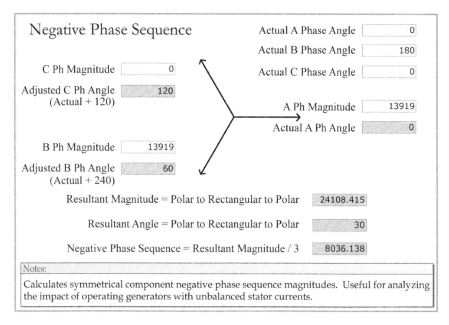

FIGURE 4.31
Negative phase sequence magnitude.

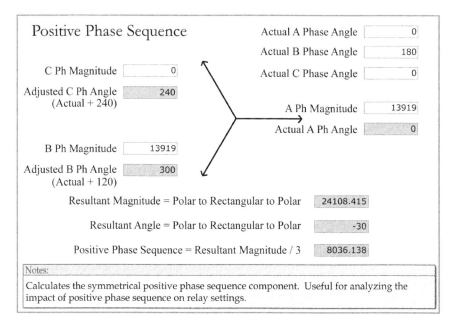

FIGURE 4.32
Positive phase sequence magnitude.

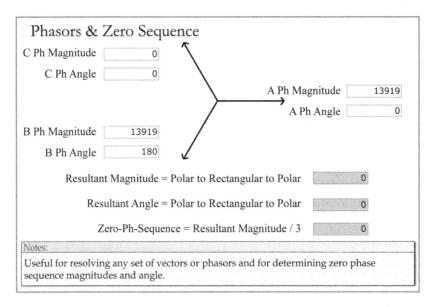

FIGURE 4.33
Zero phase sequence magnitude.

would be the symmetrical component zero phase sequence multiplied by 3, or simply the vector sum of the actual currents.

The generator withstand is based on the symmetrical component method of calculating negative phase sequence current. Figure 4.34 calculates the generator-withstand time before rotor damage can occur and discloses that the 578.6-MVA generator with a K of 10 can withstand 8036 amps, or 57.7% of the negative phase sequence I_2 stator current for 30 seconds before loss of life or rotor damage starts to occur.

The proposed negative phase sequence settings for the 578.6 MVA generator are presented in Figure 4.35. The suggested alarm and trip pickup points are 6% alarm and 8% trip. The suggested relay timing is for a generator with a K of 10 and an I^2T or K value of 9 (90% of the rotor thermal rating is depleted). These settings should not cause nuisance tripping or coordination problems on a normal utility system.

4.4.9 Inadvertent Energization (50/27)

Inadvertent energization of at-rest generating units is not all that uncommon. This can occur if unit switchyard disconnects that isolate units from energized ring buses or breaker-and-a-half configurations are inadvertently closed, or if the unit breakers are closed when the isolating disconnects are closed. Some utilities install midposition limit switches that will trip the associated high-voltage breakers when the disconnect swing hits the midpoint of its travel. Although a properly designed sync-check relay

Generator Protection

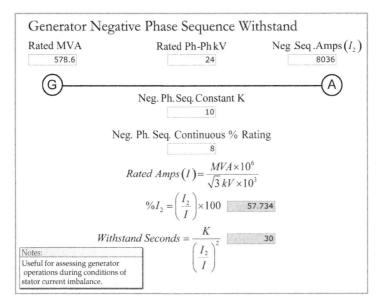

FIGURE 4.34
Negative phase sequence withstand.

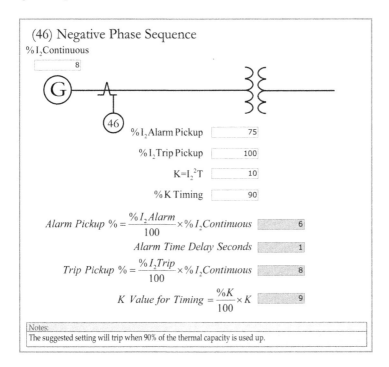

FIGURE 4.35
Negative phase sequence relay settings.

scheme should prevent inadvertent closures of generator breakers or unit switchyard breakers, one way or another, it seems to happen in the industry. If the inadvertent energization lasts long enough, catastrophic generator damage can occur. Inadvertent energization can also occur through the UAT, but the current is much lower because of the relatively high auxiliary transformer impedance, and the auxiliary transformer overcurrent relays should be able to mitigate damage to the unit. During major outages, there is also concern that protective functions that might limit the duration may be out of service for relay testing purposes, or that the potential transformers could be racked out or otherwise disabled.

Basically, the generator is being started across the line as a motor; generators are not designed for this purpose, and severe overheating of stator and rotor components can occur. Additionally, the rotor wedges and amortisseur interfaces (where applicable) will not be held securely by centrifugal force, as they are during normal operation, and the loose fits can cause additional damage from arcing. Figure 4.36 approximates the expected symmetrical current from energizing an at-rest machine. The base current is derived from the transformer megavolt-amps. The three-phase short circuit current on the 22.8-kV side of the GSUT is converted to ohms, and then to per-unit ohms. The corrected generator unsaturated negative phase sequence per-unit impedance

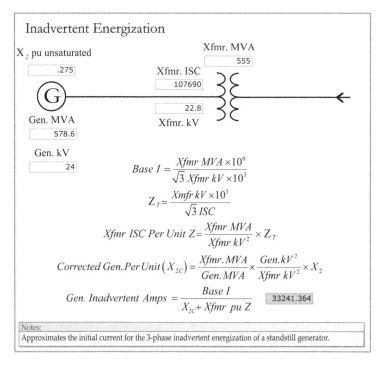

FIGURE 4.36
Inadvertent energization current.

Generator Protection

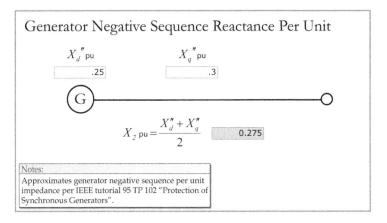

FIGURE 4.37
Negative phase sequence reactance.

X2c can then be added to the transformer's 22.8-kV short circuit per-unit impedance and divided into the base current to determine the at-rest energization current. In this example, the initial symmetrical inrush current will be around 33,241 amps, or 239% of the full-load current. In-service loss of field, reverse power or antimotoring, and generator and switchyard backup impedance relays may be able to see the event, depending on their specific settings.

If the generator negative phase sequence reactance is not available, the per-unit value can be approximated by summing the direct and quadrature per-unit subtransient reactances and dividing by 2, as shown in Figure 4.37. Newer microprocessor-based relays generally have logic functions that detect and trip for inadvertent energization conditions. This logic often involves recognizing that the unit is off-line by the absence of voltage for a specified period of time and then measuring a sudden inrush of current.

About 75% of the units that the author reviewed did not apply inadvertent energization protection. This protection is necessary to protect generator rotors from severe damage during standstill energizations. Basically, the generator is trying to act like a motor that accelerates from a locked rotor condition during across-the-line starting. This can cause severe damage to the highly stressed rotor components if the duration is long enough. Additionally, about 50% of units equipped with motor or manual operated isolating disconnects did not use midposition limit switches to trip the high-voltage breakers open in the event that an operator or electrician mistakenly tries to close the disconnect.

4.4.10 Breaker Failure (50BF)

For designs that apply a breaker at the generator bus, some of the newer digital relays provide logic elements that can be mapped to provide breaker failure protection for events that are normally isolated by opening the generator breaker. As with high-voltage switchyard breakers, three logic elements are necessary: a specified level of current, an energized breaker trip coil circuit, and a time delay that is longer than the breaker trip time with margin. Tripping would include the unit, associated high-voltage breakers, and lower-voltage auxiliary transformer breakers.

Approximately 75% of the author's surveyed units did not apply breaker failure protection for unit designs that were equipped with generator isolated phase bus breakers.

4.4.11 GSUT Instantaneous Neutral Overcurrent Ground Fault (50N)

As discussed in Chapter 2, in terms of standards, delta–wye transformers are not normally designed to withstand close-in ground faults. The windings are required to withstand the electromechanical forces exerted by three-phase and phase–phase short circuits for only 2 seconds, and not the higher current magnitudes that can be experienced during close-in radial phase-ground short circuits. As illustrated in Figure 4.38, the close-in phase-ground fault

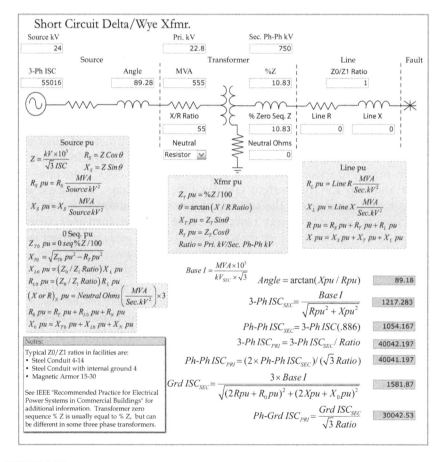

FIGURE 4.38
GSUT short circuit currents.

current is approximately 130% higher (1582/1217) than a three-phase short circuit condition. For this reason, some utilities install GSUT neutral reactors to limit the ground fault current to the three-phase magnitude. This is only a problem when a ground develops between the GSUT and the line to the high-voltage switchyard breakers prior to synchronizing.

Following synchronization and the closure of the high-voltage unit breakers, the GSUT will function as a ground bank and the total ground fault current will be much higher, but split between all three windings, thereby reducing the amount of current in each winding and the resulting electromechanical forces. An instantaneous overcurrent element can be utilized in the GSUT's neutral-CT circuit to trip the unit (deenergizing the GSUT), for high-voltage (HV) ground faults that occur before synchronizing (HV breakers open) to back up the transformer differential scheme and ensure protection of the GSUT. The element could be set for around 1 amp secondary, with a time delay of three to six cycles to ensure that the differential relays operate first. In the interest of downstream coordination, the closing of the high-voltage unit breakers should disarm this protective function.

100% of the author's reviewed units did not apply GSUT presynchronized instantaneous neutral overcurrent ground fault protection.

4.4.12 GSUT Neutral Overcurrent Breaker Pole Flashover (50PF)

With double-breaker, double-bus switchyard configurations, the high-voltage unit breakers are open and energized from their respective buses when the units are off-line. Ring bus and breaker-and-a-half configurations are also in the same operating condition after preliminary switching is completed prior to synchronizing the unit. If there is a breach in dielectric in one phase of the energized high-voltage unit breakers due to moisture, contamination, or loss of gas pressure, the resulting flashover will push phase-neutral current through the step-up transformer and cause high phase–phase currents (usually around 175% of the generator's full-load-rated amps) into the generator. Although high-voltage unit breaker pole flashover is more likely during synchronizing operations where double voltage can exist as the synchroscope rotates, a pole flashover can occur anytime the circuit breaker dielectric degrades from moisture intrusion, contamination, or a reduction in dielectric gas pressure. However, energizing an at-rest generator from a pole flashover would have the likelihood of being more damaging than a more transitory event, where the generator is at or near synchronous speed. An at-rest machine will not rotate with a single-phase energization, and centrifugal force will not be available to hold the rotor components securely, increasing the possibility of forging arc damage.

Figure 4.39 illustrates the current flow paths on both sides of the GSUT for a C-phase pole flashover event and presents an approximation expression for determining the high and low side currents into the at-rest generator.

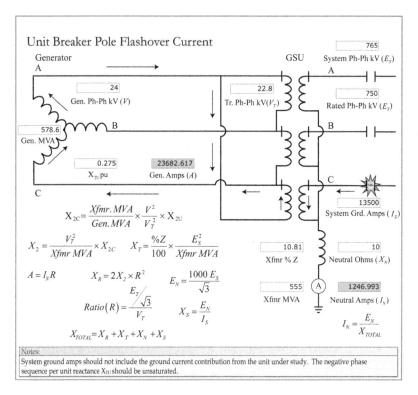

FIGURE 4.39
Pole flashover current.

First, the generator per-unit negative phase sequence unsaturated reactance (corrected to the transformer MVA and low side voltage), the transformer's %Z (for the wye-winding or 765-kV side), and the high-voltage system ground fault current are converted to ohmic values. Then double the corrected generator negative sequence ohms (accounts for current flow into both the A and C phase windings) are reflected to the system side of the GSUT. The total circuit, high-voltage side ohms would be the summation of the secondary-side reflected ohms plus the transformer high-side ohms plus the neutral reactor ohms plus the system ground fault ohms. The neutral symmetrical current can now be determined by dividing the system phase-neutral voltage by the total ohms. In this example, the transformer neutral current multiplied by the transformer winding turns ratio provides a generator side current of 23,683 amps, or approximately 170% of the rated generator amps of 13,919.

Normally, unit breaker pole flashover is detected by sensing GSUT neutral current and 52b logic that indicates that the unit breakers are open. Because the high-voltage unit breakers are already open, it is necessary to initiate breaker failure protection to isolate the faulted circuit breaker. Where two unit breakers are involved, breaker failure logic should be able to determine which breaker has the current flow and trip accordingly. The author's suggestion is to trip at 1 amp, with a six-cycle time delay to mitigate nuisance tripping from transient conditions. Pole flashover is not of major concern with generator bus breakers because the ground current would be limited by the high-impedance stator grounding scheme, and the stator ground relay should be able to detect the event.

100% of the reviewed units did not apply GSUT instantaneous neutral overcurrent protection for pole flashover protection. Not only can it be damaging to the generator, the breaker itself may be severely damaged by the high-temperature electrical arcing. If allowed to continue, the event is likely to evolve into a switchyard high-voltage ground fault.

4.4.13 Transformer GSUT Ground Bank Neutral Overcurrent (51GB)

For designs that utilize generator bus medium-voltage breakers for synchronizing, the GSUT normally remains energized from the electrical system after unit tripping to backfeed the unit auxiliary power transformer. In this mode of operation, it will also act as a system ground bank and deliver a high level of current to local high-voltage switchyard ground faults. Although this level of current is much higher than a presynchronized ground fault in the line from the unit to the yard, the electromechanical forces in the windings are much lower since the total neutral current is shared by all three phases and are normally within the transformer electromechanical force withstand of 2 seconds. However, in this mode of operation, ground bank overcurrent protection is not normally provided. Since the generator breaker is already open, there is nothing in the generator package that can protect the transformer from overcurrent conditions. A minimum trip of 125% of GSUT FLA with a time delay of 2 seconds at the unit ground current contribution to the high-voltage bus per system studies is suggested to ensure coordination with downstream system relays. If system studies are not available, suggest calculating the ground bank amps assuming an infinite source bus, and timing at 80% of that current magnitude. Figure 4.40 shows the infinite source GSUT ground bank current of 4024 amps. Because the generator isolated phase bus breaker is already open, this function needs to trip the unit high-voltage breakers in order to isolate the GSUT from the electrical system.

100% of the authors' reviewed units did not apply GSUT neutral time overcurrent protection for ground bank protection.

4.4.14 Overvoltage (59)

As the name implies, this function operates for generator 60-Hz overvoltage. Because the volts/Hz function normally covers 60-Hz overvoltage, this author does not normally put the 59 function into service.

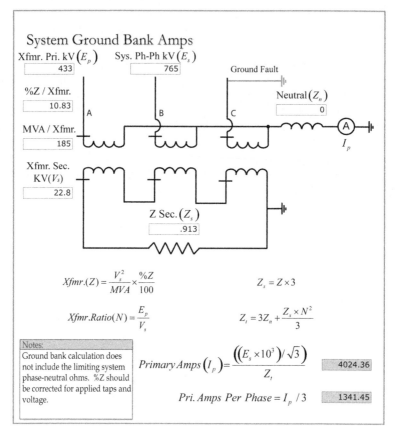

FIGURE 4.40
GSUT ground bank current.

4.4.15 Isolated Phase Bus Ground Detector (59BG)

For designs that utilize a generator bus breaker, when the unit is tripped, the generator bus breaker is opened, and the GSUT normally remains energized to backfeed the unit auxiliary power transformer. For this mode of operation, because the generator stator ground scheme is no longer active, a second ground detector (wye-grounded–broken delta) on the GSUT side of the bus is required to alert operations that there is a ground, as well as to reduce 6 per-unit arcing ground transient voltages to 2.4 per unit to substantially reduce the possibilities for development into full phase-phase or three-phase short circuits, as discussed in Chapter 3. Because of the difficulty with coordinating the bus ground detector with the generator stator ground scheme when the unit is on-line, and also because there is the possibility of nuisance tripping from blown fuses, it is suggested that the bus ground detector alarm only and not trip for ground conditions. If the generator stator ground scheme trips the unit off-line and the bus ground detector alarm does not clear, then the ground will be in or beyond the generator breaker; otherwise, it will be located on the generator side of the bus breaker. Because the grounded-wye broken delta scheme provides a ground or zero sequence path, it will also detect the third-harmonic voltage that develops when the generator is on-line, and it will also provide a ground path for a 100% stator winding subharmonic injection scheme. As mentioned in Chapter 3, the alarm-only wye–broken delta scheme is not compatible with surge capacitors; the higher voltages that develop during ground faults can fail surge capacitors thermally from the higher current flow, unless they are rated to handle the full phase to phase voltage to ground.

The bus ground detector is normally connected in a wye–broken delta configuration and can be somewhat unstable, depending on circuit impedance parameters. This is referred to as *neutral instability* or *three-phase ferroresonance*, which can blow fuses if it takes off. A blown fuse may erroneously cause the ground detector to actuate. The design details for a wye-grounded–broken delta ground detectors were covered in Chapter 3.

Ground faults on floating delta systems can also be detected with a single potential transformer that is connected from one phase to ground, as shown in Figure 4.41. If the ground occurs on the same phase, the voltage balance relay will close the moving contact to the left from the reduction in voltage, and if it occurs on another phase, the relay will close the moving contact to the right due to the increase in voltage. This single potential transformer scheme has the following disadvantages:

- It does not limit the transient voltage from arcing spitting grounds.
- It does not allow dual usage of the potential transformer secondary voltage since grounds will affect the magnitude of the voltage.

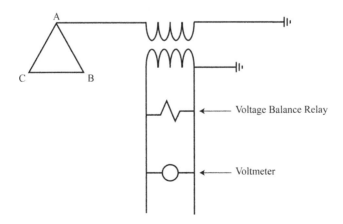

FIGURE 4.41
Single potential transformer ground detection.

- It will not mitigate high voltages that can be coupled through the inner-winding capacitance of the GSUT.
- It will likely increase the phase to neutral voltage on two phases since the potential transformer impedance pulls the capacitive neutral point toward the connected phase.

A common protection oversight with the advent of smaller combustion turbine units that were equipped with generator bus breakers was not to include a ground detector between the generator breaker and step-up transformers, which is necessary for the reasons previously outlined. This was first brought to the author's attention when a combustion turbine unit tripped off-line from its stator ground scheme and station technicians could not locate the problem. Later, an operator walking by an outdoor, single-phase potential transformer compartment for synchronizing purposes on the energized side of the bus circuit breaker heard spitting and arcing inside the compartment. The potential transformer was being fed from two different phases via vertical, 30-foot isolated phase bus tubes that connected to the horizontal main bus. The vertical sections were equipped with neoprene gasket sections for expansion and contraction, which degraded over time and permitted moisture intrusion. The station investigated the cost to bring the third phase down and to increase the size of the compartment for two different generating units for the standard three transformer ground detector. The cost came to $200,000 in 1980s dollars, which was a large expenditure for an alarm-only scheme at that time.

In thinking about the single potential transformer scheme, the author came up with a low-cost, two-transformer solution that would substantially reduce the costs without the disadvantages of the single-transformer circuit. In this scheme, illustrated in Figure 4.42, resistors are installed between two of the three phase–phase combinations to mitigate arcing transients and coupling voltages; the outside phases can still be used for synchronizing and will not be affected by system grounds, and the capacitive neutral shift will not be as large. The voltage alarm relay will operate in the same way as the single potential transformer scheme described previously. Two epoxy cast potential transformers were small enough to fit inside the existing outdoor compartment. Depending on which phase was grounded, the scheme may or may not mitigate transient voltages quite as effectively as the wye-grounded–broken delta configuration. Unlike the other schemes covered in Chapter 3, the resistors remain energized even though the system is free of grounds; basically, they act as heaters and will reduce moisture intrusion into the outdoor cabinet. While almost ideal, the cost of providing the alarm-only ground detection was substantially reduced by the two-transformer scheme and has proven to be reliable over the ensuing years.

Generator Protection

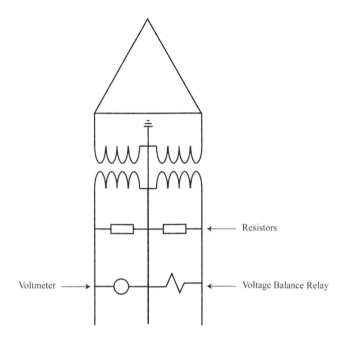

FIGURE 4.42
Two-transformer ground detection scheme.

Roughly 80% of the designs for units equipped with generator isolated-phase bus breakers during the author's reviews did not apply a second bus ground detector for the area beyond the isolated phase bus circuit breaker; and even if they did, the secondary side was not equipped with the proper secondary resistor to mitigate the high transient voltages associated with arcing spitting grounds.

4.4.16 Loss of Potential (60)

The normal practice is to prevent nuisance tripping from blown potential transformer fuse conditions. Older electromechanical relays compared one set of potential transformers to another and blocked the tripping of voltage-sensitive protective relays that might actuate for blown potential transformer fuses. Newer digital schemes look at only one set of potential transformers, and if there is a 10% reduction in voltage with no change in generator current flow, a blown fuse is assumed; in addition to alarming, the 60 function will normally block tripping of the 21, 40, 50/27, 51 V, and 78 functions. Protective functions that operate on increased voltages would not require blocking for blown fuse conditions, and the under-over frequency elements normally require a minimum voltage for measurement purposes.

4.4.17 Stator Ground (64)

Generator stator ground protection is applied to protect generator stators and all other directly connected electrical apparatus on the output side from ground faults. This can include the primary windings of the main step-up, auxiliary, excitation, and instrument transformers, associated buses and conductors, generator bus circuit breakers, and surge capacitors and arrestors (if equipped). As discussed in the "High-Impedance Grounding" section of Chapter 3, a grounding transformer is connected between the generator neutral and the station ground grid. High-impedance (resistance) grounding is preferred because generators are not typically designed to carry large unlimited close-in ground fault currents that are above the three-phase level, and there is also concern with core iron damage from ground faults involving the stator iron, which are expensive to repair. The generator stator ground overvoltage relay is connected on the secondary side of the grounding transformer. This function is often referred to as a 59G instead of 64, but the author prefers 64 to avoid confusion with other 59 elements. Relays used for this purpose are normally desensitized to the third-harmonic voltages that appear on the neutral when the machine is energized.

Figure 4.43 provides suggested settings for the stator ground overvoltage tripping relay. A default minimum trip or pickup level of 5% of the 100% ground voltage is suggested, with a time delay of 1 second for a 100% ground fault. The pickup level should be higher than typical third-harmonic voltage levels in order to mitigate any possibilities for nuisance tripping, and the timing is intended to mitigate iron damage from low-current, prolonged ground fault conditions. The stator ground scheme can also detect system ground faults on the high-voltage side of the GSUT that are coupled through the step-up transformer inner-winding capacitance, and the 1-second time delay should coordinate with the higher-voltage electrical system ground protection. The closer a ground fault is to the neutral end of the stator winding, the lower the driving voltage is. Consequently, this scheme may not have the sensitivity to detect grounds for roughly the last 10% of the neutral end of the stator winding, and it cannot provide 100% protection of the stator windings.

Generating stations built before the 1970s generally used open-delta potential transformers for measuring bus voltages. Newer station designs normally apply wye–wye potential transformers to improve the integrity of the isolated phase buses. Each potential transformer is connected phase to ground, which negates the need to bring two different phases into the same segregated potential transformer compartment, maintaining the isolation and reducing the possibility of generator phase–phase faults. However, wye–wye potential transformer connections, with both primary and secondary neutrals grounded, allow the stator ground protection scheme to detect secondary-side ground faults. Either the stator ground protection time delay may need to be increased (at the risk of iron damage) to coordinate with

Generator Protection

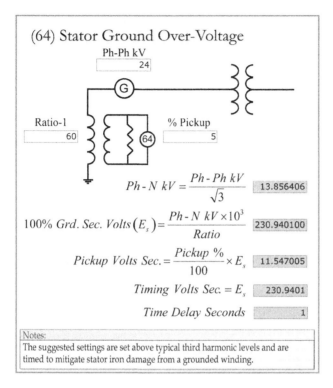

FIGURE 4.43
Conventional generator stator ground scheme protection.

potential transformer secondary-side fuses, or the grounding method at the secondary side may need to be modified. Instead of grounding the secondary neutral point, one of the phases could be grounded with the neutral floating, as illustrated in Figure 4.44. The stator ground scheme can now only see faults between the secondary neutral and the grounded phase; because the

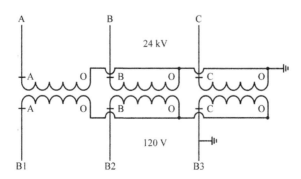

FIGURE 4.44
Modified wye–wye potential transformer grounding.

neutral wiring is only between the transformer compartments, the exposure is quite low and a fault to the neutral conductor, while possible, is unlikely. The foregoing assumes that a grounded secondary-side neutral is not required by the particular relay elements applied for generator protection, which is usually the case.

Figure 4.45 illustrates the stator winding construction for two- and four-pole, indirectly cooled, cylindrical rotor generators and shows two single-turn bars in each core iron slot. Unlike salient pole hydrogenerators, which have multiple turns in each bar and roughly 100 V of insulation between turns, cylindrical rotor machines used in fossil (oil, gas, and coal) and nuclear plants only have one turn per bar. The stator bar ground wall insulation is not graded and has the same rating at both ends of the windings of a 95-kV basic impulse level (BIL), even though the in-service voltage stress at the neutral end is dramatically lower. The total insulation between turns is roughly double; consequently, transient voltages breaking down the turn insulation and causing shorted turns is not a problem with the higher-speed cylindrical rotor designs. Accordingly, cylindrical rotor stator windings do not normally need to be protected from transient voltages with surge capacitors and arrestors. The concern

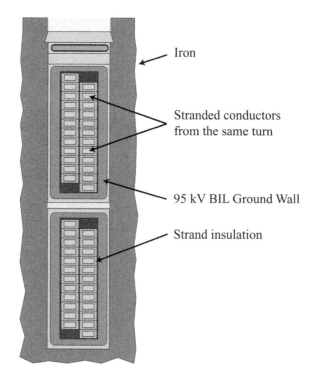

FIGURE 4.45
Cylindrical rotor generator stator coil indirectly cooled construction.

Generator Protection 209

with shorted turns, especially with salient pole generators, is that heating from the loop current could cause stator iron damage, which is expensive to repair.

As mentioned earlier, conventional stator ground protection cannot detect faults at or near the neutral end of the stator windings. Some of the unlikely (but possible) failure modes at the neutral end are:

- If the end turns are not supported or braced properly, end-turn vibration can cause cycle fatigue failures of conductor strands, which can lead to arcing and insulation failure from the heating that occurs, and eventually damage the stator iron if allowed to continue.
- If a stray conductive material bridges the core iron teeth or laminations, eddy current heating could damage the iron and nearby insulation. By the time the insulation fails from excessive temperatures, iron damage would already be expected.
- If stray magnetic material gets trapped in the magnetic field (although it is an unlikely event), it is possible for it to drill into the stator bar insulation.
- If the neutral connections are poor and have enough resistance, the insulation can fail from the overheating associated with the poor-connection watt loss.

About 50% of the surveyed units were found with conventional stator ground protection time delays that exceeded IEEE recommendations. High-impedance stator grounds can cause iron damage if the duration is long enough. IEEE recommends tripping in 1 second for 100% ground faults, as illustrated in Figure 4.43.

The next two sections will discuss two methods for providing stator ground protection for 100% of the stator winding.

4.4.18 Subharmonic (64S)

Subharmonic injection schemes, normally in the range of 15–30 Hz, are used to provide 100% stator winding protection for stator ground faults. Figure 4.46 presents a simplified subharmonic injection scheme circuit. The subharmonic frequencies are injected into the secondary side of the grounding transformer, the current is measured, and because of the transformer connection, the injected voltage is seeking a ground path on the primary side. Ground faults will cause an increase in current that is measured by an overcurrent function that can initiate a trip of the generator. Lower frequencies are used to increase the capacitive reactance impedance, thereby reducing the continuous amount of current that flows during normal operation, reducing the sizing or rating of the injection circuitry components.

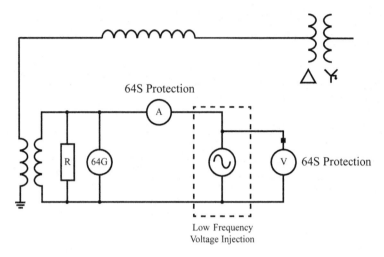

FIGURE 4.46
Typical subharmonic injection circuit.

Subharmonic injection schemes offer the following advantages:

- 100% stator winding protection from ground faults
- The ability to automatically test the generator insulation during startup

The disadvantages are

- Higher cost
- Does not normally detect stator grounding transformer primary or secondary open or shorted circuits
- May require a significant increase in the minimum trip threshold to coordinate with generator bus ground detectors

As mentioned earlier, designs that are equipped with isolated phase bus breakers or generator bus breakers, a second bus ground detector (59BG) should be installed in between the breaker and the step-up transformer. Normally, in this design, the step-up transformer remains energized to backfeed auxiliary load when the generator breaker is open. Consequently, the second bus ground detector is required to alert operations that there is a ground and to mitigate high transient voltages from arcing ground events that can develop into high-current phase–phase or three-phase short circuits, by providing a resistive current that dampens the transient. This type of ground detector provides a zero sequence or ground path for the injection scheme, and the 64S minimum trip or actuation set point may need to be increased substantially to avoid the possibility of nuisance tripping.

4.4.19 Generator Third-Harmonic Monitoring (64T)

As with everything in life, it is not possible to design and manufacture a perfect generator. Consequently, the magnetic flux will not distribute evenly in the core iron of large machines. This asymmetrical flux density will saturate some of the iron laminations and contribute to a small flattening of the sine waves associated with that area. This will not be apparent when looking at the overall 60-Hz sine waves at the generator output terminals; however, a few volts of third-harmonic voltages will be present at the secondary side of the generator stator grounding transformer. The third-harmonic voltage, 180 Hz for 60-cycle machines, timing is such that the contribution from each phase will fill in to form a continuous single-phase third-harmonic wave form. If the voltage is measured with an oscilloscope, it will appear to be a symmetrical uniform single phase 180 Hz sine wave.

Figure 4.47 represents a depiction of a three-phase machine as a single-phase, 180-Hz generator that can be used to visualize and analyze the phenomena. As you can see in the figure, the harmonic voltage from each phase will push current through the distributed capacitance on the output terminals, which are then summed as they flow through the grounding transformer to fill in the gaps and complete the circuit. It is basically a voltage divider circuit, if the capacitive reactance is lower at the output terminals; the third will normally be higher at the grounding transformer. The actual third-harmonic voltage depends on generator design details, excitation levels, and loading. The absence of a third-harmonic voltage would indicate that either the machine has a ground on the neutral end of the winding, or it has an open or short in the primary or secondary ground detection circuitry. Consequently, it is a good idea to monitor the third-harmonic voltage to prove that the ground detection scheme and generator are both healthy.

Figure 4.48 shows actual generator third-harmonic measurements for a 192-MVA 18-kV generator with a stator grounding transformer ratio of 14.4-kV /120-V or 120/1 at no load. At full load, the third-harmonic reading

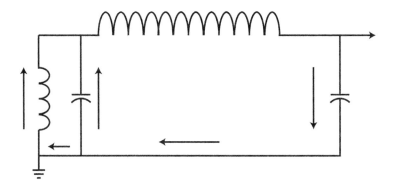

FIGURE 4.47
Generator third harmonic circuit depiction.

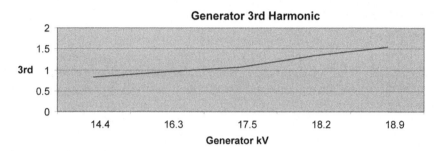

FIGURE 4.48
Generator third harmonic at no load.

will be roughly three times higher as the field is increased to take care of the internal voltage drop, or in the range of 3.0–5.0 V for the secondary side or 360–600 V on the primary side. When boosting vars or supplying part of the system var requirement, the third-harmonic voltage will be higher, and it will be lower when bucking vars or becoming a var load to the electrical system.

Third-harmonic monitoring can provide the following advantages:

- Proves that the neutral end of the stator winding is free of grounds
- Verifies that the neutral grounding conductors and primary winding of the stator grounding transformer are not open-circuited or shorted
- Proves that the secondary grounding transformer winding and associated conductors are not open-circuited or shorted
- Proves that the stator grounding transformer is racked in, if equipped
- Verifies that grounding transformer safety disconnects are closed, if equipped

Due to the robust stator winding construction of two- and four-pole generators, as well as concerns about nuisance tripping when the unit is bucking vars or operating in the lead, this author's preference for cylindrical rotor generators is to alarm only if the third-harmonic voltage drops below 50% of the value measured when the machine is operating without load at rated terminal voltage. A procedure should be in place to quickly determine if there is a problem by verifying that the excitation levels are in an acceptable range, measuring the third-harmonic voltage at the stator ground protective relay terminals, and visually inspecting the grounding transformer and associated disconnects and wiring. If the alarm appears to be legitimate, an orderly shutdown of the unit is recommended in order to complete further testing of the stator windings and circuitry. Some utilities elect to trip with their salient pole hydrogenerators because of concerns with stator winding shorted turns that can cause iron damage from the loop current heating.

Approximately 90% of the author's surveyed units did not monitor the third-harmonic voltage at the generator neutral. The loss of third-harmonic voltage indicates that there is either a ground at the neutral end of the stator winding, inappropriate excitation levels, or an open or short in the stator grounding transformer winding primary or secondary circuitry. Since it can cause nuisance tripping when operating in an underexcited mode, alarm-only operation and immediate investigation for two- and four-pole cylindrical rotor generators is suggested.

4.4.20 Out of Step (78)

Although there is concern with transient torques that can damage the turbine/generator/step-up transformer, out-of-step protection is primarily applied to protect turbine/generators from loss of synchronization or out-of-step slip frequency pulsating torques that can mechanically excite the natural frequencies of rotating systems. Each subsequent cycle can raise the shaft's natural frequency torsional levels higher, eventually resulting in cycle fatigue failures if it is allowed to continue. Slip frequencies for out-of-step events are usually in the range of 0.5–5 cycles per second.

When an electrical short circuit occurs on the electrical system, a generator's watt load is displaced with vars that feed the fault. This, in effect, unloads the machine, and the turbine speeds up. If the electrical system protection does not clear the fault quickly enough, the generator can lose synchronization (dynamic or transient instability) as the speed and associated phase angle differences increase in magnitude. The possibilities for instability are increased with the opening of line breakers that increase the power transfer impedance. Dynamic instability usually occurs when the power angle differences are 120° or greater. A three-phase fault unloads the machine the most and will have the shortest critical clearing time or cycles before it loses synchronization with the electrical system and becomes unstable—usually in the range of 6–20 cycles. At the other extreme are phase-ground faults, which only unload one phase; consequently, the critical clearing times can be three times longer in duration than three-phase events. Phase–phase fault critical clearing times are lower than phase-ground and higher than three-phase short circuits.

Electrical faults will normally move to a specific impedance point and stay there until cleared by automatic protective relaying. However, an out-of-step condition results in a traveling impedance swing or locus that moves through a supervising mho circle repeatedly until synchronization is restored or the generator is tripped off-line. Figure 4.49 illustrates typical impedance swings; the voltage at the machine terminals determines var flows during the swing, and a relatively lower generator voltage will result in leading vars into the machine. The electrical center for a typical utility system is normally in the generator or its associated step-up transformer. The center is the total impedance divided by 2, and it occurs when the two systems are 180° out; it appears as a momentary three-phase fault

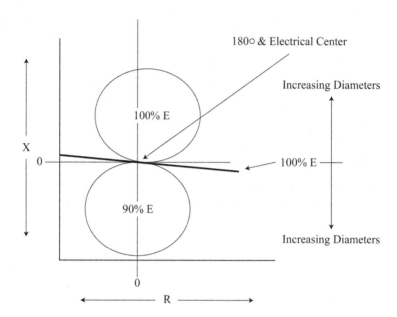

FIGURE 4.49
Out-of-step swing impedance.

to both the generator and the system, which further aggravates system integrity.

Figure 4.50 develops the settings for a single blinder scheme, which is simpler to set and more common than other designs. By convention, forward is toward the generator and reverse is toward the electrical system. The suggested forward reach for the offset mho circle is two times the generator transient reactance in order to ensure, with margin, that the generator is covered. The suggested reverse reach is two times the step-up transformer impedance to ensure that the transformer and part of the electrical system are covered. For cross-compound units or generators operated in parallel at their output terminals, the reverse reach may need to be modified to account for infeed impedance.

The right and left vertical impedance blinders are set for 120° (extrapolated power angles), or 60° on each side of a perpendicular bisect line of the vector sum of the generator, transformer, and system impedances (electrical center), as illustrated in Figure 4.51. By convention, the generator and associated step-up transformer are assumed to be purely inductive, and only the impedance angle for the system is considered. The settings are normally determined with a time-consuming graphical procedure, but the figure offers a trig method for calculating blinder settings.

With reference to Figure 4.51, the right blinder reaches very far to the left and the left blinder reaches very far to the right. Assuming that the swing is

Generator Protection 215

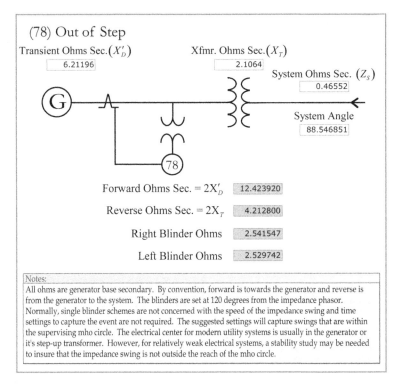

FIGURE 4.50
Out-of-step single blinder relay settings.

approaching from the right side, the left blinder will already be picked up from load current and will see the event first, followed by the supervising mho circle. As it continues to travel, it will actuate the right blinder, then drop out the left blinder, and finally exit the mho circle on the left side, completing the cycle. Tripping is normally initiated upon exiting the mho circle. If the swing originates from the left side, a reverse sequence will take place, and tripping will occur when the swing exits the supervising mho circle on the right side. Other schemes are available, but the single blinder scheme is one of the simplest, as it does not require slip frequency data or concern about the speed of the swing. Because tripping is initiated at the exit of the mho circle when the impedance is higher, the interrupting current levels for generator breakers should be favorable, depending on the breaker tripping time and the slip cycle frequency. Accordingly, a tripping time delay is not normally provided unless studies have been performed that indicate that a time delay should be provided to reduce the interrupting current of the generator breaker. This, of course, depends on the generator breaker tripping time, the electrical system configuration, generator loading, and associated inertia and impedance parameters that influence slip frequencies at the time of the event.

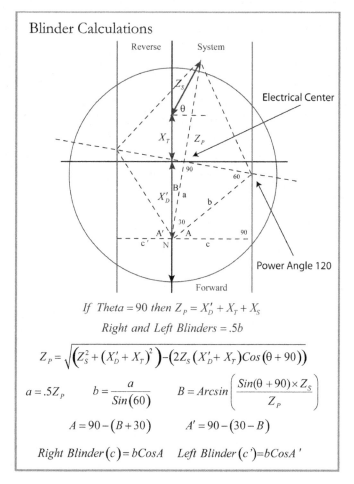

FIGURE 4.51
Out-of-step blinder calculations.

Figure 4.52 shows the calculation for determining different reach points for right or left blinders. This is useful for analyzing events and also for testing the out-of-step blinder elements.

As a matter of technical interest, a similar calculation is shown in Figure 4.53 to determine different reach points for a reactance-type impedance element. Reactance elements are sometimes used to cut off mho circles to mitigate overreaching on short transmission lines.

About 75% of the surveyed units either did not apply out-of-step protection, or it was questionable if the settings provided sufficient margin to ensure detection of all loss of synchronization events. The slip frequencies associated with out-of-step operation can repeatedly excite natural turbine component frequencies and cause cycle fatigue failures.

Generator Protection

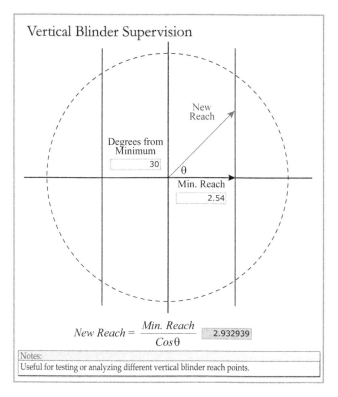

FIGURE 4.52
Vertical blinder impedance points.

4.4.21 Overfrequency and Underfrequency (81)

Generator overfrequency and underfrequency relays are normally applied to protect turbine blades from resonant frequencies during off-frequency operations. The turbine blade frequency withstand times are cumulative, and damage can occur when the withstand times are used up. The longer, low-pressure steam turbine blades are normally more sensitive to off-frequency operation. Turbine blades that are under stress (at full load) may resonate at different frequencies than those of unloaded blades, and the unit load at the time may affect the withstand time. On typical utility systems, off-frequency operation events are not numerous, and the durations are normally quite short. Although overfrequency and underfrequency resonant limitations are generally mirror images, with the exception of more severe overspeed conditions where centrifugal force becomes the major limitation, the primary concern is really underfrequency, where governor action alone cannot mitigate the occurrence. Underfrequency occurs when generation does not match the load. If the load is higher than available generation, the system frequency will decay as the turbines slow down. Utilities normally

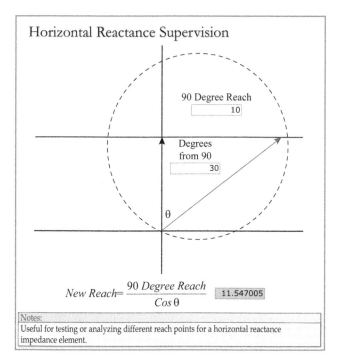

FIGURE 4.53
Horizontal reactance supervision.

have underfrequency load-shedding programs that will isolate blocks of load in an effort to rebalance available generation with the load demand.

Governor action on generating units will also open steam or fuel valves, but they have limited capability on the amount of additional load that they can pick up. Motors will also slow down; this reduces customer load demand, but it may also have a reducing impact on generating station output levels. Many generating stations are not equipped with overfrequency and underfrequency tripping and rely on operations to take generating units off-line at 57 Hz if the frequency does not show signs of recovering. By 57 Hz, the utility underfrequency load-shedding program has already done what it could to balance load, and other automatic load-shedding programs are not typically available to further reduce load. There is also a concern that the generating unit could be isolated with excessive load if other units trip off first, causing significant turbine/generator braking forces as the machines are forced to abruptly slow down.

Figure 4.54 provides preliminary settings if information is not available from the turbine manufacturer or the regional transmission authority. The preferred method is to solicit recommendations from the turbine manufacturer, who should have specific knowledge about the frequency limitations of their equipment. The settings provided in Figure 4.54 will allow a single

Generator Protection

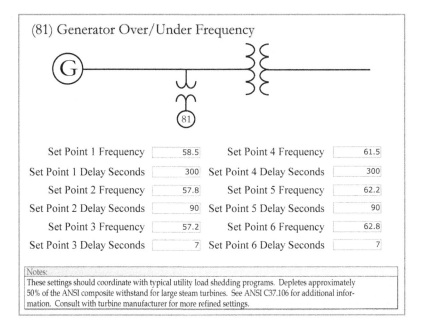

FIGURE 4.54
Overfrequency and underfrequency withstands.

event to consume approximately 50% of the American Nation Standards Institute (ANSI) composite large steam turbine accumulated withstand times before tripping the unit off-line and should coordinate with the typical utility underfrequency load-shedding programs.

Because potentially damaging events are somewhat rare and normally short in duration, and the underfrequency tripping functions need to coordinate with utility load-shedding programs and governor control responses to mitigate unnecessary tripping that can aggravate system disturbances, the suggested time delays are somewhat long in duration. The overfrequency and underfrequency settings are really a compromise between the generating station, the original equipment manufacturer (OEM), and the utility. Turbine manufacturers often take a conservative approach to ensure protection of their equipment and may deny warranty unless their recommended overfrequency and underfrequency tripping set points are implemented. This can put the generating station in conflict with the utility because the manufacturer's initial recommendations may not coordinate with the utility underfrequency load-shedding program. The utility may not allow the generation to connect to their system unless the underfrequency tripping coordinates with their load-shedding program. Sometimes an iterative process takes place before the manufacturer, generating station, and utility can all compromise on the overfrequency-underfrequency settings.

In more recent years, regional transmission authorities in the United States have recommended settings that manufacturers can use as a guide for their turbine withstands and that utilities can use for coordinating their load-shedding program.

In 2003, the Western Electricity Coordinating Council (WECC), a Western U.S. regional transmission authority, recommended not tripping faster than the following durations at the associated frequencies:

- >59.4 and <60.6—continuous
- 59.4 and 60.6—3 minutes
- 58.4 and 61.6—30 seconds
- 57.8—7.5 seconds
- 57.3—45 cycles
- 57 and 61.7—instantaneous trip

Usually, generator multifunction digital relays do not provide more than six overfrequency-underfrequency elements. Accordingly, applying the WECC shortest allowable time delays (assuming that they are required to be within the turbine manufacturer's recommendations), and trying to prioritize the six available set points, the following trip settings are suggested:

- 58.4 Hz—30 seconds
- 57.8 Hz—7.5 seconds
- 57.3 Hz—0.75 seconds
- 57.0 Hz—0.1 seconds
- 61.6 Hz—30 seconds
- 66.0 Hz—0.1 seconds

The withstand times for events higher than 58.4 or lower than 61.6 are long enough to rely on operator intervention. A short time delay of six cycles is provided at 57 Hz to try and avoid nuisance tripping from spurious events. The last or remaining element is used to back up the mechanical overspeed trips for a steam unit that is normally set for 10% overspeed. A time delay of 0.1 seconds is also provided to mitigate spurious tripping.

Roughly 80% of the surveyed units either did not apply overfrequency/ underfrequency protection or, if they did, they were not within NERC guidelines. Overfrequency/underfrequency tripping protects turbine blades from resonant frequencies that can take life away or cause cycle fatigue failures. It can also be used to back up the turbine overspeed protection and to isolate the unit from a nonrecoverable underfrequency electrical system disturbance in an effort to mitigate braking torques associated with islanding and excessive load.

Generator Protection 221

4.4.22 Lockout Relay (86)

The tripping functions discussed in this chapter normally pick up high-speed 86 lockout auxiliary relays that have isolated contacts for tripping and alarming. Several contacts are required to trip the prime mover, excitation system, unit auxiliary transformer low side breakers, and unit breakers and initiate appropriate annunciation or alarms to guide operations. Tripping should be done in a manner that accomplishes all of the following:

- Reduces overexcitation excursions
- Reduces underexcitation excursions
- Reduces overspeed excursions
- Reduces or prevents apparatus damage
- Isolates the problem
- Identifies the problem

At least two 86 lockout relays should be provided for some form of redundancy, as presented here

86 Relay 1
- Transformer differentials (87)
- Negative phase sequence (46)
- Generator stator ground (64)
- Generator loss of field (40)
- Generator reverse power (32)
- Generator out of step (78)
- Overfrequency/underfrequency (81)

86 Relay 2
- Transformer sudden pressure (63)
- Generator backup impedance (21)
- Generator volts/Hz (24)
- Transformer primary overcurrent (51)
- Transformer neutral overcurrent (51N)
- Generator 100% winding protection (64S or 64T)
- Inadvertent energization (50/27)
- Switchyard zone 1 impedance (21)
- Generator differential (87)

The rationale for the lockout relay tripping selections is that the generator backup impedance, switchyard zone 1 impedance, transformer sudden pressure, and transformer overcurrent functions all provide either full or more limited redundant protection for the transformer and unit differential functions. The negative phase sequence function can provide redundancy for transformer neutral overcurrent depending on the fault levels, and the 100% generator stator winding function backs up the conventional stator ground element. If the transformers saturate enough during an overfluxing event, the transformer differential relays could back up the generator volts/Hz function. The reverse power relay (depending on settings) may back up the inadvertent energization element, and the overall unit differential and generator backup impedance elements can back up the generator differentials.

4.4.23 Generator Differential (87)

Almost all generators 10 MVA and larger are equipped with neutral and output-side CTs for differential protection. Differential protection in general was covered in Chapter 3. In the case of generators, the relay (secondary-side) tripping thresholds at low loads are very sensitive, usually in the 200–300-milliamp range, to mitigate iron damage. The differential function cannot detect shorted stator turns because the current in equals the current out. As mentioned earlier, this is more of a concern with the slower salient pole hydromachines and shorted turns are not that likely with the two- and four-pole cylindrical rotor machines due to their robust stator coil design. The differential scheme does not normally have the sensitivity to detect ground faults on generators that are high-impedance (resistance) grounded. One concern with generator differential, because of the very-high-fault current levels, the installation of CTs with like characteristics is more important to prevent nuisance tripping from error currents during through fault conditions. Typical slopes for generator differentials are in the 10%–25% range.

Bibliography

Baker, T., *EE Helper Power Engineering Software Program* (Laguna Niguel, CA: Sumatron, Inc.), 2002.

Baker, T., *Improving BES Reliability at Large Generating Station Locations*, Texas A&M Protective Relay Conference for Protective Relay Engineers (Texas: Texas A&M University, 2014).

General Electric Corporation, *ST Generator Seminar*, Vols. 1 and 2 (Schenectady, NY, 1984).

IEEE Power Engineering Society, *IEEE Tutorial on the Protection of Synchronous Generators*, Catalog No. 95 TP 102 (Piscataway, NJ, 1995).

Klempner, G. and Kerszenbaum, I., *Handbook of Large Turbo-Generator Operation and Maintenance*, 2nd ed. (Piscataway, NJ: IEEE Press, 2008).

Klempner, G. and Kerszenbaum, I., *Large Turbo-Generators Malfunctions and Symptoms* (Boca Raton, FL: CRC Press, 2016).

Mozina, C., *100% Generator Stator Ground Fault Protection: What Works, and What's New*, International Conference of Doble Clients, Doble Engineering, 2014.

Reimert, D., *Protective Relaying for Power Generation Systems* (Boca Raton, FL: CRC Press, 2006).

Westinghouse Electric Corporation, *Applied Protective Relaying* (Newark, NJ: Silent Sentinels, 1976).

Westinghouse Electric Corporation, *Electrical Transmission and Distribution Reference Book* (East Pittsburgh, PA: Westinghouse Electric Corporation, 1964).

Westinghouse Electric Corporation, *Total Integrated Protection of Generators* (Coral Springs, FL: Tutorial, 1982).

5

Electrical Apparatus Calculations

This chapter presents miscellaneous electrical apparatus calculations that have not been covered in the preceding chapters. These calculations and the accompanying text are useful for electrical engineers engaged in support of operations, maintenance, and betterment projects for generating stations and other industrial plants. This discussion will also have more of a personal flavor, and I hope that the readers find it interesting. Additional details on operating and maintenance procedures related to this chapter are covered in Chapters 6 and 7, respectively. The following electrical apparatuses are covered in alphabetical order: buses, cable, circuit breakers, generators, metering, motors, and transformers.

Electrical apparatus standards inform the manufacturers about the requirements that they are expected to meet. The appropriate standard for existing plants is the one that was in effect at the time the equipment was designed and purchased, but that is not necessarily the current standard. Standards are not always legally mandated unless the customer refers to them in purchase order specifications.

In general, when to upgrade or replace electrical apparatus or determine the remaining life is always a very difficult question. If there are three like transformers, motors, or other types of apparatuses and two of them fail, it is much more likely that design, workmanship, and material quality are reasons to refurbish or replace the remaining like equipment with an upgraded version. If the apparatus fails an important test, such as hipot or other overvoltage testing, or has an obvious anomaly during a physical inspection, then the decision is easy. Otherwise, it is a very difficult, and it is not all that uncommon to inherit a new set of problems when replacing or upgrading equipment. Perhaps the old adage applies—"If it is not broken, don't fix it."

As technology moves forward with more sophisticated monitoring and diagnostics, the database may not be large enough or have sufficient history to ascertain with certainty what action to take. Performing a life analysis of insulation systems can be very difficult since manufacturers may use different chemical compounds or formulations for their similar products, while using identical nomenclature. The following can affect the life of electrical apparatuses:

- Ambient temperature
- Lightning
- Salt air

- Moisture
- Corrosive environments
- Contamination
- Operating practices (described in Chapter 6)
- Maintenance practices (described in Chapter 7)
- Manufacturers' material quality
- Manufacturers' fabrication quality
- Manufacturers' design margins
- Shipping damage
- Construction design margins
- Installation practices
- Protective relay applications (described in Chapters 3 and 4)
- Protective relay settings (described in Chapters 3 and 4)
- Bus transfer schemes or practices (described in Chapter 3)
- Operating voltages (described in Chapter 2)
- Transient mitigation methods (described in Chapters 3 and 5)
- Grounding methods (described in Chapter 3)
- Exposure to electrical system disturbances (Chapters 3 and 4)
- Material incompatibility (chemical interactions)

5.1 Buses

Obviously, buses are important to the operation of generating units, and particularly isolated phase buses that are designed to mitigate exposing the generator to phase–phase and three-phase short circuit conditions. In the past, isolated phase bus designs used flexible insulated bus sections for expansion and contraction, which prevented longitudinal current flow in the segregated tube sections. In more recent years, the designs allow longitudinal current flow by connecting flexible braided copper jumpers across flexible expansion and contraction sections and bonding all three phases together to ground at the generator, associated step-up transformer, and unit auxiliary transformer ends to reduce the inductive reactance and induced voltage hazards.

Practically, the aluminum tubes for each phase will need to carry the same amount of current as the main bus current-carrying conductors (almost a 1:1 ratio). Consequently, it is important to maintain the braided jumpers properly and replace those that are corroded and frayed enough to reduce their current-carrying capability. The short circuit currents and the magnetic flux density emanating from the isolated phase buses for large generating units

Electrical Apparatus Calculations 227

are much higher than any other location at a typical utility electrical system, for the following reasons:

- The load current is much higher—high-voltage megavolt-amps (MVA) at a medium-voltage level.
- The short circuit current is much higher due to the higher MVA and lower operating voltage.
- The larger spacing between phases reduces the subtractive flux from the other phases, resulting in a higher self-flux and opposing voltage.
- People can come into close proximity or in direct contact with the aluminum tubes while the unit is on-line.

Magnetic flux linking with supporting I-beam structures can induce large currents into the structure if there are paths for the current to take. Support beams are often equipped with copper bars or bands that encircle the beams to promote eddy current flow in the bands. The eddy currents in the bands will produce an opposing magnetic flux that will help reduce the current flow in the beams.

All buses need to be able to withstand short circuit currents from external or through faults without damage or loss of life. In addition to thermal concerns, they need to be supported or braced in a manner to handle the magnetic bending forces that occur from high-magnitude short circuit currents. Figure 5.1 shows the expressions for calculating peak forces for horizontal bus bars only. The calculations need to be modified for vertical bus bars to

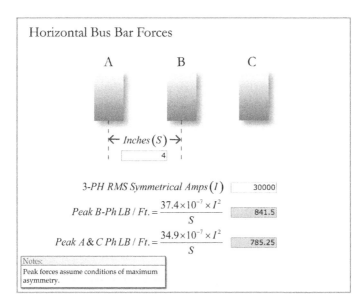

FIGURE 5.1
Horizontal bus bar forces.

consider gravitational effects, as well as for bus bars that bend in a different direction. Only the horizontal forces are considered here to provide the reader with some understanding of the magnitude of electromechanical forces. The expressions assume maximum asymmetry and determine the peak forces.

As shown in Figure 5.1, the forces for B-phase or the middle bar are higher because both of the other phases are in close proximity and act on the middle phase. The forces cause vibration at double frequency or 120 hertz (Hz) because of the change in polarity during each half-cycle. B-phase experiences 841.5 foot-pounds for each foot of unsupported length. At first blush, this does not seem like much force for such a high current level with a relatively short separation distance between bars. However, imagine an 800-pound weight swinging or oscillating at 120 cycles per second.

5.2 Cable

This section will discuss cable thermal limitations for carrying short circuit currents, as well as calculations that determine overhead and underground cable watt loss and var consumption.

5.2.1 Short Circuit Withstand Seconds

The cable withstand calculation is used to verify that protective relay systems will clear through faults or external short circuits before damage occurs to cables. Figure 5.2 determines the number of seconds and cycles that a cable can withstand a high current level before loss of life. For example, you would not want to replace the cables if a 4-kilovolt (kV) motor developed a short circuit condition. Normally, it is assumed that the cables are at rated temperature when the fault occurs. The calculation then determines the number of seconds and cycles that it will take to reach a conductor temperature of 250°C, which is the annealing point of copper and considered the maximum temporary withstand temperature for copper conductors. As you can see in the figure, 2/0, 90°C cable can withstand 30,000 amps for only 0.1 seconds or six cycles before reaching 250°C. Because instantaneous fault interruption is usually considered to be around six cycles, and 30,000 amps is a typical 4-kV three-phase fault current level, 2/0 cable would not provide sufficient design margin. Consequently, many engineering firms will not use cable that is smaller than 4/0 for 4-kV applications in large generating stations.

5.2.2 Short Circuit Fusion Seconds

Figure 5.3 covers the amount of time that a cable can carry a high level of current before the copper conductors fuse open. One useful application for this

Electrical Apparatus Calculations

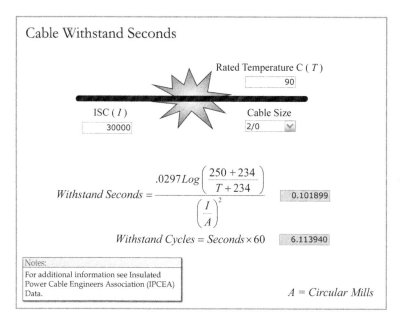

FIGURE 5.2
Cable short circuit withstand seconds.

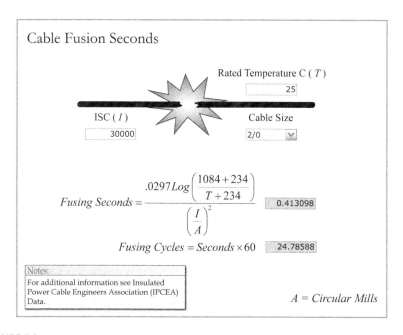

FIGURE 5.3
Cable short circuit fusion seconds.

calculation is to determine how long a personal ground will last during an inadvertent energization. This is important because safety grounds cannot protect personnel from an inadvertent energization unless they are able to last longer than it takes for the protection to automatically deenergize the circuit. In this case, the personal grounds are assumed to be at ambient temperatures initially, or 25°C. The calculation then determines the number of seconds and cycles that it will take to reach a conductor temperature of 1084°C, which is the fusion or melting point for copper. As shown in the figure, 2/0 cable can withstand 30,000 amps for only 0.413 seconds, or 24.79 cycles, before the conductors fuse open and can no longer provide protection. This limitation may be of concern for switchgear personal safety grounds, as transformer secondary-side overcurrent protection usually takes longer than 0.4 seconds to interrupt short circuit conditions. Consequently, larger personal ground conductors or parallel cables may be required in order to provide adequate safety protection.

5.2.3 Cable Line Loss

The line loss calculation is used to determine the amount of watt loss and var consumption of underground and overhead cables. It is useful for economic and voltage analysis of overhead/underground lines and for verifying the accuracy of revenue meters that are compensated for line losses to upstream or downstream points of delivery. Usually, a conductor temperature of 50°C is assumed for determining the resistance of overhead conductors for revenue metering applications.

The calculation procedures are shown in Figure 5.4. Basically, the watt loss for balanced conditions can be determined by squaring the current in one phase, multiplying it by the resistance of one phase, and then multiplying it by 3 to include the losses from the other phases. The net var consumption can be determined with a somewhat similar procedure. First, the squared line current is multiplied by the inductive reactance ohms for one phase, and then by 3 to obtain the total inductive vars. Then, the square of the phase-neutral voltage is divided by the capacitive reactance per phase and then multiplied by 3 again, to determine the total capacitive vars. Finally, the total or net vars are determined by subtracting the capacitive vars from the inductive vars. A negative result means that the line is predominantly capacitive at the specified line current. Predominantly capacitive vars will cause a voltage rise in the line that can be calculated according to the formulas presented in Chapter 2.

The author worked with a high-end metering manufacturer, in the development of an Independent System Operator (ISO) revenue meter for generating station deregulation applications in California during 1996–1998. This was probably the first line-compensated meter in the United States to consider the capacitive component of overhead transmission lines since prior meters only considered the inductive portion. Unloaded transmission lines are capacitive in nature; as the lines are loaded, they will hit a unity power factor (PF)

Electrical Apparatus Calculations

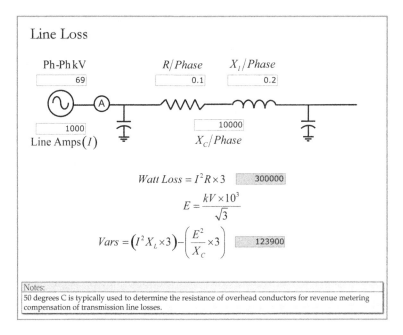

FIGURE 5.4
Cable line loss.

point and then become inductive with further loading. During that time, major utilities in California agreed to set the overhead aluminum conductor steel-reinforced (ACSR) cable temperature at 50°C as a practical matter for determining cable resistance and represents approximately 75% of the line ampacity. In reality, in addition to loading, the conductor temperatures are influenced by variations in ambient temperatures due to weather patterns and differences in wind flows and sun incidences over the length of the line.

5.3 Switchgear Circuit Breakers

Switchgear circuit breakers represent the greatest electrical hazard to station personnel. During the 1990s, this author assembled a task force to assess the present condition and remaining life of switchgear air circuit breakers at eight different generating station locations for a large utility. Also participating in the review were an apparatus engineer who specialized in switchgear circuit breakers, and an electrician training instructor who was quite experienced in the routine and overhaul maintenance of air circuit breakers. Altogether, 17 circuit breakers from eight different plants were randomly selected for detailed inspection—1 13.8-kV, 8 4.3-kV, and 8 480-volt (V) breakers. Altogether, the review disclosed that almost half the breakers had the following serious problems:

- 1 megohm across the open poles of a 13.8-kV breaker (arc chute contamination).
- Moisture/salt air-induced carbon tracking within 1 inch of two phases beneath the bottom insulated phase spacer board for a 4.3-kV vertical lift circuit breaker.
- The anti-pump feature was found jumpered out on a 4.3-kV breaker.
- The cubicle or cell safety interlocks that prevent operators from racking a closed 480-V breaker in or out did not function properly on five different breakers.

The review also revealed that the breakers would probably outlast everything else in the plants if they were maintained properly; suggested maintenance procedures or practices for switchgear circuit breakers are presented in Chapter 7. Of particular importance is the adjustment of air circuit breaker main contacts for bus source and tie breakers. Feeder breakers are normally operated well below their continuous current rating and can usually tolerate maladjusted main current-carrying contacts. However, source and tie breakers are often loaded at or close to their rated current, and consequently, they are very intolerant of improper adjustments of their main current-carrying contacts, which can lead to three-phase short circuit conditions as the dielectric between phases is breached by molten metal splatter, gassing, and ionization from overheating.

5.3.1 Alternating Current (AC) Hipot Testing

The routine alternating current (AC) overvoltage or hipot testing of medium-voltage switchgear circuit breaker insulation systems is strongly recommended based on typical failure modes, the inability of a lower-voltage megger insulation test to always identify the problem area, the low cost for the test equipment, the low cost to repair a failure, and finally and most important, the potential hazard to operating personnel who rack the breaker into the cubicle while standing in front of it, which energizes approximately half the breaker. AC testing is preferred over direct current (DC) because of the physical geometries of the equipment. Switchgear circuit breakers have very little capacitance and, consequently, AC hipot sets can be very small in size and capacity and relatively inexpensive as well. Insulation failures are usually the result of surface tracking; thus, they can be repaired easily. Because the consequence of a test failure is relatively minor, electricians can perform the testing when they overhaul the breaker. By comparison, the DC hipot testing of medium-voltage motors, and medium-to-large generators are normally performed by highly trained engineers or technicians and often witnessed by supervisors or engineers because of the high cost of failure.

Electrical Apparatus Calculations

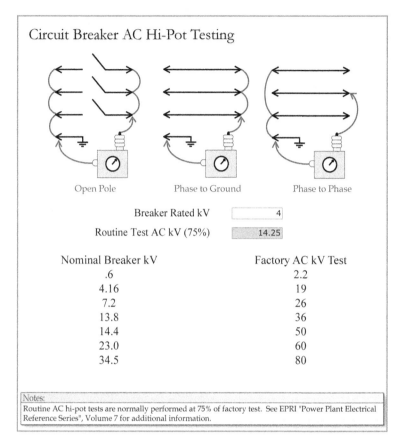

FIGURE 5.5
Circuit breaker AC hipot testing.

Figure 5.5 illustrates the three overvoltage tests that should be performed on switchgear circuit breakers during routine maintenance and provides the recommended routine test values. An open-pole test is performed to verify that the arc chutes are not overly contaminated from interrupting arc by-products, and also to test the movable contact insulation to ground in the open position. The phase–phase test verifies the integrity of the phase–phase insulation, and the phase-ground test proves that the ground insulation level with the breaker closed is acceptable for continued service.

5.3.2 Circuit Breaker Duty

Switchgear circuit breakers have two ratings for short circuit currents. The *interrupting rating* is the maximum asymmetrical current that a circuit breaker can interrupt without suffering damage, and the *momentary rating* is the maximum asymmetrical current that the breaker can close into and mechanically latch. In either case, the resulting high-temperature arcing, if long enough in duration, will cause severe damage to the circuit breaker.

Figure 5.6 presents the procedure for calculating the breaker interrupting and momentary asymmetrical currents for breakers manufactured during the mid-1960s to 2000. Verification on which standard was used by the manufacturer should be made for breakers built near the ends of this time span. In recent years, the standards have been leaning more toward vacuum breaker technology than air circuit breakers, and they are also being merged with European standards. As illustrated in the figure, current from the source transformer and motors will flow through the breaker. If the breaker in question feeds a motor, the contribution from that motor does not need to be considered. Normally, a load breaker that feeds a downstream transformer would represent the worst-case position in the study, or the smallest motor if transformer feeders are not provided in the design. The motor contribution is breaker specific, and the source breakers do not see the motor contribution from the bus that they are feeding.

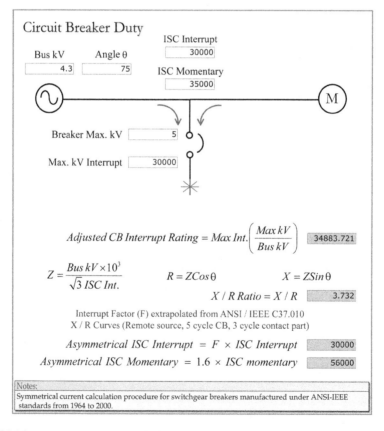

FIGURE 5.6
Circuit breaker duty.

Electrical Apparatus Calculations 235

The source transformer and motor symmetrical short circuit contributions and the summing of currents from two sources were discussed in Chapter 2. Prior to beginning this calculation, the transformer's three-phase, symmetrical short circuit current is paralleled with the motor symmetrical interrupting current, and then again with the motor symmetrical momentary current to determine the total (transformer + motor) interrupting and momentary symmetrical values, respectively.

Normally, the circuit breaker maximum interrupting current is given at the maximum rated kilovolts. If the bus is operating at a lower kV, you can take credit for it and increase the interrupting rating by a factor of maximum kV/bus kV. Then the total interrupting short circuit impedance is calculated, and the X/R ratio is determined from the angle. The X/R ratio is used to estimate the asymmetrical interrupting current decrement at the time that the circuit breaker contacts part.

Figure 5.6 assumes that the breaker trips open in five cycles and the contacts part in three cycles. If the contacts take longer to part, the asymmetrical current may be lower; and conversely if they part faster, the current may be higher. If the circuit breaker does not appear to have sufficient margin and the breaker trip and contact part times are different, a more precise interrupt factor should be extrapolated from the ANSI/IEEE X/R curves presented in the C37.010 standard. The momentary rating assumes asymmetrical currents of 1.6 per unit, or 160%, as the event is instantaneous and there is no time for decay. The ANSI/IEEE remote source indicated in the figure means that the short circuit impedance is high enough that the generator current decrement does not need to be included. Generator decrement does not need to be considered if the fault impedance is larger than the remote generator impedance by a factor of 1.5 or greater.

When the available short circuit duty is too high for the circuit breaker ratings, there are a number of mitigation methods. From the least expensive to the most, these methods include using motor feeder cables in the calculation; using actual locked rotor currents instead of maximum nameplate values; limiting the available ground current (if that is the problem area); adding a few cycles of time delay to instantaneous overcurrent circuit breaker tripping (if asymmetrical interrupting current is the problem); relocating some motors to other buses; upgrading selected feeder breakers; and installing phase reactors to reduce the short circuit current.

5.4 Generators

5.4.1 Generator Stator Acceptance DC Hipot

Insulation systems for medium-voltage generator stators are normally DC overvoltage or hipot tested at acceptance values after shipment or installation

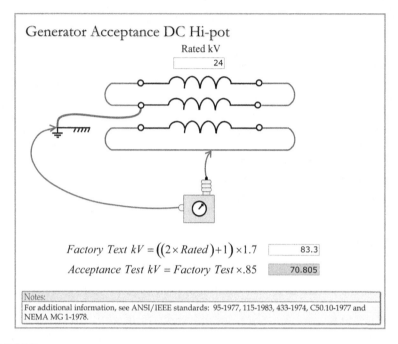

FIGURE 5.7
Generator stator acceptance DC overvoltage testing.

(or following rewinds when another party is responsible for the cost of repair), and also to prove that the apparatus is fit for service. DC is used instead of AC because of the amount of vars needed to support the insulation capacitance. The relatively high AC var load in comparison to DC fosters the need for a high-capacity, large AC hipot test set that is relatively expensive and more difficult to transport to the site. Normally, one phase is tested at a time when the other two phases are grounded. The ends of the coils are connected to reduce transient voltages in case the test is suddenly interrupted.

Figure 5.7 provides recommendations on DC acceptance test kV levels. Numerous standards address this matter, as shown in the "Notes" section of the figure. The AC factory test is performed at twice-rated phase-phase kilovolts plus 1 kV. Acceptance tests are normally performed at 85% of the factory value. The AC value is then multiplied by 1.7 to convert it to an equivalent DC magnitude. This is not a conversion from root mean square (RMS) to peak; it was determined by an industry consensus that a 1.7 factor provides an equivalent DC searching level.

5.4.2 Generator Stator Routine DC Hipot

Insulation systems for generator stators are normally routine DC overvoltage or hipot tested during unit overhauls or major outages to provide a measure

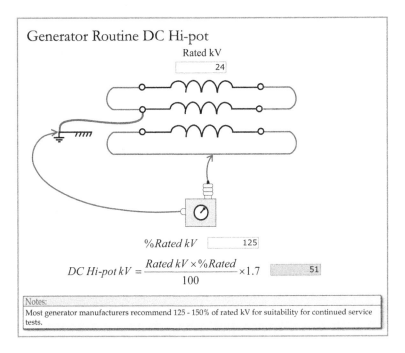

FIGURE 5.8
Generator stator routine DC overvoltage testing.

of confidence that the insulation will perform its intended job until the next overhaul.

Figure 5.8 provides recommended kV values for the routine DC overvoltage of testing of generators. Most manufacturers feel that a test level of 125%–150% of the phase-phase voltage is high enough to predict a future, although not defined, life. A value of 125% of the rated phase–phase voltage multiplied by 1.7 for a DC equivalency is suggested, as it is more defendable if the insulation system fails to carry the minimum test voltage. A turn failure in the stator slot areas normally requires a partial or full rewind. If you elect not to overvoltage-test during a major outage and the insulation fails shortly after returning the unit to service, you may not be in a defendable position.

5.4.3 Generator Rotor Acceptance DC Hipot

Insulation systems for large generator, cylindrical, rotor field windings are normally DC overvoltage or hipot tested at acceptance values after shipment to the site or following onsite rewinds, when another party is responsible for the cost of repair should it fail, and also to prove that the apparatus is fit for service.

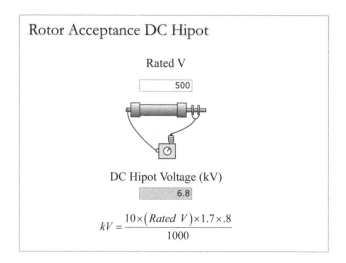

FIGURE 5.9
Generator rotor acceptance DC overvoltage testing.

Figure 5.9 provides an illustration of the test and the suggested DC hipot test values. Most manufacturers suggest testing at 80% of the factory test value, with both ends of the winding shorted together to mitigate transient voltages in case the test is suddenly interrupted. As the figure shows, a 500-V field winding would be acceptance-tested at 6.8-kV DC.

5.4.4 Generator Rotor First-Year Warranty DC Hipot

Insulation systems for large generator, cylindrical, rotor field windings are normally DC overvoltage or hipot tested while still under warranty after completing 1 year of service, when another party is responsible for the cost of repair should it fail, and also to prove that the apparatus is fit for continued service.

Figure 5.10 provides an illustration of the test and the suggested DC hipot test values. Most manufacturers suggest testing at 60% of the factory test value, with both ends of the winding shorted together to mitigate transient voltages in case the test is suddenly interrupted. As you can see in the figure, a 500-V field winding would be tested at 5.1-kV DC after completing 1 year of service.

5.4.5 Rotor Routine Overhaul Insulation Testing

Cylindrical rotor generator field windings are routinely tested during unit overhauls. Normally, a polarization index (PI) of 2 is required (measure of moisture or contamination) and is calculated by dividing the 10-minute megohm reading by the 1-minute measurement. On brush machines, the exposed insulation for the collector or slip rings usually have a hard buildup

Electrical Apparatus Calculations 239

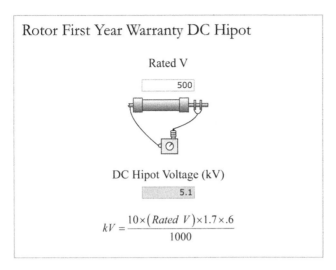

FIGURE 5.10
Generator rotor first-year-warranty DC overvoltage testing.

of oil/carbon from the bearings and brushes, and extensive cleaning of the area may be required before getting a satisfactory PI.

Figure 5.11 provides an illustration of the test and suggests using a 1000-V megger or insulation test instrument. As you can see in the figure, the minimum megohms are 50. This is higher than the standard nonoverhaul requirement, which is only 9 megohms per kilovolt, or 4.5 megohms, and can be performed with a 500-V test instrument.

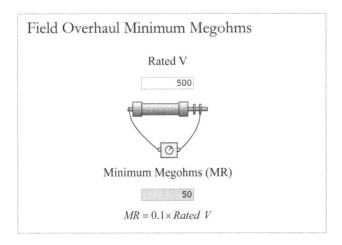

FIGURE 5.11
Generator rotor overhaul insulation testing.

5.4.6 Generator Temperature

Figures 5.12 and 5.13 present the formulas for determining temperature based on copper resistance, as well as resistance based on temperature. The formulas will work on any copper winding.

Generator and larger medium-voltage motor stators normally have 10-ohm copper resistance temperature detectors (RTDs) embedded in their stator windings and in gas passages to monitor stator temperatures. Usually, the RTD will have an ohmic value of 10 at 25°C. Where applied, these devices can drive recorders, protective relays, or distributed control systems (DCSs) and alarm or trip if the temperature reaches the set point value. As you can see in Figure 5.12, a resistance value of 10.5 ohms would indicate a temperature of 38°C. Figure 5.13 provides the inverse expression (or °C) to copper ohms.

A similar approach can be used for determining the temperature of a generator's rotating copper field winding that is equipped with collector rings and brushes. The resistance of the field can be determined by dividing the field voltage (allowing brush voltage drop) by the field current. This is one of the most important measurements for monitoring the health of the generator and is often overlooked when replacing recorders with DCS measurements. If the temperature suddenly increases, it can indicate brush-rigging problems that can add resistance to the circuit or a bad connection in the axial studs or main leads to the field windings. A sudden drop in temperature could be an indication of shorted field turns.

The field temperature measurements can also help operations in determining the appropriate action for field ground alarms. An internal generator

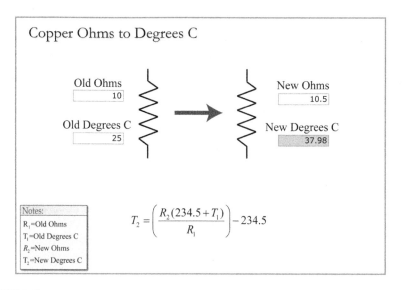

FIGURE 5.12
Copper ohms to degrees Celsius.

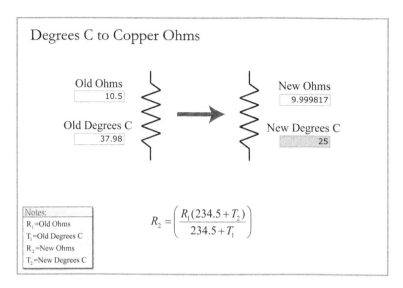

FIGURE 5.13
Degrees Celsius to copper ohms.

field ground could be caused by high-temperature electrical arcing, insulation damage, contamination, elongated end turn conductors, brush-rigging problems, or other anomalies in the generator field. If a second ground occurs inside or outside the field in the opposite polarity, DC short circuit currents will flow that can damage the rotor forging, retaining rings, collector rings, journals, and bearings. For that reason, manufacturers normally recommend automatic tripping for field ground events. Most generating stations usually elect not to trip automatically because experience has shown that the ground is usually located in the peripheral excitation equipment, not necessarily in the generator rotor; consequently, the risk or exposure to the rotor may be low. However, if the field ground is coincident with step changes in either generator bearing vibration or field temperature, then the probability of the ground being in the rotor is quite high, and the unit should be quickly removed from service to mitigate the possibility of rotor damage.

Hall-effect devices (DC clamp-on ammeters) can also be used to determine if the ground is toward the generator or toward the excitation system for generators equipped with collector rings and brushes. The author had the original idea for this concept during the 1990s, and the utility developed electronic instrumentation and clamping sensors with large-enough flexible laminated iron diameters to fit around the DC cables that feed the generator brush rigging. If the ground was toward the generator, the instrument could measure the differential DC current that flowed to ground with the generator energized and the normal ground detector out of service.

The instrument inserted two resistors between the polarities that were grounded at their common point to develop the ground current. Some of the larger generators were equipped with DC buses that fed the brush rigging. In this case, the spacing between the bus polarities was too large, and the hall-effect sensors saturated because the flux had not fully canceled out. My good friend, Dr. Isidor Kerszenbaum, suggested cutting the bus open and inserting a coaxial bus. The coaxial section was fabricated from smaller- and larger-diameter copper pipes. The smaller diameter was inserted inside the larger diameter and insulated with an epoxy compound. The hall-effect sensor and flexible laminated iron were then circumferentially wrapped around the larger-diameter outer coaxial pipe, and this design successfully eliminated the prior saturation problem.

5.4.7 Cylindrical Rotor Shorted Turns

Shorted turns in generator, cylindrical rotor field windings can contribute to vibration problems due to rotor thermal bending from the uneven heating associated with nonsymmetrical DC current flow and watt losses in the windings. Shorted turns can also cause unbalanced magnetic flux in the air gap, which can also aggravate vibration problems. Basically, balance problems from rotor bending will develop fundamental 60-Hz signatures and magnetic problems contribute to double frequency or 120-Hz vibration.

Since vibration signature analyses for rotor shorted turn problems is not always an exact science, it is desirable to have confirming data from other testing before proceeding with very costly disassemblies and repairs of large machines. Additional tests for confirming the existence of shorted turns in generator rotor field windings are commonly performed before committing to expensive repairs. At this time, the following four test procedures are generally used in the industry to help verify whether generator field windings have shorted turns:

- *Thermal stability testing*: Involves changing generator-operating parameters (watts, vars, and cooling) and recording and analyzing the impact on rotor vibration signatures.
- *Field current open circuit testing*: If accurate previbration electrical measurements have been recorded, precise field current measurements can be compared to the historical data; if there is a shorted turn, the field current will be higher for an identical terminal voltage. If the machine is loaded, it is much more difficult to compare since var flow is dependent on system voltage levels.
- *Flux probe analysis*: Utilizes an installed air gap probe to measure and analyze the magnetic flux from each rotor slot as it passes the location of the sensor. Some generators are permanently equipped with flux probes, but many are not. Installing the probe normally requires a unit outage, especially with hydrogen-cooled machines.
- *Repetitive surge oscilloscope (RSO) testing*: Will be discussed in more detail in the following text.

Electrical Apparatus Calculations 243

It should be noted that these analyses (vibration analysis, thermal stability, field current open circuit, flux probe analysis, and RSO testing) by themselves do not provide absolute certainty that there is a shorted turn problem in the generator rotor. However, when confirmed by other testing, the probability that the field winding is the cause of the vibration problem increases significantly. Shorted turn anomalies can be masked if they are near the center of the winding or otherwise balanced, if there are multiple shorted turns, if they are intermittent, if there are grounds, and if there are other contributors to the overall machine vibration levels.

RSO testing has some advantages over other testing, in that it can be used periodically during rewinds to verify that windings are free of shorts on both at-rest and spinning deenergized rotor windings. In 2004, Isidor Kerszenbaum and the author cooperated in the design for an RSO test instrument. Previously, Kerszenbaum had been the lead engineer for an Electric Power Research Institute (EPRI) project on RSO testing. The Generator Rotor Shorted Turn Analyzer produces a succession of step-shaped, low-voltage pulses. The pulses are introduced simultaneously to the DC rotor winding field winding from both ends, as shown in Figure 5.14.

If no discontinuities or anomalies are present in the winding (due to grounds or shorted turns), both traces will be nearly identical, and if inverted and summed, a single trace will be displayed as a horizontal straight line, with a minor blip at the origin and an almost imperceptible ripple. Any significant discontinuities arising from a fault will be shown as an irregularity on the summed trace. By estimating the location of the anomaly on the screen, an inference can be made as to the approximate location of the fault. For instance, large irregularities near the origin of the trace are attributed to faults close to either end of the winding. However, the inductance of a cylindrical rotor field winding is not linear since it is different in the retaining ring areas.

Figure 5.15 shows the wave forms captured with a digital scope when there is a shorted turn problem. As you can see, the two top positive waveforms represent the input channels, and the bottom trace represents the summed trace with a more sensitive scaling factor. With digital scopes, the traces can

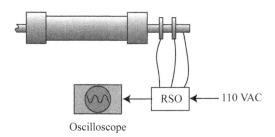

FIGURE 5.14
Typical RSO test connections.

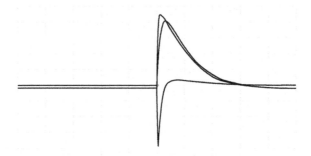

FIGURE 5.15
Typical RSO test waveforms.

be displayed in three different colors. RSO testing does not work on the slower salient pole hydromachines because the inductance is too high and the signal cannot penetrate far enough into the winding. Additionally, salient pole generator field windings do not provide a uniform plane between each conductor and ground.

5.4.8 X/R Ratio

Large generators are usually assumed to be purely inductive. However, if a more refined study is desired, the generator's X/R ratio needs to be known so that a more precise short circuit angle can be calculated. Generators 40 MVA and larger will typically have X/R ratios that range from 40 to 120, and smaller 1–39 MVA machines usually range from 10 to 39. Smaller X/R ratios have lower short circuit angles, which cause asymmetrical currents to decay more rapidly.

Figure 5.16 illustrates a procedure for calculating generator X/R ratios for the 60-Hz generator in Chapter 4. Three parameters are needed for the calculation: the direct axis saturated subtransient reactance (X_d''), the saturated negative phase sequence reactance (X_2), and the generator armature time constant (Ta_3). The effective R can then be calculated by using the formula presented in the figure. The calculated R can then be backed out of the subtransient reactance in order to determine the X/R ratio.

5.5 Metering

The author's first utility engineering job was in revenue metering, and following an early retirement in 1996 due to the deregulation of generation in California, he was called back to be the lead engineer for a fast-track metering

Electrical Apparatus Calculations

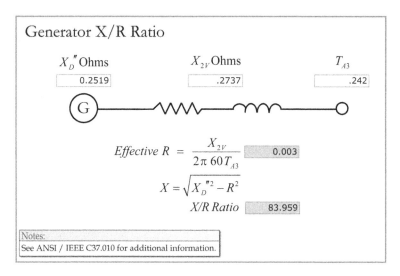

FIGURE 5.16
Generator X/R ratio.

project involved with installing Independent System Operator (ISO) revenue metering for 52 different generating units (mostly multi-unit fossil, nuclear, and hydro).

ISO guidelines at the time indicated that each generating station had to deliver its output power to the electrical system. If the local generating station high-voltage switchyard had transmission lines that went to different substations, the local switchyard was considered the point of delivery. If the transmission lines went to a single remote substation, then the lines were considered part of the generating station package and the ISO point of delivery was the remote substation. Under prior Federal Energy Regulatory Commission (FERC) guidelines, generator step-up transformers (GSUT) were in the transmission account, and consequently typical generating station metering was designed to measure the watt and var outputs of the generator only.

The vast majority of locations were not equipped with revenue metering instrument transformers that could measure watts and vars at the new point of delivery. The cost to install new high-voltage revenue metering class current and voltage instrument transformers would be substantial, and there were concerns that the expenditures might not be defendable at California Public Utilities Commission (CPUC) hearings. It was decided to use existing revenue metering class instrument transformers that were typically installed at the generator output terminals and also for the reserve and unit auxiliary transformers that feed the plant auxiliary power system requirements. This meant that the metering had to compensate for transformer losses and var consumption, net out auxiliary load, and (in some cases) compensate for transmission line watt losses and var consumption to the point of delivery as discussed in the section entitled "Line Loss," earlier in this chapter. Actually, compensated metering can be more accurate due to the undermetering problem for conventional metering. For example, if a transformer is energized without load, the meters may not be able to measure the no-load var consumption and watt losses accurately at low current levels (0.5% of full load or less). However, with transformer-compensated metering, since the calculations are based on voltage magnitudes rather than current, the watt and vars can be measured fairly accurately.

Although this was the most complicated revenue metering project ever attempted at that point in time, the cost ended up being one-tenth that of high-voltage metering at the point of delivery. There were three high-end revenue metering manufacturers in the United States at the time, but none had meters with the full capabilities that were required. Process Systems, in North Carolina, was the successful bidder, and numerous trips were made to the company's facilities to explain the features that were required to their software engineers.

As mentioned in Section 5.2, it became the first U.S. meter to compensate for transmission lines properly by subtracting the capacitive vars. It was also the first meter to include a transducer card that could provide 4–20-ma signals to represent the compensated watt and var flows to the point of delivery. Normally, plant operators are blind to switchyard power flows, and assume that they are delivering vars to the system. In most cases, the plants were found to be var-deficient and required system vars to feed the full requirement of their GSUTs. Process Systems was successful, and their meter initially became the only revenue meter approved for large generation by the California ISO. Eventually, larger metering companies acquired their design.

As mentioned before, meetings were held with the other major utilities in California to resolve how to handle the temperature issue with overhead transmission conductors. Other issues were also looming. Transmission losses are exponential, which means that whoever comes on last causes the greater amount of losses per megawatt. Meetings were also held on how to distribute the transmission losses to the various units at a generating station, and it was finally decided to net the transmission watt losses and var consumption against a single unit. A priority of units at each particular location was established, and the full transmission losses were netted against a running unit using pulses from one meter to another. For example, if unit 1 were running, the pulses could be sent to that meter; if not, they could be directed to the unit 2 meter; and so forth. In a similar manner, the auxiliary transformer load could also be netted against its own unit using pulses.

If more than 1 meter or service is involved, the pulses need to be totalized in order to determine coincident demand. In the past, this was accomplished with pulse-metering totalizers that were designed to accept parallel pulses and output them in a serial fashion as if there were only 1 meter. The author developed and received patents for a pulse metering totalizer in the late 1970s using complementary metal oxide semiconductor (CMOS) logic chips. The speed of the output pulses needs to be controlled so that they do not exceed the capability of the end recording device, and it is also preferable that the maximum output rate be distributed among the input channels in any percentage; in other words, if the other channels do not have pulses, the full output pulse rate can be accepted by the active input channel or by any proportion of pulses that is possible. In more recent years, with the advent of microprocessor technology, high-end revenue meters can handle more complex calculations and now can totalize pulses internally without the need for external devices, although the external type metering totalizers can still be found in service at older plants.

5.5.1 Theory

Watts and vars can be measured with three elements, as presented in Figure 5.17. Basically, under balanced conditions, the three-phase MVA can be determined by measuring the MVA in one transformer and multiplying by 3. The three-phase MVA multiplied by the cosine of the PF angle will yield the total megawatts, and the sine of the angle will provide the total megavolt ampere reactive (MVAR) consumption. However, if a neutral conductor is not provided for load purposes, three-phase watts and vars can be measured accurately with only two elements.

Figure 5.18 illustrates the voltage and current connections for three-phase, two-element metering. The connection is not phase rotation sensitive and requires two phase currents that are 120° apart, with the voltage or potentials for each element lying inside their associated current by 30°. Blondel's theorem indicates that two watt elements can accurately measure any level of unbalanced phase load and PF, so long as there is no neutral current flowing. For example, assuming a three-phase balanced load of 10 amps and 100 V at unity PF, each element would measure 10 × 100 × 0.866 (cosine 30) or 866 watts. The total for both elements would be 866 + 866, or 1732 watts (the same as the standard power formula). If the PF or load angle is changed from zero to 30° lagging, the top element would measure 500 watts and the bottom 1000 watts or 1500 watts total (the same as standard power formula).

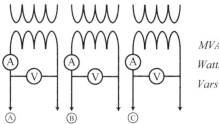

$$MVA = (E_A \times I_A) + (E_B \times I_B) + (E_C \times I_C)$$
$$Watts = MVA \times Cos(Angle)$$
$$Vars = MVA \times Sin(Angle)$$

FIGURE 5.17
Three-element metering.

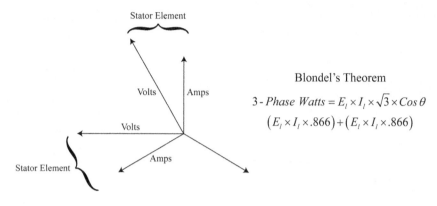

FIGURE 5.18
Two-element metering.

5.5.2 Watt Demand

Watt demand is the maximum or peak watts consumed over a selected time interval. The peak value is stored until it is replaced with a higher peak from a different time interval, or it is reset to zero for the next revenue period (usually monthly). Most commonly, demand intervals of 15, 30, or 60 minutes are used for revenue metering applications. Normally, demands are corrected to a 60-minute or 1-hour period. For example, to make a 15-minute measurement equivalent to a 1-hour period, it would be multiplied by 4. At that point, watts and watt-hours would have the same value. In other words, if a load of 100 watts were held constant for 1 hour, it would consume 100 watt-hours. Customers with demand readings will usually have an extra utility demand charge, above the kilowatt-hour (kWh) consumption charge, for the peak demand measurement. The rationale for the demand charge is that the utility is required to size their equipment to carry the peak demand. Depending on tariff details, the monthly demand charges often exceed the total energy charges for kilowatt-hours.

Figure 5.19 illustrates isolated contact pulses coming from a meter. One pulse is normally represented by a transition of the form C output contacts. The contact transitions are usually provided by isolated mercury wetted reed relays or electronic solid-state relays. The pulses could be used for driving a data acquisition system, load management equipment, or the plant DCS system. For revenue metering applications, the utility or metering engineer will provide the amount of kilowatt-hour energy in one pulse or transition. If only half the output contacts (form A) are used, then a contact open is a pulse and a contact close is another. Form C or three-wire pulses were designed for increased integrity. Normally, three-wire pulses are conditioned by a reset-set (RS) flip-flop logic that will not allow a pulse to register until one contact opens and the other contact closes. This prevents false registration from bouncing contacts and induced voltages. To condition form A contacts,

Electrical Apparatus Calculations

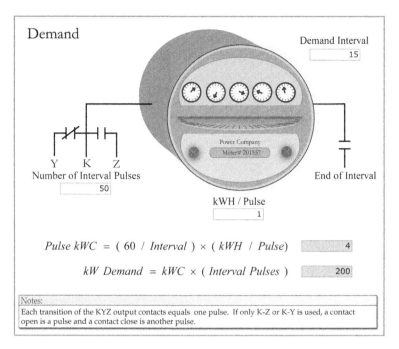

FIGURE 5.19
Watt demand.

a time delay of 20–30 milliseconds (ms) is usually provided. Revenue meters are typically equipped with end-of-interval contacts. They are normally provided in a form A contact configuration (single contact) and momentarily close at the start of each new demand interval. If the utility furnishes customers with the end-of-interval contact, the customer can tell precisely when the demand period starts and ends and can synchronize their load management systems with the revenue meter. As can be seen from the example calculations in Figure 5.19, if 50 1-kWh pulses were received in 15 minutes, each pulse would have a demand constant of 4 and the calculated demand for the 15-minute period would be 200 kW.

5.5.3 Watts

Figure 5.20 covers a procedure for determining real-time watts from an induction disk watt-hour meter if the instrument transformer ratios, the secondary kilowatt-hour value, and the number of revolutions over a specified number of seconds are known. This, of course, assumes that the load is held constant over the measuring period. The secondary Kh is the amount of secondary watt-hours for one revolution of the meter disk. Modern digital watt-hour meters will normally display the instantaneous watt and var measurements.

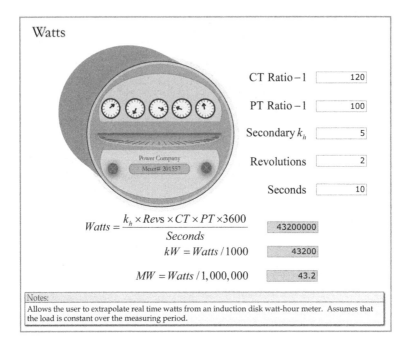

FIGURE 5.20
Watts.

5.6 Motors

The first polyphase round rotor motor was invented by Nikola Tesla in 1888, followed by the first squirrel cage rotor design in 1889 by Mikhail Debrovolsky (German). Of course, generation at fossil and nuclear plants is not possible without a significant amount of medium- and low-voltage motor loads to drive the various fans, compressors, and pumps required for the process. This section will describe the calculations for stator insulation testing, locked rotor amps, unbalanced voltages, X/R ratios, and voltage drop. Additionally, detailed information is provided on motor switching transients and motor reliability studies.

5.6.1 Motor Insulation Resistance

Institute of Electrical and Electronics Engineers (IEEE) and other industry standards and recommendations indicate that the minimum insulation megohms to ground for motors is 1 megohm per rated phase–phase kilovolt, plus 1 megohm at 40°C. The insulation megohms are determined by testing and need to be corrected for any temperature deviations from the reference temperature of 40°C. The actual correction factors are different depending on the class or type of insulation. In general, the higher the temperature, the lower the insulation resistance is because more free electrons are available for conduction.

One can roughly calculate the minimum megohms as three times higher, at 25°C (typical ambient testing temperature), or 3 megohms per kilovolt for a three-phase test; motors are normally tested to ground without breaking internal phase connections, as presented in Figure 5.21. In the interest of simplicity, for a single-phase test, one would expect the insulation resistance to be three times higher, or 9 megohms per kilovolt, which is conservative since there is also phase-to-phase insulation resistance, which slightly reduces the actual factor, especially if the untested phases are grounded. If you are testing too much equipment at the same time, you may need to disconnect or isolate apparatus units in order to meet the minimum requirement.

Although each type of electrical apparatus (buses, cables, circuit breakers, generators, and transformers) has its own minimum megohms, the industry in general uses the motor standard for the routine insulation testing of all apparatuses. Experience has shown that 3 megohms per kilovolt will not be a problem for energizing any apparatus unless there are other contributing factors. General practice is to use a 1000- or 500-V DC test instrument. Here, 1000 V is basically an overvoltage or hipot test for low-voltage equipment

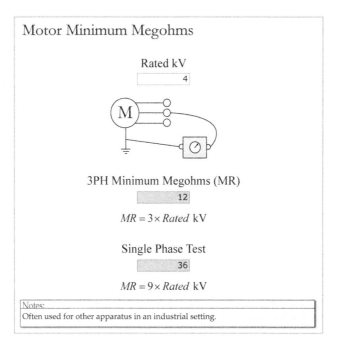

FIGURE 5.21
Motor insulation resistance.

(480-V apparatus), and consideration should be given to using a 500-V test during routine or short outages to avoid extending the outage for cleaning and dryout, and 1000-V test for overhauls and longer outages, where a higher-proof test for an undefined future life is desired.

Particularly concerning is the field practice of using 1 megohm per rated phase to phase kilovolt (eliminating the plus 1 kV) and not correcting for temperature differences. The author was in proximity to a transformer cooling cabinet when it was energized from a motor control center (MCC) feeder that exploded at 1 megohm per kilovolt. Another case was a 4-kV FD fan motor that failed when it was energized at 1 megohm per kV and resulted in a cascading series of events and four fires.

5.6.2 Acceptance DC Hipot

Insulation systems for medium-voltage motor stators are normally DC overvoltage or hipot tested at acceptance values to set after shipment, new installation, or repair warranty responsibilities, and also to prove that the apparatus is fit for service. DC is used instead of AC because of the amount of vars needed to support the insulation capacitance. The relatively high AC var load in comparison to DC necessitates the need for a higher-capacity AC hipot test set that is relatively expensive and more difficult to transport to the site. Normally, all three phases are jumpered together to reduce the possibility of damaging transient voltages if the test is suddenly interrupted.

Figure 5.22 provides recommendations on DC acceptance test kilovolt levels. National Electric Manufacturers Association (NEMA) standard MG

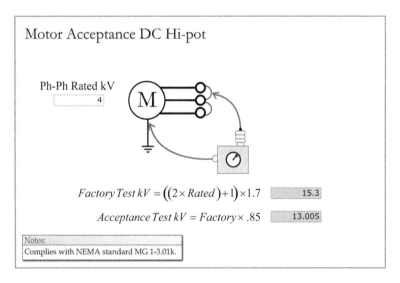

FIGURE 5.22
Motor acceptance DC overvoltage testing.

Electrical Apparatus Calculations

1–3.01 K addresses this matter, as shown in the "Notes" section of the figure. As with generators, the AC factory test is performed at twice-rated phase–phase kilovolts plus 1 kV. Acceptance tests are normally performed at 85% of the factory value. The AC value is then multiplied by 1.7 to convert it to an equivalent DC magnitude.

5.6.3 Routine DC Hipot

Figure 5.23 presents recommendations on DC routine test kilovolt levels for motors. There is also an industry consensus that a test level of 125%–150% of the phase–phase voltage is high enough to indicate an unknown future life (hopefully until the next major outage). As with generators, a value of 125% of the rated phase–phase voltage multiplied by 1.7 for an equivalent DC magnitude is suggested because it is more defendable should the insulation system fail to carry the minimum recommended test voltage. A failure in the stator slot areas normally requires a partial or full rewind. If you elect not to overvoltage-test during a major outage and the insulation fails shortly after returning the unit to service, you may not be in a defendable position, especially if it causes a complete unit outage or a reduction in the rated output megawatts of the generating station.

5.6.4 Locked Rotor Amps

Motors manufactured according to NEMA standards normally have a starting kilovolt-amp (kVA) per horsepower (HP) code letter stamped on the nameplate, which is used to determine locked rotor amps. Tables that are

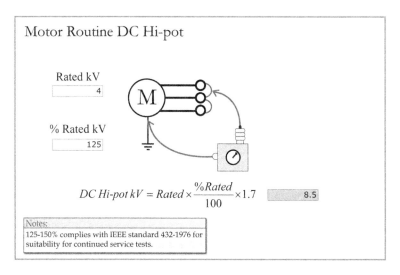

FIGURE 5.23
Motor routine DC overvoltage testing.

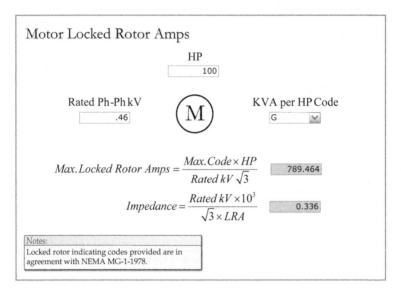

FIGURE 5.24
Motor locked rotor amps.

provided in NEMA MG-1-1978 indicate a range of kVA/HP factors for each code letter. Figure 5.24 provides a procedure for calculating locked rotor amps if the code letter is known. In the interest of providing some margin in the calculations, the maximum kVA/HP factor is normally assumed for most applications.

5.6.5 Unbalanced Voltages

Motors are very sensitive to the unbalance of supply voltages. Even a 1% or 2% imbalance can cause overheating and shorten the life of the motor. The unbalance causes negative sequence current flow in the rotor, at twice the supply frequency. The negative phase sequence currents increase rotor and stator winding losses and cause a small component of negative torque. Unbalanced resistances and reactance along current paths and in the various connections, external and internal to the motor, also may cause a voltage unbalance, with the potential of reducing the expected life of the motor.

Figure 5.25 demonstrates a NEMA motor standard procedure for calculating the percentage of voltage unbalance and determining a derating factor for the motor. The procedure is particularly useful where the angles are not known and a negative phase sequence calculation would be difficult to perform. The generally recognized upper limit for running a motor with unbalanced voltages is 5%. As the figure shows, the average voltage is 484, the maximum deviation from average is 24, and the voltage unbalance is almost 5%. The voltage unbalance derating factors are shown in the "Notes" section

Electrical Apparatus Calculations

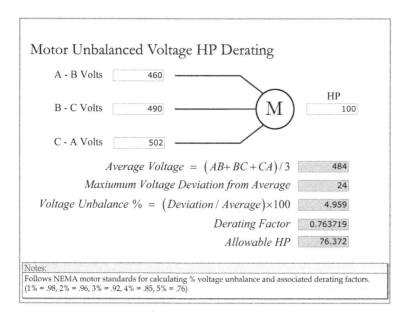

FIGURE 5.25
Motor HP derating for supply voltage unbalances.

of the figure in 1% steps to 5%. The respective HP derating factors are 0.98, 0.96, 0.92, 0.85, and 0.76, respectively.

5.6.6 X/R Ratio

For more refined motor studies, the X/R ratio needs to be determined. Figure 5.26 approximates the X/R ratio for a typical 100 HP, 1,755 RPM, at

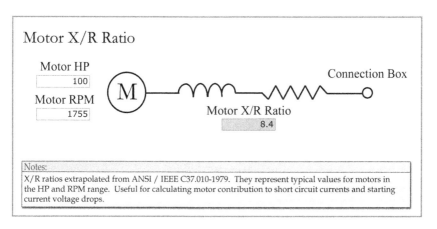

FIGURE 5.26
Motor X/R ratio.

locked rotor amps. ANSI/IEEE Standard C37.010-1979 presents curves for estimating typical X/R ratios based on motor HP and RPM values. The X/R ratio is useful for calculating motor starting current voltage drops and for determining motor short circuit contribution angles.

5.6.7 Switching Transients

During the late 1980s, EPRI contracted with Ontario Hydro and Rensselaer Polytechnic to complete a study of medium-voltage motor turn to turn insulation and possible damage from switching transients. The study was particularly thorough and considered surge capacitors, cable shielding, the location of the feeder breaker in the line-up, power supply grounding impedance, HP, nominal voltage, the length of cable, the type of cable, the type of circuit breaker, and the number of bus loads for medium-voltage motors. With the exception of power supply grounding, circuit breaker type, and location, all the foregoing parameters affected the maximum per-unit transient magnitude, the rise time or both.

In Figure 5.27, six turns are illustrated in both the top and the bottom bars. The turn-to-turn insulation capability of form wound medium-voltage stator coils is much less than the ground wall insulation since the operating voltage is distributed across many turns; it is generally in the range of 10–100 V. However, circuit breaker switching can produce high-frequency, high-voltage transients that can be damaging to the turn to turn insulation.

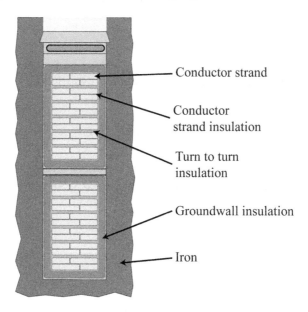

FIGURE 5.27
Motor form wound coils.

Figure 5.28 shows a simplified model of a stator winding with distributed capacitance. Because Xl increases with high frequencies and Xc decreases, the capacitance shunts the transient voltage to ground across the first few turns, which dramatically increases the turn-to-turn voltage magnitudes on those turns. Generally, surge capacitors do not decrease the magnitude, but they increase the rise time, which lowers the frequency and distributes the transient voltage across a greater number of turns, thereby reducing the turn-to-turn voltage. Most generating stations are reluctant to apply surge capacitors because they represent a stored energy device, have a higher failure rate, are not compatible with alarm-only ground schemes unless each capacitor is continuously rated to handle the full phase–phase voltage, and hamper efforts to test motor insulation systems.

Depending on the parameters discussed previously, the maximum transients were both measured and simulated during circuit breaker second-pole closing, and not interruption. The highest measured switching transient was 4.6 per unit, and the simulated transients were typically around 5.1 per unit, with rise times of 125 nanoseconds (ns) depending on the selected parameters. The per-unit values are in relationship to nominal phase-neutral voltages. The greatest impact for reducing transient magnitudes was cable shield grounding. The standard practice is to ground the shield only at the switchgear end. However, the EPRI study disclosed that grounding the shield only at the motor end, without the surge capacitors (as shown in Figure 5.28), would reduce 5.1 transients to 2.3 per-unit, with an 85-ns rise time. Intuitively, the 125-ns rise time probably limits the voltage to the first few turns anyway, and reducing the rise time further probably does not have that much of an impact, but reducing the magnitude by more than 50% should have a much greater impact depending on circuit details.

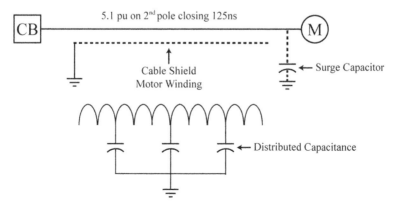

FIGURE 5.28
Motor-switching transients.

5.6.8 Reliability

Also during the 1980s, EPRI contracted with General Electric to complete another study, on motor failure rates and mechanisms. It focused on generating station switchgear-fed motors, 100 HP and larger. Altogether, they looked at 56 electric utilities and 132 generating units and discovered that there was a 12:1 ratio when plant failure rates were compared. The worst locations had 9.3% failures per year, and the better locations had 0.8% failures per year.

The motor failure mechanisms, identified in the report by percentage, were the following:

- Bearing-related (41%)
- Stator-related (37%)
- Rotor-related (10%)
- Other (12%)

Although environmental factors differ from site to site, there are a number of opportunities to improve the life of motors, reduce failure damage, and improve operating productivity at any location. Many of these points have been covered in this and the preceding chapters, and good operating and maintenance practices remain to be discussed in Chapters 6 and 7. Opportunities to extend motor life are listed here, along with the chapter in which they are discussed:

- Optimizing operating voltages—Chapter 2
- Motor overcurrent protection—Chapter 3
- Ground protection—Chapter 3
- Arcing ground transient voltage mitigation—Chapter 3
- Bus transfer schemes—Chapter 3
- Insulation testing—Chapters 5 and 7
- Motor switching transient mitigation—Chapter 5
- Operating practices—Chapter 6
- Maintenance practices—Chapter 7

5.6.9 Voltage Drop

Figure 5.29 illustrates a procedure for calculating motor starting and running voltage drops. First, the source R and X and cable R and X values are determined. Then, the motor R and X for starting and running conditions are determined. Finally, the total starting and running amps are calculated by dividing the phase-neutral voltage by the total circuit starting and running impedances, respectively. The phase-neutral voltages at the motor terminals for starting and running operations are determined by multiplying

Electrical Apparatus Calculations 259

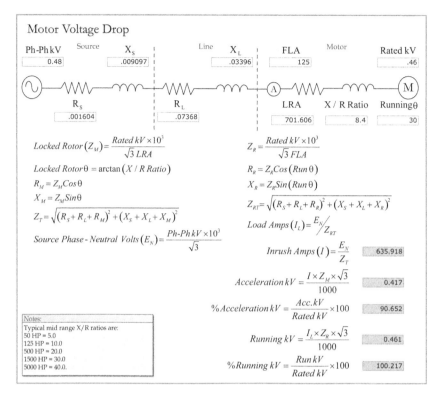

FIGURE 5.29
Motor 4/0 voltage drop.

the calculated amps for each operational mode by the calculated motor impedances. The phase–phase voltages are then determined by multiplying the phase-neutral voltages by the square root of 3 (or 1.732).

This calculation is more important for 480-V motor applications since the cable impedance has a much greater impact on low-voltage motors than medium-voltage motors. Motors are normally specified or guaranteed to accelerate at either 80% or 85% voltage; this author prefers the 80% requirement for larger, switchgear-fed, 480-V motors for greater margins. The longer the cable run, the greater is the concern that the motor may not be able to accelerate the load. Many engineering companies tend to oversize the cables to reduce the risk, and they also may get into an area of diminishing returns. When the cable size is increased for smaller cables that are mostly resistive in nature, it has a more dramatic effect on reducing voltage drop, but the benefit diminishes as the cable size gets larger, and the cable becomes predominantly inductive in nature. Increasing the cable size reduces the inductive reactance only slightly, and several cable size increases may be required just to gain a small advantage in voltage drop. At that point, it is usually better to have more than one conductor per phase. Installing two conductors per phase is approximately equivalent to cutting the distance of the run in half. Increasing the size of conductors to reduce the watt loss does not justify the expenditure.

Figure 5.30 represents a 100-HP motor for a primary fuel oil pump that had been previously retired in place to accommodate a new MCC in the tank farm area. The original 750-kcm cables for the motor were utilized to feed the new MCC. Years later, the station wanted to reactivate the motor and only had 2-inch spare conduit for the 1200-foot run. A 2-inch conduit is not large enough to accept three 750-kcm cables, and the station requested assistance. In running the calculations, the author determined that 4/0, which was seven sizes smaller than the original cables, was acceptable and could fit into the existing 2-inch spare conduit. Figure 5.30 shows the same calculation, except that the 4/0 cables have been replaced with the original 750-kcm cables for comparison purposes.

As you can see, by increasing the size of the cable by seven sizes, the acceleration voltage increased by only 3% (93.5 versus 90.7). If you compare the changes in cable impedance parameters, you will find that the resistance decreased dramatically, but the inductive reactance hardly changed at all.

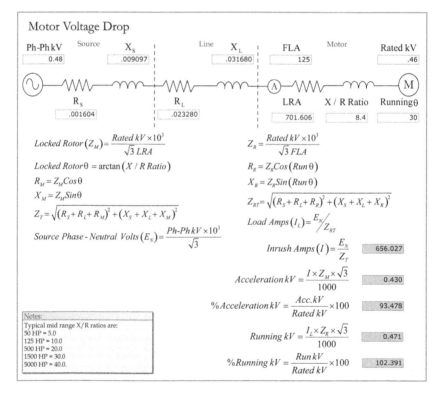

FIGURE 5.30
Motor 750 voltage drop.

Electrical Apparatus Calculations

After reactivation of the motor with the 4/0 cables, it accelerated to full speed in approximately 2 seconds. Since the 480-V buses normally run higher than 480, there may be additional margins in the calculations presented.

5.7 Transformers

The economic transmission and distribution of AC electrical power would not be possible without transformers. This section will cover power transformer loss and var consumption calculations, which are handy for verifying that transformer compensated metering is performing properly with in-service voltages and currents applied, as discussed in the "Metering" section earlier in this chapter, and finally the calculations for determining a transformer's X/R ratio for more refined studies are discussed.

5.7.1 Power Transformer Losses

Power transformers can be visualized or represented as having equivalent parallel and series legs, as illustrated in Figure 5.31. The parallel leg consists of a resistor and inductor in parallel that accounts for the no-load watt loss and var consumption, and the series leg has a resistor and inductor in series that represent the full load watt loss and var consumption.

The no-load watt losses are commonly called *iron losses* because the majority of them are associated with hysteresis and eddy current losses in transformer core iron laminations. The hysteresis loss in AC electrical apparatus cores normally predominates, usually accounting for roughly two-thirds of the total no-load watt loss. Although the no-load watt losses can be reduced by adding silicon to the alloy to reduce the realignment or flip energy, carbon

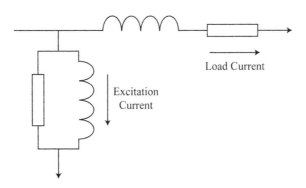

FIGURE 5.31
Transformer loss model.

steel generally provides a better magnetic path. The magnetic path can also be improved by using grain-oriented steel, which also reduces the hysteresis loss. The remaining no-load losses (roughly one-third) are reduced by laminating the iron into sheets in the direction of the magnetic flux to impede the flow of eddy currents. This elongates the annular current path and thereby increases the impedance.

The full-load watt losses are commonly referred to as *copper losses* because the majority of them are caused by resistance in the primary and secondary windings. However, eddy currents that flow in the tank and supporting structures cause an additional amount of full-load watt losses. These are commonly called *stray losses* and are usually around 10% of the copper losses. Sometimes the term *copper loss* refers to both copper and stray losses. Another term, *impedance loss,* is sometimes used to indicate the total copper loss plus stray loss.

Figure 5.32 covers a procedure for determining three-phase transformer losses at different loads and system voltages if the no-load and full-load watt losses are known. The procedure can be used to determine watt and var deliveries on the secondary side for revenue metering applications or to perform economic evaluations of transformer losses.

To determine the no-load watt loss and var consumption, first calculate the no-load impedance per phase from the exciting current. Then derive the equivalent no-load phase resistance that is required to produce the no-load watt loss. Next, determine the amount of parallel inductance by backing out the equivalent resistance from the calculated no-load impedance using the conductance, susceptance, and admittance parallel impedance method.

Now the actual no-load watt loss and var consumption at the actual applied voltage can easily be determined. The per-phase values can be calculated by squaring the applied phase-neutral voltage and dividing by the equivalent resistance and then inductive ohms in turn. Multiplying these values by 3 will yield the three-phase, no-load watt loss and var consumption.

To determine the actual load watt loss and var consumption, first calculate the rated full load amps. Then, divide the full-load watt losses by the rated amps squared and then multiplied by 3 to determine the equivalent per-phase series resistance. Correct the transformer %Z for applied taps and voltages and then convert from percent impedance to ohms. Next, determine the amount of X in the series circuit by backing out the equivalent resistance from the impedance ohms. Now the actual load watt loss and var consumption can easily be determined. The per-phase values for load watt loss and var consumption can be calculated by squaring the actual primary current and multiplying it by the phase resistance and the phase inductance ohms, respectively. Multiply these values by 3 to obtain the three-phase load watt loss and var consumption, respectively. The total watt loss will be the sum of the no-load and load-loss watts, and the total var consumption will be the sum of the no-load and load vars.

Electrical Apparatus Calculations

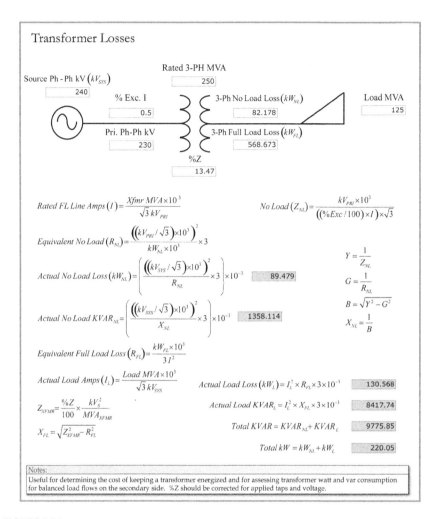

FIGURE 5.32
Transformer losses.

5.7.2 Power Transformer X/R Ratio

If a three-phase transformer's full-load losses (copper + stray) are known, a more refined X/R ratio during three-phase short circuit conditions can be calculated. The copper loss is a direct result of the resistance in the primary and secondary windings. The stray losses are caused by eddy currents in the tank and support structure areas and usually account for around 10% of the total load losses.

Figure 5.33 presents a procedure for determining the X/R ratio. First, the full-load current and impedance on the primary side is calculated. Then, the per-phase resistance is determined by dividing the full load losses by

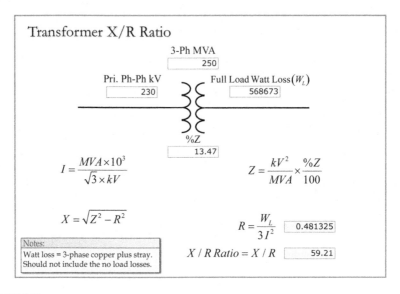

FIGURE 5.33
Transformer X/R ratio.

the full load amps squared and then multiplied by 3. Finally, the X component is determined by backing out the calculated R from the calculated Z.

Bibliography

Baker, T., *EE Helper Power Engineering Software Program* (Laguna Niguel, CA: Sumatron, Inc., 2002).

Baker, T., *RSO Shorted Turn Analyzer Instruction Manual* (Laguna Niguel, CA: Sumatron, Inc., 2004).

Electric Power Research Institute (EPRI), *Assessment of the Reliability of Motors in Utility Applications, 83 SM 476-9* (Palo Alto, CA: EPRI, 1983).

Electric Power Research Institute (EPRI), *Power Plant Electrical Reference Series*, Vols. 1–16 (Palo Alto, CA: Stone & Webster Engineering Corporation, 1987).

Electric Power Research Institute (EPRI), *Turn Insulation Capability of Large AC Motors*, EL5-5862 Main Rpt. Vol. 1 (Palo Alto, CA: EPRI, 1988).

General Electric Corporation, *ST Generator Seminar*, Vols. 1 and 2 (Schenectady, NY: General Electric Corporation, 1984).

Klempner, G. and Kerszenbaum, I., *Handbook of Large Turbo-Generator Operation and Maintenance* (Piscataway, NJ: IEEE Press, 2008).

Klempner, G. and Kerszenbaum, I., *Large Turbo-Generators Malfunctions and Symptoms* (Boca Rotan, FL: CRC Press, 2016).

Westinghouse Electric Corporation, *Electrical Transmission and Distribution Reference Book* (East Pittsburgh, PA, 1964).

6

Electrical Operating Guidelines

The guidelines presented in this chapter are based on more than 50 years of electrical experience in generating stations. The author has developed, reviewed, and/or revised operating guidelines for large industry, utilities and generation companies in the past. They represent a procedural effort to improve the availability of generating stations and large industrial facilities and reduce hazards to plant personnel. Some of the information was developed following operating errors in stations, and some to try to mitigate outages or equipment failures reported by the industry. However, there are a number of areas where the guidelines may need to be revised or upgraded by a generating station to meet its particular requirements. Some of these areas are presented here

- The guidelines are general in nature, and manufacturers' recommendations for their equipment may take precedence over the guidelines.
- There may be government regulatory agencies whose recommendations have priority over the guidelines, such as the Federal Energy Regulatory Commission (FERC), North American Electric Reliability Corporation (NERC), Occupational Safety and Health Administration (OSHA), and National Electrical Code (NEC).
- The guidelines assume that the operators are well trained and qualified to perform the suggested tasks.
- The station may have site-specific environmental conditions that need to be accounted for in the guidelines.

The guidelines here represent a good starting document. However, they should be treated as a work in progress that reflects the ever-changing government regulations, as well as site-specific and industry experience in general.

6.1 Operation of Large Generators

6.1.1 Purpose

The purpose of this guideline is to ensure the continuing operational integrity of generators. Operating conditions that have forced units off-line or have damaged or shortened the life of turbine/generator components in the past are highlighted in the guideline to prevent recurrences in the future.

6.1.2 Startup Operation

In addition to monitoring the various generator support systems for cooling and lubrication, electrical parameters, temperatures, and vibration, inattention to the following areas has caused problems in the past:

- At no time should excitation interlocks or protective relay functions be bypassed or disabled for the purpose of energizing a generator's direct current (DC) field winding.
- For generators requiring field prewarming, the manufacturer's instructions and established procedures should be followed relative to the allowable field currents.
- A generator field should not be applied or maintained at turbine speeds above or below that recommended by the manufacturer. On cross-compound units where a field is applied at low speeds or while on turning gear, extreme caution must be exercised. Should either or both shafts come to a stop, the field current should immediately be removed to prevent overheating damage to the collector or slip rings.
- After the field breaker is closed, the generator field indications should be closely monitored. If a rapid abnormal increase occurs in field current, terminal voltage, or both, immediately open the field breaker and inspect the related equipment for proper working condition before reestablishing a field.
- During off-line or unloaded conditions, at no time should the field current be greater than 105% of that normally required to obtain rated terminal voltage at rated speed in an unloaded condition. Typically, turbo-generators are designed to withstand a full load field with no load on the machine for only 12 seconds; after that, severe damage can occur to the stator core iron laminations.
- When synchronizing a generator to the system, the synchroscope should be rotating less than one revolution every 20 seconds. Phase angle differences should be minimized and no more than 5° out of phase when the circuit breaker contacts close. Phase angle differences as little as 12° can develop shaft torques as high as 150% of full load and damage shaft couplings and other turbine and generator components. Manufacturers usually recommend limiting maximum phase angle differences to 10°. It is also desirable that incoming and running voltages are matched as closely as possible to minimize reactive power flow to or from

Electrical Operating Guidelines 267

the electrical system. In general, the voltages should be matched within 2% at the time of synchronization with a slightly higher generator voltage to ensure var flow into the system instead of the generator. The speed of the turbine should be slightly greater than synchronous speed prior to breaker closure to help ensure that the unit will not be in motoring condition following connection to the electrical system.

NOTE: Under no circumstances should operators allow a unit to be synchronized using the sync-check relay as the breaker-closing device (i.e., holding a circuit breaker control switch in the closed position and allowing the sync-check relay to close the breaker). Some sync-check relays can fail in a contact closed state, allowing the circuit breaker to be closed at any time.

6.1.3 Shutdown Operation

Normally, units are removed from service through operator initiation of distributed control system (DCS) commands or turbine trip buttons that shut down the prime mover. Closure of steam or fuel valves will then initiate antimotoring or reverse power control circuits, that is, isolate the unit electrically by opening the generator circuit breakers, field breakers, and, depending on the design, unit auxiliary transformer (UAT) low-side breakers. If limit switch circuitry or antimotoring/reverse power relays fail to operate properly, the unit may stay electrically connected to the system in a motoring condition. If excitation is maintained, this condition is not harmful to the generator. However, the low pressure (LP) turbine blades may overheat from windage or air impingement. On steam units, the low pressure turbine blades are affected the most, with typical withstands of 10 minutes before damage. The unit can be safely removed from service with the following operating steps:

- Verify that there is no steam flow or fuel flow in the case of combustion turbine units to ensure that the unit will not overspeed when the generator circuit breakers are opened.
- Transfer the unit auxiliary power to the alternate source if opening the unit breakers will deenergize the UAT.
- Reduce or adjust the generator's output voltage (voltage regulator) until the field current is at the no-load value, and transfer from automatic voltage regulator (AVR) mode to the manual mode of operation.
- Open the generator circuit breakers.
- Open the generator field breaker.

6.1.4 On-line Operation

When the generator is on-line, particular attention should be paid to the following:

- Generators should be operated within their capability curves, which limit loading and field current (as related to var input and output) at various levels of hydrogen pressure. Operation beyond these limits will result in overheating and loss of life of various generator components.
- The generator stator and rotor operating temperatures should be closely monitored.
- Proper water flow to the hydrogen or air coolers (as pertinent) must be maintained. As little as 10% of rated load can severely overheat a hydrogen-cooled generator that does not have water flow to the hydrogen coolers.
- If conditions require operation with the voltage regulator on manual, the unit should be off automatic generator control (AGC) with the turbine at blocked load. Operators should keep a close watch on var output to ensure that the unit remains in a boost mode (supplying vars to the system). With the AVR out of service, changes in system voltages can cause the unit to buck vars and trip on loss of excitation.
- Units that are designed with stator ground voltmeters or protective relays that display stator ground voltages should be monitored by operations periodically. A healthy generator will produce a few volts of third-harmonic voltage during normal operation. The exact level is generator-specific and depends on the amount of real and reactive power that the generator is carrying. The absence of voltage when the generator is on-line indicates that either the stator ground scheme has an open circuit (safety switch open, grounding transformer racked out, or open conductor) or the generator has a stator ground near the neutral end where the conventional ground protection does not have the sensitivity to detect the problem. Loss of the third-harmonic voltage when the unit is on-line should be quickly investigated and resolved to ensure that the generator is properly protected and to minimize damage in the case of an actual ground condition. Some units are equipped with third-harmonic stator ground relays that will automatically alarm or trip the unit off-line if the normal third-harmonic voltage decays.

NOTE: Operating with a significantly reduced field (buck vars) can lower the third-harmonic levels to the extent that it may actuate the loss of field or third-harmonic tripping functions (if equipped).

6.1.5 System Separation

If the unit is separated from the system during system trouble, close attention to the field current and generator terminal voltage must be maintained, particularly when the unit is not operating at rated speed, to prevent catastrophic volts per hertz (volts/Hz) damage to the generator stator core iron. Reduction in speed can result in overexcitation (volts/Hz) as the automatic voltage regulator attempts to maintain rated voltage. At no time should the field current of an unloaded machine exceed 105% percent of that required to produce rated terminal voltage, at rated speed, with no load.

In general, generating units should not try to pick up blocks of load that exceed 10% of the machine rating. Attempts to do so may cause a collapse of the generator voltage and the loss of the unit.

Many generating units are not equipped with out-of-step relaying. If the watt/var outputs are swinging wildly, the unit may have loss synchronization with the electrical system. Loss-of-excitation relays may or may not operate for this condition. After quickly eliminating the AVR, power system stabilizer (PSS) (if equipped), process control, and governor instability as problem areas, quickly remove the unit from service to reduce the possibility of damage to the turbo/generator system.

6.1.6 Field Grounds

An internal generator field ground could be caused by high-temperature electrical arcing, insulation damage, elongated-end turn conductors, brush-rigging problems, or other anomalies in the generator field. If a second ground occurs inside or outside the field in the opposite polarity, DC short circuit currents will flow that can damage the rotor forging, retaining rings, wedges, collector rings, journals, and bearings. For that reason, manufacturers normally recommend automatic tripping for field ground events. Generating stations usually elect not to trip automatically because experience has shown that the ground is usually located in the peripheral excitation equipment and not in the generator rotor and, consequently, risk or exposure to the rotor is quite low. However, if the field ground is coincident with a notable step change in either generator's bearing vibration or field temperature, then the probability of the ground being in the rotor is quite high and the unit should be quickly removed from service to mitigate the possibility of rotor damage.

Each station should have detailed operating instructions that address locating and isolating field grounds for each type of excitation system. Timely and thorough investigations should be performed to identify and isolate the source of the field ground. Considering the numerous components associated with the excitation system, it is more than possible that the ground is outside the generator rotor and can be corrected without removing the unit from service. The following investigation steps are recommended to try and isolate the location of field grounds:

- Operators momentarily transfer from automatic voltage regulation to manual.
- Operators pull fuses to noncritical loads (i.e., monitoring transducers, DCS inputs, and field temperature recorders).
- Technicians verify that the field ground detector relay is operating properly.
- Technicians/electricians inspect the brush-rigging and excitation system.
- Technicians lift wires (where possible with the unit running) to further isolate excitation system components.

If the ground cannot be isolated, the unit should be quickly removed from service for further investigation.

6.1.7 Voltage Regulators

AVR instability is evident when the regulator output meter and the unit var meter swing between buck and boost. The swings may increase in magnitude. If this occurs, the voltage regulator should be transferred to the manual mode of operation. Electrical system disturbances are typically indicated by large initial voltage regulator output and unit var swings, which dampen out in a few seconds.

When transferring between AVR control and manual operation and vice versa, verify that the regulator output is null or at zero differential. Improper transfers can expose the generator to overexcitation (volts/Hz) or underexcitation (loss-of-excitation) and trip the unit off-line.

6.1.8 Moisture Intrusion

Even small amounts of moisture inside generators can result in reduced dielectric capability, stress corrosion pitting of retaining rings, and lead carbonate production and plating of the machine surfaces. The following recommendations should be adhered to in order to maintain generators in a dry condition:

- The backup seal oil supply from the bearing lube oil system has a higher moisture content and should be used only in emergency situations.
- Hydrogen dryers must be maintained properly.
- Hydrogen dryer desiccant should be monitored and replaced or regenerated as required.

Electrical Operating Guidelines 271

- Hydrogen cooler and inner cooled stator coil water leaks should be reported and repaired in a timely manner.
- Moisture monitors and detectors should be routinely checked for proper operation and calibration.
- Space heaters, when installed, should be verified as working when the unit is off-line.

Generators equipped with nonmagnetic 18-5 retaining-rings must be maintained in a dry condition to reduce the possibility of stress-corrosion pitting and cracking damage to the rings. These units should be taken off-line at the first indications of moisture intrusion (maximum dew point of 30°F) and any required repairs completed before returning the unit to service.

6.1.9 Routine Operator Inspections

To maintain the operating integrity of generators, the following checks should be routinely performed by operations personnel when making their daily inspection rounds:

- Check hydrogen purity levels (where applicable), and adjust gas flow to the purity monitor as required.
- Check that the seal oil system is operating properly. Verify that the proper pressure differential between the seal oil and hydrogen gas systems is maintained (where applicable).
- Check that the hydrogen dryers are in service (where applicable) and operating properly, and check that the desiccant is in good condition.
- Check the liquid detectors for accumulation of water or oil.
- Verify proper water flow to hydrogen or stator coolers (where applicable).
- On generators equipped with water-cooled stator coils, verify proper flow, conductivity, and differential pressure between the water and the hydrogen gas systems.
- Check the stator, gas, and field temperatures.
- Check the bearing vibration levels.
- Check the generator stator ground scheme for proper residual or third-harmonic voltages.
- Check the brush rigging (where applicable) for broken, vibrating, and arcing brushes.
- When the rotor is stopped or on turning gear, the brush-rigging area should be checked periodically for hydrogen leakage (where applicable). Hydrogen gas can leak through the bore conductors and accumulate in the brush-rigging areas when the unit is off-line.

- Check the shaft grounding brushes or braids to verify physical integrity. In those units, where the grounding brushes or braids are not visually accessible, please refer to maintenance guideline for periodic maintenance.
- Verify that the field winding ground fault detection system is operational.

Anomalies found during the routine inspections should be monitored and work orders prepared to resolve problems noted.

6.1.10 Generator Protection

6.1.10.1 Differential (87)

Generator differential relays compare the secondary currents from current transformers (CTs) installed on the neutral end of the generator windings to current transformers installed on the output side of the generator or the output side of the generator circuit breaker (if equipped). If an internal phase-phase or three-phase fault occurs between the neutral and output CTs, the current flows will not balance and the differential relay will instantaneously actuate to trip the unit off-line. In general, these relays will not be able to detect ground faults or shorted turns in the stator windings.

NOTE: The unit should not be reenergized following a generator differential trip until the cause of the relay operation can be determined and resolved by engineers or technicians.

6.1.10.2 Stator Ground (64) or (59G)

Generator stator ground schemes protect the generator, isolated-phase buses, generator bus circuit breaker (if included), arrestors and surge capacitors (if included), and the primary windings of potential, auxiliary, and main step-up transformers from breakdowns in the insulation system to ground. Typically, the stator ground relays are set to operate in 1 second for a 100% ground fault condition. The relay senses voltage on the secondary side of the generator stator-grounding transformer, which is connected between the generator neutral and the station ground grid. Under normal operation, the relay will see a few volts of third-harmonic voltage (180 Hz for 60-Hz machines), due to the nonsymmetry and partial saturation of the stator core iron, and zero volts at normal system frequencies (60 Hz). The stator ground relays are usually desensitized at 180 Hz and are normally set to operate for a 5%–10% ground (depending on the sensitivity of the particular relay) at normal system frequencies. Therefore, they do not have the sensitivity to provide protection for 100% of the stator winding.

Tripping schemes are available (64TH) that will remove the unit from service on loss of third-harmonic voltage to provide full winding protection from ground faults. At minimum, this voltage should be monitored by operations daily to verify that the grounding scheme does not have an open circuit and is healthy, and there are no grounds at the neutral end of the generator stator winding where a conventional relay does not have the sensitivity to detect grounds for the last 10% of the winding. In general, when a protective relay measures the third-harmonic voltage, tripping is preferred for salient pole hydromachines, but alarm-only operation is preferred for cylindrical rotor machines because of the robust construction of the higher-speed generators.

Subharmonic injection 100% winding protection schemes (64S) are also available to detect grounds at the neutral end of the stator winding. These schemes inject a subharmonic frequency and have the capability to detect grounds for 100% of the winding.

NOTE: The unit should not be reenergized following a (64) generator stator ground, (64TH) third-harmonic, or (64S) subharmonic trip until the cause of the relay operation can be determined and resolved by station engineers or technicians. For loss of third-harmonic, alarm-only schemes, there should be a timely investigation of the reasons for the loss, and the unit should be quickly removed for service if it appears to be legitimate.

6.1.10.3 Bus Ground Detectors (59BG)

Units equipped with generator medium-voltage isolated phase bus circuit breakers require a second ground detector scheme to protect the generator circuit breaker, isolated phase bus, and connected transformers on the downstream side of the generator circuit breaker when it is open. This scheme normally uses a wye–broken delta connection with a voltage relay installed on the secondary side to sense ground conditions. Under normal conditions, when the unit is running, this scheme will detect a few volts of third-harmonic voltage and zero volts at running frequency. However, the scheme can become unstable under certain conditions (neutral instability or three-phase ferroresonance), causing blown fuses in the ground-detecting transformers. A blown fuse may cause the ground detector relay to actuate. Also, coordination between the generator stator and bus ground detector schemes is difficult and, depending on the design, the station may not be able to quickly ascertain which side of the generator circuit breaker has the ground condition.

Preferred designs for the bus ground detector is alarm only operation for bus grounds or alarm when the generator circuit breaker is closed, and may trip with a short time delay when the generator circuit breaker is open. This prevents the unit from tripping for blown fuse conditions and allows operators to quickly determine which side of the generator circuit breaker has the ground condition. With these designs, the generator stator ground protection will trip the unit for a ground on either side of the generator circuit breaker when the unit is on-line. If the ground is on the generator side of the circuit breaker, the ground will be isolated after tripping. Otherwise, the bus ground detector will alarm or trip the unit main step-up transformer after generator tripping. For alarm-only schemes, the bus ground should be immediately investigated by operations to determine if it is valid or not. If it appears to be valid (generator stator ground protection actuated first, or the fuses are intact, the generator step up transformer (GSUT) should be quickly deenergized to reduce the possibility of the low-level ground fault current escalating into a damaging high-current double line to ground, phase–phase, or three-phase short circuit.

NOTE: The main transformer should not be reenergized following a generator bus ground event until the cause of the relay operation can be determined and resolved by engineers or technicians.

6.1.10.4 Loss of Excitation (40)

Synchronous generators are not designed to be operated without DC excitation. Unlike induction machines, the rotating fields are not capable of continuously handling the circulating currents that can flow in the rotor forging, wedges, amortisseur windings, and retaining rings, during underexcited or loss-of-field operation. Consequently, loss-of-excitation relays are normally included in the generator protection package to protect the rotor from damage during underexcitation operating conditions. Impedance-type relays are normally used to automatically trip the unit with a short time delay whenever the var flow into the machine is excessive. Minimum excitation limiters in the AVR should be set to prevent loss-of-excitation relaying whenever the voltage regulator is in the automatic mode of operation.

NOTE: The machine should not be resynchronized to the system following a loss-of-excitation trip until an investigation has been completed to determine the cause of the relay operation.

However, considering the complexity of modern excitation systems, unexplained events are not that uncommon. Engineers or technicians should inspect the physical excitation system, verify calibration of the loss-of-excitation relays, check the DC resistance of the generator field windings, and review any available data acquisition monitoring that would verify the operating condition. The unit can then be started for testing, and proper operation of the excitation system can be ascertained by operations before synchronizing the unit to the system.

6.1.10.5 Overexcitation (24)

Overexcitation (volts/Hz) relays are applied to protect the generator from excessive field current and overfluxing of the generator stator core iron. Typically, generators are designed to handle a full load field with no load on the machine for only 12 seconds before the stator iron laminations become overheated and damaged. Normally, the relays are set to trip the unit in 45 seconds at 110% volts/Hz and 2 seconds at 118% volts/Hz. The term *volts/Hz* is used to cover operation below normal system frequencies (60 Hz), where generators and transformers can no longer withstand rated voltages. Generators are continuously rated for operation at 105% voltage and transformers normally for continuous operation at 110% voltage. Consequently, the generators are usually the weak link, and safe operation for generators, in most cases, will automatically protect unit transformers that are connected to generator buses.

NOTE: The unit should not be resynchronized to the system following a volts/Hz trip until an investigation has been completed to determine the cause of the relay operation.

6.1.10.6 Reverse Power (32)

Reverse power or antimotoring relays are often applied for control purposes and for protective relaying. In the control mode, they are typically used to automatically remove units from service during planned shutdowns and to ensure that prime movers have no output before isolating units electrically to prevent overspeed conditions. In the protection mode, they are used to protect turbine blades from air impingement overheating and sometimes to protect combustion turbine units from flameout conditions. The reverse power or antimotoring protective relays should have enough of a time delay before tripping to allow synchronizing excursions (20 seconds is suggested). Motoring is not damaging to generators, so long as proper excitation levels are maintained. In steam turbines, the low-pressure turbine blades will overheat from windage. Typically, steam turbine blades can withstand motoring conditions for 10 minutes before damage. In hydroturbines, motoring may cause water cavitations. Combustion turbines also consume a fair amount of watts when in a motoring mode of operation.

NOTE: Following a reverse power or antimotoring protection trip, operations should determine if the trip was caused by control instability by reviewing recorder or DCS trending of unit megawatt outputs. If control instability is evident, engineers or technicians should investigate and resolve the problem before resynchronizing the unit. If control instability is not evident, engineers or technicians should check the calibration of the reverse power or antimotoring relay before returning the unit to service.

6.1.10.7 Negative Phase Sequence (46)

Unbalanced phase current flow in generator stators cause double-frequency reverse rotation negative-phase sequence currents to circulate in the rotor body that can damage the forging, wedges, amortisseur windings, and retaining rings. Many generators designed according to American National Standards Institute (ANSI) standards are capable of continuously carrying

8%–10% negative phase sequence current. Depending on the design details of the generator (indirectly or directly cooled), with two phases at rated current and zero current in the third phase; the machine may be able to carry this unbalance for only 30 seconds before damage can occur to the rotor components. Accordingly, negative phase sequence protection is necessary to protect generator rotors from the heating effect caused by the twice-frequency circulating rotor current during all possible operating conditions, including phase-to-phase and phase-to-ground faults on the high-voltage transmission system.

Some negative sequence overcurrent relays provide an alarm function with a pickup value set somewhere below the trip point. This alerts the unit operator to a negative sequence condition prior to an actual trip. If the negative sequence alarm is initiated, the operator should take the following actions:

- Notify the transmission dispatcher of the negative sequence condition and find out if there are any electrical problems on the transmission system. When a negative phase sequence alarm is activated, operators should also check the phase currents for balance. In addition to offsite causes for unbalance, open conductors, disconnect poles, or circuit breaker poles at the site can cause unbalanced conditions. If no abnormalities exist, notify the energy control center that the generator output will be reduced until the alarm clears.
- The generator should be taken off AGC, and the unit output reduced until the alarm clears.
- Engineers or technicians should verify calibration of the negative phase sequence relay and review any data acquisition monitoring devices (protective relay digital storage or DCS trends) to verify that the unit operated with a significant current unbalance.
- If the alarm is coincident with any electrical switching in the switchyard or within the plant, the switching should be reversed, and if the alarm clears, the apparatus in question should be investigated for proper operation.

NOTE: Following a negative phase sequence trip, the unit should not be returned to service until the cause is determined and resolved by engineers or technicians.

6.1.10.8 Backup Impedance (21) or Voltage Restraint Overcurrent (51V)

Backup impedance relays are normally provided to backup generator and transformer differential relays and to protect generators from overheating from prolonged three-phase balanced electrical system disturbances that do not clear properly from downstream electrical protective devices. Zone 1 backup protection will usually operate in around 0.1 seconds to back up the differentials, and zone 2 backup protection will operate in approximately

Electrical Operating Guidelines 277

2 seconds at 150% of rating to protect the machine and associated step-up transformer from overheating. Although generators should not be operated outside their capability curves, industry standards require a capability of 130% of rated stator current for 1 minute. Some units may be equipped with a voltage restraint or controlled overcurrent relay that basically performs the same function as the backup impedance relay.

NOTE: Following a backup impedance or voltage restraint overcurrent relay trip, the unit should not be returned to service until an investigation is completed by engineers or technicians and the problem is resolved.

6.1.10.9 Out of Step (78)

Out-of-step or loss-of-synchronization protection is primarily applied to protect turbine/generators from slip frequency power swings that can mechanically resonate and cause cycle fatigue failures of turbo-generator components. Out-of-step events can be characterized by wildly swinging watt and var indications and are usually caused by close-in electrical system faults that do not clear fast enough. When a close-in electrical system fault occurs, the watt load on a generator is displaced by vars. This basically unloads the machine and it speeds up. If the system relays do not clear the fault quickly enough (typically around 6–20 cycles), the speed of the turbine/generator system will increase past the point of recovery and the unit will lose synchronism with the system. In this case, the unit will go in and out of phase and the output indications will swing wildly. The out-of-step event may have been caused by an electrical system fault that did not clear promptly because of circuit breaker or protective relay anomalies. Operations should communicate with the energy control center to see if that is the case.

NOTE: Following an out-of-step relay trip, the unit should not be returned to service until engineers or technicians investigate and resolve the problem.

6.1.10.10 Overfrequency and Underfrequency (81)

Almost always, the overfrequency or underfrequency protection of the unit is there to protect the turbine before it protects the generator. Turbines blades (especially if they resonate at the particular speed or frequency) can experience a permanent reduction in future life. The damage done to turbine blading during off-frequency resonant conditions is accumulative and dependent on the duration of the event. Operators should be aware of the frequency limitation for their particular turbines and should limit excursion durations that are outside the manufacturers' recommended limits.

NOTE: If a unit trips by the operation of an overfrequency or underfrequency relay, the unit should not be returned to service until the system frequency stabilizes within acceptable limits.

6.1.10.11 Sync Check (25)

When synchronizing a generator to the system, the synchroscope should be rotating less than one revolution every 20 seconds. Phase angle differences should be minimized and ideally be no more than 5° out of phase when the circuit breaker contacts close. Phase angle differences as little as 12° can develop shaft torques as high as 150% of full load and damage shaft couplings and other turbine and generator components. Manufacturers usually recommend limiting maximum phase angle differences to 10° (including breaker closing time). It is also desirable that incoming and running voltages are matched as closely as possible to minimize reactive power flow to or from the electrical system. In general, the voltages should be matched within 2% at the time of synchronization with the generator a little higher to ensure var flow into the system. The speed of the turbine should be slightly greater than synchronous speed prior to breaker closure to help ensure that the unit will not be in a motoring condition following connection to the electrical system to mitigate possibilities for tripping on reverse power protection.

Slow breaker protection is now available in some of the newer technology relays. This function will operate breaker failure protection to isolate the generator if the breaker is slow to close for mechanical reasons. A typical breaker control circuit seals in when the close signal is dispatched and the close cannot be directly aborted. The breaker trip coil cannot be activated until the breaker actually closes; consequently, the only way to prevent an out-of-phase closure during a slow breaker-closing event is to operate breaker failure tripping that opens adjacent breakers to isolate the failed breaker.

NOTE: Following slow breaker tripping, the unit should not be returned to service until an investigation is completed by engineers or technicians and the problem is resolved.

6.1.10.12 Inadvertent Energization (50/27)

This scheme protects an off-line at-rest generator from an inadvertent closure of a unit breaker. The logic recognizes that the unit is off-line and will trip the unit breakers instantaneously if there is a sudden application of voltage and a corresponding high level of current. Generators are not designed to be started as induction motors, and severe generator damage can be expected if the event is long enough in duration.

NOTE: Following an inadvertent energization trip, the unit should not be returned to service until an investigation is completed by engineers or technicians and the problem is resolved.

6.1.10.13 Pole Flashover (50NF)

For designs that utilize the switchyard breakers for synchronizing, pole flashover logic will protect generators from damaging current flow if one pole or phase flashes over. In general, circuit breakers are at greater risk during synchronizing operations because the voltage across the poles can double during revolutions of the synchroscope. Basically, if the scheme detects current flow in the main step-up transformer neutral, and if the logic indicates that the unit breakers are open, it will actuate breaker failure tripping to isolate the faulted breaker. During a pole flashover event, the generator may experience high levels of phase-to-phase current, which could severely damage the generator, as well as the circuit breaker if allowed to continue.

NOTE: Following pole flashover tripping, the breaker should not be returned to service until an investigation is completed by engineers or technicians and the problem is resolved.

6.1.10.14 Main and Auxiliary Transformer Differential (87)

Transformer differential relays compare the current magnitudes and associated phase angles on the primary side of the transformer to the currents on the secondary side. Under normal conditions, the primary and secondary currents should balance out after considering the winding ratios. If they are out of balance enough, an electrical fault is assumed and the protection will actuate instantaneously to remove the transformer from service.

NOTE: Following a differential protection operation, the transformer should not be returned to service until an investigation is completed by engineers or technicians and the problem is resolved.

6.1.10.15 Feeder Differential (87)

Some designs utilize a separate differential scheme to protect the line from the unit main step-up transformer, reserve or startup auxiliary power transformer to the switchyard, or both. If the line currents at each end are not in agreement, a fault is assumed and the protection will instantaneously actuate to remove the line from service.

NOTE: Following a line differential protection operation, the line should not be reenergized until an investigation is completed by engineers or technicians and the problem is resolved.

6.1.10.16 Overall Unit Differential (87)

In lieu of installing an extra feeder differential, most generating stations apply an overall or unit differential that compares the current flow in the generator neutral side to the current flowing in the unit high-voltage switchyard circuit breakers. Accordingly, the overall scheme will detect faults in the generator, isolated phase bus, main step-up transformers, unit line to the switchyard, and unit switchyard circuit breakers. If, after considering the main step-up transformer and current transformer ratios, the currents are out of balance, the relay assumes a fault and will actuate instantaneously to trip the unit off-line.

NOTE: The normal practice for switchyard ring bus or breaker-and-a-half configurations is to open unit disconnects after tripping and reclose the unit switchyard breakers to tie the buses together to improve the integrity of the switchyard. For this case, the unit differential relay scheme is still protecting a switchyard bus section and breakers, even though the unit is off-line. Accordingly, anytime the unit breaker(s) are closed and energized, the unit differential protection must remain in service. Following a unit differential protection operation, the unit and its associated line to the switchyard should not be reenergized until an investigation is completed by engineers or technicians and the problem is resolved.

6.1.10.17 Unit Switchyard Disconnect Position Switch (33M)

High-voltage switchyards that are in a ring bus or breaker and one half configuration, will have manual or motor-operated (MOD) disconnect switches that can be opened to isolate generating units from the high-voltage yard. This allows operations to reclose the unit connected high-voltage breakers to restore the integrity of the switchyard buses without energizing the units. An inadvertent closure of the disconnect switch would energize an at-rest generator and could cause severe damage to the machine if the duration is long enough. This can easily be avoided by installing a mid-position limit switch that will trip the unit breakers open when the swing actuates the limit switch.

6.1.10.18 Auxiliary and Main Transformer Sudden Pressure (63)

Large power transformers are normally equipped with either gas- or oil-type sudden pressure relays to quickly detect internal transformers faults in order to mitigate the possibility of tank ruptures and fire in mineral oil transformers.

NOTE: Transformer sudden pressure relays must be taken out of service before work is performed that may affect the internal pressure of the transformer (for example, adding nitrogen gas, adjusting gas regulators, or any work that would allow the gas to vent to the atmosphere). Following a sudden pressure protection operation, the transformer should not be returned to service until an investigation is completed by engineers or technicians and the problem is resolved.

6.1.10.18.1 Zone 1 Impedance (21)

Some designs apply a zone 1 impedance relay that looks from the high-voltage switchyard into the generating unit to provide backup protection for the unit differentials, the high-voltage line feeder differential relay, and/or

Electrical Operating Guidelines 281

the high-voltage transformer differential schemes. This protection can detect electrical short circuits in the unit breakers, the line from the switchyard to the main step-up transformer, the main step-up transformer, the isolated phase bus, and the generator, and almost instantaneously actuate to isolate those faults. In the case of reserve or startup auxiliary power transformers fed from the switchyard, it will back up feeder and transformer differential relays and protect the switchyard breakers, the line to the transformer, and the transformer itself from short circuit conditions.

NOTE: Following a zone 1 protection operation, the unit or transformer should not be returned to service until an investigation is completed by engineers or technicians and the problem is resolved.

6.1.10.18.2 *Breaker Failure (50BF)*

Breaker failure protection is normally provided for switchyard circuit breakers, and with newer technology, digital relay breaker failure protection can be used for breakers on the isolated phase bus. Basically, if the breaker does not open in a specified amount of time after a trip signal is applied and a high level of current is flowing through the breaker, the breaker failure relay logic will automatically open adjacent breakers in order to isolate the stuck breaker condition.

NOTE: Following a breaker failure protection operation, the unit and its associated line to the switchyard should not be reenergized until an investigation is completed by engineers or technicians and the problem is resolved.

6.1.10.18.3 *Transformer Overcurrent (51)*

Auxiliary power transformers (generator bus fed or system fed) are normally equipped with primary-side overcurrent relays to back up the differential protection and, more important, to provide breaker failure protection in case the low-side breaker fails to operate to clear a faulted bus. For those cases, more upstream protection (of a generator or high-voltage electrical system) will not normally have the sensitivity to detect the faults on the secondary-side buses.

NOTE: Following a unit or reserve auxiliary transformer high-side overcurrent protection operation, the transformer should not be reenergized until an investigation is completed by engineers or technicians and the problem is resolved.

6.1.10.19 **DC Low-Voltage (27DC)**

Critical protection DC circuit voltage levels should be monitored to ensure that the protection can operate properly. Whenever the DC voltage level is at or below the set point threshold, an alarm should be initiated.

NOTE: Following a DC low-voltage alarm, plant operators, electricians, and technicians should immediately investigate and resolve the problem or remove the protected equipment from service.

6.1.10.20 DC High-Voltage (59DC)

Critical protection DC circuit voltage levels should be monitored to ensure that the protection can operate properly. Whenever the DC voltage level is at or above the set point threshold, an alarm will be initiated. This would normally indicate that the battery charger failed in a raise voltage direction.

NOTE: Following a DC high-voltage alarm, plant operators, electricians, and technicians should immediately investigate and resolve the problem or remove the protected equipment from service.

Many generating stations at this time are not equipped with all of the foregoing protective functions; some of these functions are available only in the newer technology multifunction digital relays. However, many stations will be upgraded with the newer technology relays in the future and some of the foregoing functions may be added at that time.

6.2 Operation of Large Power Transformers

6.2.1 Purpose

The purpose of this guideline is to provide suggested procedures and schedules for the routine operator inspection of power transformers (500 KVA and larger). Good operating practices are important in order to obtain the best possible service and performance. Every transformer failure represents a potential hazard to personnel and other equipment in the plant and can cause a forced outage of the unit.

6.2.2 Operator Inspections

It is the responsibility of operations to establish and maintain scheduled routine inspections of all large transformers. In general, performing one inspection per day is recommended.

The following daily checks are recommended:

- Oil leaks (where applicable).
- Proper nitrogen pressure (where applicable).
- Abnormal noise.
- Proper oil level in tanks and electrical bushings (where applicable).
- Explosion relief device semaphores or targets (where applicable).
- Proper oil pump operation (where applicable).

Electrical Operating Guidelines 283

- Proper fan operation (where applicable).
- Condition of high- and low voltage side overhead-connected bushings, insulators, and lightning arrestors (where applicable). Check the bushings and insulators for external contamination, white banding (if silicone coated), audible spitting, corona activity, or unusual radio interference.
- If switching provisions are provided, change cooling fans and circulating pumps to ensure equal running hours (usually monthly).
- Inspection of security fences, doors and gates, locks, warning signs, and so on, for integrity/operation.
- Silica gel breather desiccant crystal color (where applicable).
- Top oil-temperature gauge values. If the drag-hand is above normal range, log the value and reset the maximum pointer afterward. Investigate why the maximum temperature value was reached.
- Winding temperature gauge values (where applicable). If the drag-hand is above normal range, log the value and reset the maximum pointer afterward. Investigate why the maximum temperature value was reached.
- Neutral grounding resistors/reactors (where applicable). Look for corrosion and oil leaks, and check the condition of associated bushings and insulators.

When abnormalities are found that could cause a major failure of the transformer or personnel hazard, steps should be taken to remove the equipment from service quickly. Less serious conditions that require maintenance or repair should be identified on a maintenance order and remedied at the first practical opportunity.

6.2.3 Sudden Pressure Relays

Sudden pressure relays should be taken out of service whenever work is to be performed that may affect the internal transformer pressure. Examples of when sudden pressure relays should be made nonautomatic include the following:

- Adding nitrogen gas to the transformer
- Adjusting the nitrogen gas regulator
- Replacing the nitrogen gas bottles
- Removing the transformer top filter press inlet plug
- Adding oil to the transformer
- Performing work that may allow the nitrogen to vent or escape to the atmosphere
- Performing work on the sudden pressure relay or associated circuitry

6.2.4 Transformer Differential or Sudden Pressure Relay Operations

After a transformer's sudden pressure or differential operation, the transformer should be thoroughly tested prior to reenergization unless it can be determined that the protection operation was due to a calibration or setting error, improper current transformer circuitry, or a failed relay.

Where the integrity of the main step-up transformer or auxiliary transformer remains suspect because of questionable repairs or unknown reasons for the relay operation, the transformer should be energized once for testing. Energizing the GSUT from the system (preferably with a dedicated transmission line) with the generator isolated via open isolated phase bus links or generator breaker has the following advantages:

- A soft energization from the generator may not provide enough fault current to actuate the protection quickly, and the damage to the transformer may be greater.
- Energizing the transformer from the generator potentially exposes the generator to a fault.

6.2.5 Emergency Cooling and Loading

Running service water on the exterior surfaces of transformers to reduce operating temperatures is a practice that should be performed only during emergencies; it is not generally recommended for the following reasons:

- Service water contains a significant amount of dissolved solids that can form a temperature insulation barrier on the heat exchangers and other exterior surfaces, forcing the transformer to operate continually at higher temperatures.
- Service water is corrosive to transformer heat exchangers and other accessories.

When it appears necessary to use service water, a transformer loadability study can be performed to see if the transformer can operate at higher temperatures, negating the need for the service water. In general, transformers can be loaded beyond their nameplate ratings for short durations, with no or very little loss of life. Many transformers can carry 140% of the full load for 2 hours with very little loss of life. However, the permissible short time loadability varies depending on the design details of the specific transformer. An engineering study should be completed before transformers are operated above their rated load to substantiate that the loss of life is acceptable and that the associated buses, cables, and breakers can also carry the increased load.

6.2.6 Oil Pump Operation

Electrostatic voltage transfer is the phenomenon by which a charge develops between the oil and the insulation systems of a transformer when the transformer is deenergized and oil is circulated by the oil pumps.

Running oil pumps on deenergized transformers should be limited to 10 minutes. This is to prevent the failure of transformers from electrostatic charging and subsequent tracking of the insulation systems. If additional operation of the pumps is required, a minimum wait time of 2 hours should be observed before operating them again.

6.3 Operation of Large Electric Motors

6.3.1 Purpose

The purpose of this guideline is to provide suggested procedures and schedules for the routine operator inspection of switchgear-fed motors (100 HP and larger). Good operating practices are important to obtain the best service and performance. Every motor failure represents a potential hazard to other equipment in the plant and may cause a forced outage of the unit.

6.3.2 Operator Inspections

It is the responsibility of operating personnel to establish and maintain scheduled routine inspections of all large motors. In general, an inspection frequency of once per day is recommended.

The following checks are recommended for *running* motors:

- Abnormal noise levels
- Increase in bearing vibrations
- Increased operating temperatures of bearings or stator
- Status of lubricating oil (i.e., contamination, proper levels, temperatures, and pressures)
- Cleanliness of exterior surfaces
- Cleanliness of air filters
- Obstruction of air intake

The following checks are recommended for deenergized *standby* motors:

- Lubricating the oil system in proper standby condition
- Motor heaters in service (where applicable)

- Power and control circuits in proper standby condition
- Cleanliness of exterior surfaces
- Cleanliness of air filters
- Obstructions that would interfere with the starting of the motor

Maintenance orders should be prepared for any anomalies found during the foregoing inspections.

6.3.3 Starting Duty

Generally, induction motors should be limited to the following starts:

1. Two starts in succession, coasting to rest in between, with the motor initially at ambient temperature.
2. After two successive starts, a minimum 1-hour cooling period should be provided before a third start. The motor can be at rest or running during this cooling period.

The preceding information is to be used only as a guide. Manufacturers' specifications may be more stringent or liberal and take precedence over this general guideline.

6.3.4 Heaters

Many large motors are equipped with heaters to prevent moisture buildup in the windings. The heaters should automatically switch on when the motors are deenergized. Motor heater circuits should be monitored by ammeters or current-driven sensor light-emitting diodes (LEDs) to provide indication that the heaters are functioning properly. Operations should routinely verify that the heaters are working when the motors are deenergized. If it is found that the heaters are not working properly, the motor stator insulation should be tested with a megohm tester or megger, prior to returning it to service.

6.3.5 Protection

Operation of motor protective relays should not be taken lightly. The following investigation steps should be completed before allowing restarts of motors following protective relay operations.

6.3.5.1 Instantaneous Phase Overcurrent Tripping

Instantaneous phase overcurrent relay minimum trip points should be set well above the surge and locked rotor values for the motor. An instantaneous

Electrical Operating Guidelines 287

target and corresponding trip indicates that either the electrical protection malfunctioned or was set improperly, or else a permanent ground (solidly grounded systems only) phase-to-phase or three-phase electrical fault exists in the cable or motor. The motor should not be reenergized; this is to avoid aggravating damages and overstressing the plant electrical system until the cause of the relay operation can be determined and repaired, or until reasonable testing to prove the electrical integrity of the motor or cables has been completed.

6.3.5.2 Time-Phase Overcurrent Tripping

Time-phase overcurrent targets and corresponding trips are usually caused by mechanical problems in the motor or driven equipment, control anomalies, or malfunctioning or improperly set protective relays.

If the motor trips on time overcurrent during *starting* (rotor normally thermally stressed), perform the following actions:

- Verify that the motor was unloaded properly during the starting cycle.
- Visually inspect the bearings and lubricating systems for both the motor and the driven equipment.
- Rotate the motor and driven equipment (where practical) to verify mechanical freedom.

If no mechanical cause for the relay operation can be determined, complete maintenance testing (which will be discussed in Chapter 7).

If the reason for the trip cannot be determined after testing and allowing a minimum 1-hour cooling period, restart the motor for test with an operations or maintenance person at the motor location (at a safe distance) to prove rotational capability.

If the motor trips on time overcurrent while *running* (stator may be thermally stressed):

- Review bearing temperatures (where possible).
- Visually inspect the motor and driven equipment.
- Review associated operating parameters for acceptable ranges prior to the trip.
- Perform the maintenance testing discussed in Chapter 7.

If no cause for the relay operation can be found after completing the recommended testing and allowing a minimum 1-hour cooling period, restart the motor for testing and closely monitor the running amperes.

6.3.5.3 Feeder Ground Tripping

Feeder ground targets and corresponding tripping would not occur unless there is a protective relay malfunction or permanent cable or motor insulation single phase-to-ground failure. The motor should not be reenergized until the cause of the failure can be determined and repaired or until reasonable testing to prove electrical insulation integrity has been completed.

6.4 Operation of Auxiliary System Switchgear

6.4.1 Purpose

The purpose of this guideline is to provide suggested practices for the operation and inspection of medium-voltage [2–13.8-kilovolt (kV)] circuit breakers and contactors, and low-voltage [200–480-volt (V)] draw-out switchgear circuit breakers.

Good operating practices are critical to obtain the best service and performance from plant equipment and to ensure a safe environment for plant personnel.

6.4.2 Operator Inspections

It is the responsibility of operating personnel to establish and maintain scheduled routine inspections of all plant switchgear. Circuit breakers, medium-voltage contactors, buses, and main unit 480-V switchgear breakers must be kept clean and dry to reduce the possibility of insulation failures that can result in explosions and fire. In general, performing one inspection per day is recommended.

The following daily checks and inspections of switchgear locations are recommended:

- Dropped or actuated protective relay targets. Any targets found should be reset and recorded in the control room log book.
- Audible noise from electrical arcing.
- An unusual smell from overheated or burning insulation.
- Moisture intrusion (e.g., roof leaks or water on the floor).
- Status lamps and semaphores are working properly.
- Pressurizing room fans and dampers are functioning properly to mitigate moisture intrusion and other contamination.
- Switchgear room doors closed properly to mitigate contamination.
- Switchgear cubicle doors are closed to mitigate contamination.

Electrical Operating Guidelines 289

- Panels for accessing breaker racking mechanisms, cable terminations, and other purposes are closed to mitigate contamination.
- Breakers and contactors are kept in their respective cubicles or in special enclosures (usually equipped with heaters) designed to keep the equipment clean and dry.
- Switchgear room lighting is functioning properly.
- Cubicle labeling is consistent with plant policy and accurately describes the source, tie, and feeder positions.
- Rack-in tools and protective safety gear are stored and maintained properly.
- Housekeeping is performed often enough to keep the room clean and orderly.

Maintenance orders should be prepared for any anomalies found during the foregoing inspection process.

6.4.3 Protection

Protective relays are coordinated in such a way that only those circuit breakers or contactors that need to be operated to isolate faults are automatically tripped. This allows the maximum amount of equipment to remain in service and reduces the impact to on-line generating units. It also provides an indication as to the location of the electrical fault.

Electrical faults in transformers, motors, buses, cables, circuit breakers, and contactors are permanent in nature, and protective relay operations must be thoroughly investigated before reenergizing the equipment. Electrical short circuits are usually in the range of 15,000–45,000 amps, depending on the size and impedance of the source transformer. Reenergizing faulted electrical apparatus always results in more extensive damage and commonly causes fires because of the stored heat from the first event. Therefore, the equipment should not be reenergized until a thorough investigation is completed by engineering, technicians, and maintenance personnel. Source and tie breaker overcurrent relays should be set high enough to handle bus transfer conditions where all connected motors may be in an inrush or starting condition. Consequently, these relays do not always provide ideal thermal overload protection, and operations must rely on transformer temperature alarms, ammeters, or DCS alarms for that purpose. Accordingly, the overcurrent tripping of a source or tie circuit breaker usually indicates a short circuit, not an overload condition.

6.4.3.1 *Load Feeder Overcurrent Protection*

Load feeders are equipped with fast-acting instantaneous overcurrent elements (e.g., fuses for contactors) that will clear short circuits in the cables or load (e.g., motors or transformers) by isolating the faulted circuits before source and tie breaker overcurrent relays can operate.

6.4.3.2 Load Feeder Ground Protection

Designs that limit the ground fault current (usually around 1000 amps) apply separate ground relays that will actuate for ground faults only. These relays trip with very short time delays to isolate the grounded feeders before source or tie circuit breaker ground relays can operate.

6.4.3.3 Source and Tie Overcurrent Protection

Source and tie breakers are not equipped with instantaneous tripping elements. They rely on time delay to achieve fault coordination with downstream buses and loads. Typically, these relays are timed at maximum three-phase short circuit current levels and are set to operate in 0.4–0.8 seconds. Normally, the relays have an inverse time characteristic, and lower current levels will increase the time delay for all relays correspondingly. Typically, the tie breaker to another bus will be set to operate in around 0.4 seconds and the source transformer low-side breaker in around 0.8 seconds. Designs using two tie breakers in series often have the same relay settings on each breaker, and either one or both tie breakers may operate for faults downstream of both tie breakers. The source and tie overcurrent relays protect the buses and all breakers that are racked into the buses. They also provide backup protection if a feeder breaker fails to clear a faulted cable or load.

6.4.3.4 High-Side Source Transformer Overcurrent Protection

The source transformer high-side overcurrent relays are normally set to operate for maximum three-phase short circuits on the low-voltage side in around 1.2 seconds. This provides enough of a time delay to coordinate with low-voltage or secondary-side overcurrent relays. The relays usually have an inverse time characteristic, and lower current levels will increase the time to trip correspondingly. The source transformer high-side overcurrent relays assume that the fault is in the transformer, low-voltage, side-connecting buses or cables, or in the low-voltage circuit breaker and will trip all equipment necessary to isolate the fault. In the case of UATs, which are usually equipped with differential protection, the high-side overcurrent relays will also provide a complete electrical trip of the unit and main step-up transformer. The high-side overcurrent relays also provide breaker failure or stuck breaker protection if the low-side breaker fails to interrupt a fault on the associated bus.

6.4.3.5 Source and Tie Residual Ground Protection

Designs that limit the ground fault current (usually around 1000 amps) apply separate ground relays that will actuate for ground faults only.

Source and tie breaker ground relays are not equipped with instantaneous tripping elements. They rely on time delay to achieve fault coordination with downstream buses and loads. Typically, these relays are timed at maximum ground fault current levels and are set to operate in 0.7–1.1 seconds. Normally, the relays have an inverse time characteristic, and lower current levels will increase the time delay for all relays correspondingly. Typically, the tie breaker to another bus will be set to operate for 100% ground faults in around 0.7 seconds and the source transformer low-side breaker in around 1.1 seconds. Designs using two tie breakers in series usually have the same relay settings on each breaker, and either one or both tie breaker ground relays may operate for ground faults downstream of both tie breakers. The source and tie breaker ground relays protect the buses and all breakers that are racked into the buses. They also provide backup protection if a feeder breaker fails to clear a grounded cable or load.

6.4.3.6 Source Transformer Neutral Ground Protection

Designs that limit the ground fault current (usually around 1,000 amps) apply separate ground relays that sense the ground current flowing in the transformer neutral. Only ground faults will actuate these relays. The source transformer neutral ground relay is normally set to operate for maximum ground faults in around 1.5 seconds. This provides enough of a time delay to coordinate with the source and tie breaker ground relays. The relay usually has an inverse time characteristic, and lower current levels will increase the tripping time correspondingly. Transformer differential protection may not operate for ground faults in this zone because the limited amount of ground current in this type of scheme may not be high enough to operate the relays. The neutral ground relay is designed to isolate ground faults on the low-voltage or secondary side of the source transformer. This includes the transformer's low-voltage winding, low-voltage circuit breaker, and connecting buses or cables. The neutral ground relay will also back up the low-side breaker if it fails to clear a ground fault condition.

6.4.3.7 Alarm-Only Ground Schemes

These schemes limit the ground fault current to just a few amps. Typical values are 1.1 amps for 480-V systems and 3.4 amps for 4-kV systems. On wye-connected source transformers, the neutral is normally grounded through a grounding transformer. On delta-connected source transformers, the ground detector current is usually supplied by three transformers connected grounded wye on the primary and broken delta on the secondary side. In both cases, voltage relays are applied on the secondary sides of the grounding transformers to alarm for ground conditions. In the later scheme, blown primary fuses to the ground detector transformers can cause an alarm condition. Both relay schemes provide alarms (typically 10% and higher) for all grounded apparatus on the particular electrical system, including the source transformer's low-voltage or secondary windings and all connected buses, cables, breakers, potential transformers; and loads.

When one phase of a three-phase system has a 100% ground, the voltage to ground on the other two phases increases by a factor of 1.73. This means that the electrical insulation for all equipment on the system has to withstand a much higher voltage to ground when one of the other phases is grounded. This voltage is normally above the continuous insulation rating of installed cables and other electrical apparatus. Consequently, plant operators are expected to isolate and clear ground faults within a relatively short time period. If the ground fault is not cleared in a timely manner, it may develop into a high-current short circuit condition if another phase fails to ground from the higher operating voltage to ground. Because short circuits often result in explosions, fire, and loss of generation, it is in the interest of operations to have a predeveloped plan for quickly isolating grounds. This usually involves switching off nonessential loads, transferring to a different source transformer, and reducing generation to a point where other loads can be taken out of service.

6.4.4 Switchgear Bus Transfers

6.4.4.1 Paralleling Two Sources

Paralleling two different sources is the preferred method of transferring from one source to another. This method is not stressful to motors, is bumpless, and does not jeopardize a running unit. However, in most designs, the amount of short circuit current available during the parallel exceeds the interrupting capability of feeder breakers. Source and tie breakers will not be affected, but feeder breakers may not be able to clear close-in faults and the breakers may be destroyed in the process. Consequently, the duration of parallel should be kept to a minimum (i.e., a few seconds) to reduce the exposure time and likelihood of a feeder fault occurring. Parallel operations should not be performed if the voltage phase angles between two systems are out of phase by 10° or more, as extrapolated from a synch-scope. Depending on the impedances of the transformers involved, phase angle differences as little as 10° can cause more than rated current to flow during parallel operation. The higher current may cause the operation of source or tie overcurrent relays and the resulting loss of the bus and generation. Typically, this is more of a problem when the generating unit feeds one system (transmission) and the reserve, or startup transformer is fed from a different system (subtransmission). Reducing generation will normally bring the phase angles closer together as the generator's power angle reduces from the lower load.

6.4.4.2 Drop Pickup Transfers

Drop pickup or switch time transfer schemes are potentially damaging to motors and may cause the loss of a running unit or interruption of an operating process if the new source breaker fails to close after the prior source breaker trips open. When a bus is deenergized, the connected motors act like generators and provide a residual voltage to the bus. This voltage typically collapses in approximately 1 second. However, drop pickup transfers are much faster than 1 second, and the residual voltage can add to the new source voltage. If the vector sum of the two voltages exceeds 133% of the rated voltage of the motors connected to the bus, the motors may be damaged or their remaining life reduced.

6.4.4.3 Automatic Bus Transfer Schemes

Automatic bus transfer schemes are normally designed to reduce motor stress during transfer conditions and to coordinate with fault relays. Coordination with overcurrent relaying is achieved by initiating the transfer after the source circuit breaker trips open. If overcurrent relays trip the source breaker (indicating a bus fault), the automatic transfer will be blocked. Additionally, these schemes normally apply residual voltage, and sometimes high-speed synchronizing check relays that will not allow transfers unless the vector sum of the residual and new source voltages is below 133%. Usually these schemes will time out if the transfer is blocked by 86 lockout relays. However, if that is not the case, operations should verify that the automatic transfer scheme is disarmed before resetting the bus 86 lockout relays.

Bibliography

Baker, T., *Electrical Operating Guidelines* (Laguna Niguel, CA: Sumatron, Inc., 2003).
EPRI, *Power Plant Electrical Reference Series*, Vols. 1–16 (Palo Alto, CA: Stone & Webster Engineering Corporation, 1987).
Klempner, G. and Kerszenbaum, I., *Handbook of Large Turbo-Generator Operation and Maintenance* (Piscataway, NJ: IEEE Press, 2008).
Klempner, G. and Kerszenbaum, I., *Large Turbo-Generators Malfunctions and Symptoms* (Boca Raton, FL: CRC Press, 2017).
Westinghouse Electric Corporation, *Applied Protective Relaying* (Newark, NJ: Silent Sentinels, 1976).

7
Electrical Maintenance Guidelines

The maintenance guidelines given in this chapter are based on more than 50 years of electrical maintenance and engineering experience in generating stations. They represent a procedural effort to improve the availability of generating stations and large industrial facilities and reduce hazards to plant personnel. Some of this information was developed following actual failures in stations that the author was overseeing, and some to try to mitigate equipment failures reported by the industry in general. However, there are a number of areas where the guidelines may need to be revised by a particular generating station to meet its specific requirements. Some of these areas are presented here:

- The guidelines are general in nature, and manufacturers' recommendations for their equipment take precedence over the guidelines.
- There may be government regulatory agencies whose recommendations have priority over the guidelines, such as the Federal Energy Regulatory Commission (FERC), North American Electric Reliability Corporation (NERC), Occupational Safety and Health Administration (OSHA), and National Electrical Code (NEC).
- The guidelines assume that electricians, technicians, engineers, and contractors are well trained and qualified to perform the suggested tasks and all appropriate safety measures are taken.
- The station may have unusual environmental conditions that need to be accounted for in the guidelines.

The guidelines here represent a good starting document. However, they should be treated as a work in progress that reflects the ever-changing government regulations, site-specific experience, and industry experience in general.

7.1 Generator Electrical Maintenance

7.1.1 Purpose

The purpose of this guideline is to provide suggested procedures and schedules for the inspection, care, and electrical maintenance of generators.

The guideline is aimed at electrical specialists and does not cover generator support systems that are normally maintained by mechanical crafts, including seal oil systems, seals, bearings, blowers, rotor balancing, water coolers, gas and air cooling systems, leak detectors, and stator inner-cooled water systems.

7.1.2 Routine On-line Slip-Ring Brush-Rigging Inspections

NOTE: The rotor ground fault protection should be taken out of service whenever maintenance is performed on the brush rigging to avoid nuisance alarms or trips and to reduce the voltage reference to ground. Low-voltage rubber gloves should be worn prior to touching any energized conductors. The ground detector should be placed back into service as soon as the brush-rigging inspection is completed. All activities are to be coordinated with control room personnel.

Station maintenance should be performed on excitation system brush rigging in the following circumstances:

- Before a unit is returned to service after an electrical fault that results in a unit trip
- At least once per week on in-service and standby units (on turning gear), preferably before the weekend on Friday

NOTE: Information should be recorded about the condition of the brushes and brush holders and general visual findings during the inspection. An equipment-specific form should be prepared for each machine that accommodates the number of brushes and other design details. The as-found information should be entered while performing the inspection steps delineated next.

The following checks should be performed during the inspection process:

- Check brush leads (pigtails) for looseness, heating (discolored leads), and frayed wires.
- Check for proper and uniform brush tension. (Once a month, a spring tension scale should be used to spot check and verify the recommended tension.)
- Check for brush wear. Brushes should be replaced when 75% of their useful life has been expended. If spring tension becomes less than recommended and the springs are in good condition, the brushes may have to be replaced earlier.

NOTE: In general, no more than 20% of the brushes on a polarity should be replaced at one time with the unit in service. Ample time should be allowed for the new brushes to seat properly before replacing additional brushes.

Electrical Maintenance Guidelines

Under no circumstances should brushes of different grades or manufacturers be mixed on the same polarity. If special jigs for bedding the brushes are available at the site, up to 50% of brushes can be replaced at one time, if properly bedded.

- Lift each brush approximately a quarter-inch and gently allow it to return to the operating position under normal spring pressure. Remove brushes showing a tendency to stick or bind and correct the cause.
- Pull out a single brush from each polarity and observe its face for etching and excess powdering.
- Feel the top of each brush and check for running smoothness.
- Check for sparking or arcing.
- Inspect the brushes for mechanical wear, cracks, and hammerhead or anvil deformations on the metal pigtail plate.
- Inspect the integrity of springs and clips.
- Using dry air [not to exceed 35 pounds per square inch (psi)], remove accumulated carbon dust from the brush rigging. A filter mask should be worn during this process to mitigate inhalation of carbon fibers.
- Spot-check 10% of the brushes on each polarity with a direct current (DC) clamp-on ammeter for uniformity in current loading.
- The appearance of rings should be clean, smooth, and highly polished.
- A dark color does not necessarily indicate a problem. Look for streaking, threading, grooving, and a poor film. Collector ring-polishing or brush-seating procedures may be accomplished by using canvas or brush seating stones when necessary. A stroboscope will go a long way in aiding the visual inspection of the collector rings with the unit in operation.

NOTE: Grinding of collector rings to remove metal should not be performed on on-line energized units (excitation applied to the field) by plant personnel. However, contractors with a proven record, using specially designed stones and a vacuum system to remove the particulate and following proper procedures, have been successfully used to grind the collector rings on in-service generators.

- Check the brush compartment for cleanliness, loose parts, contamination by oil or dirt, or other conditions that may suggest a source of trouble. If an air filter is installed, check for cleanliness of the filter and proper seating of the filter, and change or clean as required.

- Problems found during the foregoing inspection steps should be corrected to the extent possible with the unit in service. Needed maintenance or repairs requiring a unit outage should be identified on a maintenance order.

7.1.3 Inspection of Rotor Grounding Brushes and Bearing Insulation

If accessible with the machine on-line, at the same time that the excitation brush rigging is inspected, the condition of the turbine rotor grounding brushes or braids should be checked. The integrity of the grounding elements is essential for minimizing shaft voltages and associated shaft or bearing currents.

On some designs with pedestal bearings and sandwich-insulation construction, it is possible to measure the insulation resistance with the unit on-line. When possible, this insulation should be tested for proper values during the inspection. According to the literature, it should fall somewhere between 100 kΩ and 10 MΩ or more.

In some pedestal-bearing arrangements, it is possible to assess the integrity of the insulation by measuring the voltage across the bearing insulation with a high-impedance voltmeter. If the grounding brush is lifted during the test, the shaft voltage is impressed on the bearing insulation. Therefore, if the reading is zero, the integrity of the insulation is compromised. Experience will show what voltage magnitudes to expect for a particular machine. It is not uncommon that shaft voltages are as high as 120 volts (V).

NOTE: On some designs, the bearing insulation can be measured only during the assembly of the mechanical components. Accordingly, it is important that the measurements be completed at the appropriate time or reassembly sequence. Refer to the original equipment manufacturer (OEM) for specific recommendations for the testing of the bearing insulation.

7.1.4 Routine Unit Outages

The following maintenance is recommended during unit outages:

- Brushes should be lifted on generator collector rings whenever units are off-turning gear over periods in excess of 24 hours to 1 week to mitigate brush imprinting (depending on the local environmental conditions and the plant's experience). Experience has shown that brush imprinting may result in ring polishing or grinding requirements that shorten the life of the rings.
- Inspect the rings for condition and protect them from the elements to prevent physical damage and oxidization when applicable. A polyvinyl chloride (PVC) film and corrugated cardboard overwrap can provide a protective cover.

Electrical Maintenance Guidelines 299

- At the beginning of each shutdown and weekly thereafter, test for hydrogen leaks (if applicable) around the collector rings and outboard radial terminal studs.

During routine shutdowns, or at approximately 3-month intervals, accomplish the following maintenance:

- Thoroughly clean the brush rigging and slip rings.
- Check all brush faces and change any brushes that are less than one-third their original length. If more than 20% of the brushes require replacement, they must be sandpapered to fit the contour of the ring.
- Thoroughly check the brush rigging for loose parts.
- Check the brush-holder for collector ring clearance, which should be uniform and within the manufacturer's tolerances (typical value is 3/16 of an inch from brush-holder to ring). Brushes should not be allowed to ride closer than 1/8 inch from the edge of the collector rings, and the average center line of the brushes in the axial direction should coincide with the center line of the ring when the machine is at operating temperature.
- The cooling grooves for generator collector rings should be inspected for proper depth. If carbon dust is found in the grooves (often it combines with oil vapor escaping through adjacent bearing seals to form a hard compound), carefully remove it to reinstate the cooling capability of the grooves.
- With the unit off-turning gear and the brushes lifted, and after completing a thorough cleaning of the insulating material beneath the collector rings, megger the field and perform DC resistance and alternating current (AC) impedance testing of the field winding. The resistance measurements should be corrected to ambient temperature and compared to previous data by an experienced person. Impedance and resistance changes caused by rotor shorted turn and connection problems may be slight.

7.1.5 Overhauls

Generators are usually refurbished whenever their respective turbines, whether high pressure (HP) or low pressure (LP), are overhauled. The following maintenance is recommended during major overhaul outages:

- Normally, experienced personnel, contractors, or both will remove the generator rotor from the bore to facilitate thorough stator and rotor inspections. Care must be taken to mitigate moisture and contaminant intrusion of both the rotor and stator during the outage. It is customary to enclose the removed rotor in a canvas tent and provide canvas flaps that cover the bore openings at each end of the machine for environmental protection. Additionally, where temperature and humidity could be a problem, heated dry air should be blown through the stator bore and the temporary rotor enclosure to maintain these components in a dry condition. Alternatively, recirculating dehumidifiers could be used for long maintenance periods.

- The rotor and stator inspections should be completed by engineers or specialists with prior experience in inspecting generators.

- The outside diameter (OD) surfaces of 18-5 retaining rings should be nondestructive examination (NDE) tested for stress corrosion pitting and cracking during major overhauls. When there is evidence of abnormal moisture intrusion or moisture-induced pitting on the OD of 18-5 rings, they should be removed to facilitate NDE inspection of the inside surfaces. The OD of 18-18 retaining rings should also be NDE-tested during major overhauls. If any abnormal cracking or anomalies are found on the OD surface areas, the rings in question should be removed and NDE testing of the inside surfaces performed. Recently, 18-18 rings have been found to be susceptible to stress corrosion cracking when contaminated with chlorides. Thus, they need to be kept away from such contamination and cannot be assumed to be free of stress corrosion-induced cracking.

- Generators with 18-5 retaining rings should be equipped with continuous dew point monitoring systems and alarms during overhauls to ensure that moisture intrusion problems are quickly identified, recorded, and mitigated.

- The generator rotor bore at the excitation end of hydrogen-cooled machines should be pressure-tested with clean dry air or an approved inert gas at 125% of the rated hydrogen pressure during overhauls. The pressure is normally applied to the bore plug opening. The bore should be able to hold capped (isolated from air/gas supply) pressure for 2 hours without an indication of a leak. Older machines may be equipped with hydrogen seals on the in-board radial conductor studs only, and the collector ring area or outboard radial studs will need to be sealed off with a can or other means to facilitate the pressure testing. It is prudent to stress-check the integrity of the rotor bore plug itself, which in some cases has been found to be deficient. This plug often is forgotten during inspections, but it has the potential to cause hydrogen leaks, so it should be inspected (and replaced if necessary) during major overhauls.

- All generator-monitoring instrumentation should be calibrated during overhauls and functionally tested at 2- to 3-year intervals between overhauls.

- Brushless excitation system diodes and associated fuses should be thoroughly cleaned and inspected during overhauls and at 2- to 3-year

Electrical Maintenance Guidelines 301

intervals between overhauls. Diodes and fuses requiring replacement should be replaced with identical manufacturer's models and weight measured before and during replacement to ensure that rotor balance is not affected. Where identical diodes or fuses cannot be supplied, the complete set should be replaced with a new matching set.
- Electric insulation tests would normally be performed on stator and rotor windings at the beginning of the outage, when the unit is still assembled and warm (under hydrogen pressure when applicable). This will allow more time to complete any required repairs during the outage window and ensure that the machine is in a dry condition for the testing.
- When applicable, generator heaters should be left switched on to avoid moisture ingress. The space heaters must be switched off whenever personnel are performing inspections or working in the interior stator areas.
- If applicable, lead-box drain-lines should be inspected for blockage and cleaned if necessary. Liquid level-detection systems should be checked for proper alarms and site glasses cleaned.

7.1.6 Vibration

Shorted turns in generator two- and four-pole cylindrical rotor field windings can contribute to vibration problems due to rotor thermal bending from uneven heating associated with the nonsymmetrical DC current flow and watt losses in the field windings. Shorted turns also can cause unbalanced magnetic flux in the air gap, which can aggravate vibration problems.

Since vibration signature analyses for generator rotor shorted-turn problems is not always an exact science, it is desirable to have confirming data from other testing before proceeding with very costly disassembly and repairs of large machines. Additional tests to confirm the existence of shorted turns in generator rotor-field windings are commonly performed before committing to expensive repairs. At this time, the following four test procedures are generally used in the industry to help verify whether generator field windings have shorted turns:

- *Thermal stability testing*: Involves changing generator-operating parameters (watts, vars, and cooling) and recording and analyzing the impact on rotor vibration signatures.
- *Flux probe analysis*: Utilizes an installed air gap probe to measure and analyze the magnetic flux from each rotor slot as it passes the location of the sensor. Some generators are permanently equipped with flux probes and many are not. Installing the probe normally requires a unit outage, especially with hydrogen-cooled machines.

- *Repetitive surge oscilloscope (RSO) testing*: Involves applying a succession of step-shaped low-voltage pulses that are applied simultaneously to each end of the field winding. The resulting reflected waveforms can be viewed on dual-channel analog or digital scope screens as two separate waveforms, or after one of them is inverted, and both summed as a single trace. If no discontinuities are present in the winding (due to grounds or shorted turns), both traces will be nearly identical, and if inverted and summed, a single trace will be displayed as a horizontal, straight line. Any significant discontinuity arising from a shorted turn will be shown as an irregularity or anomaly on the summed trace.
- *Field current open circuit testing*: If accurate previbration electrical measurements have been recorded, precise field current measurements can be compared to the historical data. If there is a shorted turn, the field current will be higher for an identical terminal voltage. If the machine is loaded, it is much more difficult to compare, since var flow is dependent on system voltage levels.

It should be noted that the foregoing tests do not provide absolute certainty by themselves that there is a shorted-turn problem in the generator rotor. However, when confirmed by other testing, the probability that field winding is the cause of the vibration problem increases significantly. Shorted-turn anomalies can be masked if they are near the center of the winding or otherwise balanced, if there are multiple shorted turns, if they are intermittent due to centrifugal force or thermal expansion and contraction, if there are grounds, and if there are other contributors to the overall machine vibration levels.

RSO testing has some advantages over other testing, in that it can also be used periodically during rewinding to verify that windings are free of shorts on both at-rest and spinning deenergized rotor windings.

7.2 Transformer Electrical Maintenance

7.2.1 Purpose

The purpose of this guideline is to provide suggested procedures and schedules for the inspection, care, and maintenance of oil-filled and dry-type power transformers 500 kVA and larger, including onsite spare units.

7.2.2 Inspections

The systems and components identified here should be inspected by maintenance personnel on in-service and spare transformers at 2–3-year intervals,

Electrical Maintenance Guidelines

or at an interval specified by the manufacturer if it happens to be shorter than 2 years.

- Check heat exchangers and associated components for oil leaks, rust, exterior paint condition, cleanliness of airway passages, and deterioration of cooling fins.
- Check main and conservator oil tanks for oil leaks, correct level, rust, and exterior paint condition. Ensure sight glasses are clean and oil levels are clearly visible.
- Control cabinets should be cleaned, wiring checked for tightness, and motor contactor contacts inspected, cleaned, and dressed as required. Fan and pump loads should be megger-tested and started to prove proper fan rotation and operation of oil flow indicators. Molded-case circuit breakers should be exercised several times to prevent frozen mechanisms. Motor load current and phase balance should be checked with a clamp-on ammeter. Heaters in the control panel should be checked for proper operation. Care should be taken not to enable trip functions during this inspection.
- Each transformer alarm should be functionally tested to prove proper operation. Care should be taken not to activate a trip function during the alarm testing.
- Check high- and low-voltage bushings and insulators for physical condition and cleanliness of porcelain or coatings. Check the bushing oil level, where applicable, and for leaking flanges. If the inspection identifies the need for adding oil during the next outage, care should be taken to use fresh, clean, and dry oil of identical or equivalent characteristics.
- Where applicable, verify that a positive nitrogen pressure blanket is maintained within the limits specified by the manufacturer. Typically, low-pressure alarms are set for 0.5 psi, and normal operating pressure is around 2 psi.
- Perform a thermograph test of transformer and associated external electrical connections. The test should be performed when the transformer's load is greater than 25% of full load and has been over that value for at least 1 hour. Prior to the thermography testing of the cooling system, make sure that all cooling pumps and fans are running for at least 1 hour (this may require switching the cooling system to manual operation).
- Transformer oil leaks should be cleaned during the inspection and mitigated to the extent possible by retorqueing bolts, tightening valve packing, applying epoxy or sealants, and gasketing leaking bolts. Care should be taken to select sealants that do not adversely affect the transformer's oil by reacting with it or contaminating it.
- Desiccant breathers should be inspected and refurbished as required.

Required corrections that can be performed with the transformer in service should be accomplished during the inspection. Maintenance or repairs requiring a transformer outage should be identified on a maintenance order.

7.2.3 Transformer Testing

The following tests should be performed on unit-related transformers during major turbine overhauls (HP turbine for cross-compound units): Testing of transformers that serve more than one unit should be performed during major turbine overhauls of the lowest-number unit. The following tests should also be performed whenever new or spare transformers are placed into service:

- Power factor (PF) test (Doble) all unit transformer windings, including bushings and lightning arrestors (except a transformer with secondaries rated less than 2.4 kV).
- Resistance-test all high- and low-voltage side windings (include all tap positions to check for pyrolytic-carbon buildup on the tap changer contact surfaces).
- Test the insulation (megger) of each winding to ground and between windings.
- Test the winding turns-ratios, polarity, and excitation current (except for a transformer with a secondary winding that is rated less than 2.4 kV).
- Test the winding impedance (except a transformer with a secondary winding rated less than 2.4 kV).
- Test and calibrate all transformer monitoring instrumentation.

NOTE: The preceding tests, for the most part, require the transformer be disconnected from external conductors and buses. Before reconnecting, care should be taken to inspect the condition of mating surfaces (clean and resilver plate where required); use proper contact anti-oxidation grease; torque bolts to the proper value for the particular softer mating surface material; and verify that proper joint connection is achieved. Particular care should be taken to prevent oxidization when terminating aluminum connections. Normally, after cleaning, deoxide grease is applied and then file-carded to dislodge any oxides that might have formed.

7.2.4 Avoiding Pyrolytic Growth in Tap Changers

Pyrolytic growth is the phenomenon by which oil breaks down to an amorphous hard carbon deposit through the effects of the high electric-field strength around contact blocks. Once started, it grows along the contours of the field. This growth can force the contacts apart, causing a high contact resistance that can lead to gassing and failure. To mitigate this phenomenon, the following maintenance activities are recommended:

- *No-load tap changers*: At least once every 2 years, or more often where experience indicates that it is necessary. No-load and off-circuit tap changers should be operated through the entire range of taps no less than six times. The transformer should be deenergized during this operation. Care must be taken to return the tap to the proper operational position.
- *Load tap changers*: If possible, once a month or whenever an outage allows it, load tap changers should be operated through the entire range of taps. The transformer can be energized, but preferably with the low-voltage breaker open to ensure that secondary system voltages do not become too high or low.

7.2.5 Internal Inspection

Internal inspections are normally performed only when it is absolutely necessary to drain the oil to perform a maintenance repair or to investigate an anomaly. The inspection should be completed by engineers or specialists with prior experience in inspecting transformers, as follows:

- Conduct internal inspection of the transformer for abnormal conditions or deterioration of windings, core, core-bolt tightness, insulation, blocking (winding clamping mechanisms), shields, connections, leads, bushings, and tap changer (as required). Look for arcing, carbon tracking, burning, or excessive heating discoloration and debris in the tank. Perform a core ground test, where feasible. Micro-ohm (ductor test) suspected bolted joints.
- Keep exposure of core and windings to air to a minimum. When core/windings are to be exposed to air for a significant time (i.e., overnight), then an external dry air or dry nitrogen source should be connected to ensure positive pressure in the tank.

NOTE: Care must be taken not to damage continuous dissolved gas analysis (DGA) monitoring systems during the draining and refilling of transformers (particularly the application of vacuum). Follow the manufacturer's instructions to prevent damage to the DGA equipment during the foregoing maintenance/inspection operations.

Vacuum-drying and oil-filling (with degasified/filtered oil) should be performed based on the manufacturer's recommendations, test data, parametric data, and/or as indicated by the internal condition of the transformer. After dryout of the core and windings, the transformer should be filled with degasified oil, followed with a tail vacuum and running of the oil-circulating pumps to remove trapped air from the pumps and core/coil assembly. Vacuum should be broken with nitrogen. These activities should be performed only by qualified personnel.

7.2.6 Electrostatic Voltage Transfer

Electrostatic voltage transfer is the phenomenon by which a charge develops between the oil and the insulation systems of a transformer, when the transformer is deenergized and oil is circulated by the oil pumps.

- Running oil pumps on deenergized transformers should be limited to 10 minutes. This is to prevent the failure of transformers from electrostatic charging and subsequent tracking of the insulation systems. If additional operation of the pumps is required, a minimum 2-hour wait should be followed.
- For oil settling before energization, the oil should be allowed to settle for at least 18 hours after completing the filling process, before an energization of the transformer is attempted.

7.2.7 DGA

When performed properly and in a timely manner, DGA of transformer mineral oil can provide an advance warning about a deteriorating condition in a transformer before a catastrophic failure occurs. It can indicate the present condition of a transformer by analyzing the gas contents. A long-term assessment of a transformer can be obtained by trending the results of DGA on a specific unit. Samples should be taken annually and sent to an experienced lab for analysis. In addition to DGA, the sample should be tested for dielectric quality, moisture content, and aging parameters.

The following gases are normally analyzed for DGA purposes:

- Hydrogen (H_2)—(<101 PPM)—Possible corona
- Carbon dioxide (CO_2)—(<121 PPM)—Possible decomposition of cellulose
- Ethylene (C_2H_4)—(<51 PPM)—Possible low-energy spark or local overheating
- Ethane (C_2H_6)—(<66 PPM)—Possible low-energy spark or local overheating
- Acetylene (C_2H_2)—(<36 PPM)—Possible arcing
- Methane (CH_4)—(<2500 PPM)—Possible low-energy spark or local overheating
- Carbon monoxide (CO)—(<351 PPM)—Possible decomposition of cellulose

Electrical Maintenance Guidelines 307

The foregoing parts per million (PPM) levels are suggested for alarm limit purposes. Transformers that have gas concentrations at or higher than the alarm threshold should be subjected to further tests and investigations and, at a minimum, DGA-tested at shorter intervals for trending purposes and assessment of developing problems. See IEEE Standard C57.104 for additional information.

7.2.8 Dielectric Breakdown Test

A dielectric breakdown test of the oil should be completed during the DGA. The dielectric breakdown indicates the capacity of the oil to resist electric voltage. Its capability depends mainly on the presence of physical contaminants (i.e., undissolved water, fibers, etc.). The test is normally performed with a 0.040 gap per ASTM D-1816 and the breakdown values should not be less than the following:

- Apparatus <69 kV—20 kV minimum
- Apparatus 69–288 kV—24 kV minimum
- Apparatus >288 kV—28 kV minimum

7.2.9 Insulators and Bushings

Each generating station should establish appropriate intervals for washing, cleaning, or greasing or coating transformer and neutral-reactor bushings, arrestors, and insulators based on the environmental conditions present at the particular location.

Locations in a mildly contaminating area will most likely be able to pressure-wash with nonconductive water at established intervals based on experience at the particular location. Hot or energized washing in general is not recommended for substations, transformers, and their associated dead-end structures because of the flushing of contamination onto lower-level bushings and insulators that can cause power arcs and electrical failures.

Silicone greases that are manufactured for this purpose are suggested for high contamination areas (plants close to the ocean or high industrial contamination areas). It is expected that most high contamination locations will need to grease insulators at 2-year intervals. White dry bands will form on the insulators and bushings when the grease is near the end of its useful life. At the first indication of white banding, arrangements should be made to remove clean, and apply new grease to the porcelain surfaces. The grease should be applied at a minimum thickness of 1/8-inch to prevent premature white banding. Over time, the surface contamination migrates to the interior areas which refreshes the exterior. Longer-duration RTV like materials are also available for this purpose. With these materials, the corona activity due to surface contamination will pop off the contamination and thereby refresh the exterior surfaces of the insulator or bushing. These materials are applied once and should last for 3–10 years or longer. At the end of the material's useful life, it requires scrubbing or hydroblasting for removal before another coat can be applied. Spare transformer bushings should be stored in either a vertical position or in a horizontal position with the top end at least 18 inch (46 cm) higher than the bottom end. In either of these positions, the insulation structure of the bushing remains under oil.

7.2.10 Sudden-Pressure Relays

Transformer gas or oil sudden-pressure relays must have their trips disabled whenever work is to be performed that may affect the internal transformer's pressure, including adding nitrogen gas, adjusting gas regulator, or work that permits gas venting to the atmosphere. This should be coordinated with the control room to ensure that the relay is returned to service correctly after the maintenance work is completed.

7.2.11 Spare Transformer Maintenance

Nitrogen-blanketed spare transformers should be kept fully assembled with nitrogen pressure maintained above the oil level. Current transformer (CT) terminals not used (including hot-spot temperature detectors) are to be short circuited, and transformer-bushing terminals should be protected from the weather to prevent corrosion. Paint should be maintained on the external surfaces to prevent corrosion. For spare transformers, oil samples should be taken every 2 years for the purpose of checking for water content and dielectric quality.

Control cabinet heaters should be temporarily wired and energized to keep control cabinet interiors dry. Temporary power to cooling circuits is also advantageous to allow momentary periodic cycling of the cooling system to exercise the fans and pumps. As mentioned before, the pumps on deenergized transformers should not be run longer than 10 minutes to avoid the development of electro-static charges.

7.2.12 Phasing Test

Whenever a unit main transformer is replaced with a spare or a new transformer, a thorough phasing test should be performed before completing the connections to the generator. The transformer should be energized from the switchyard with the generator isolated-phase-bus links open (where applicable). Undisturbed generator bus potential transformers (PTs), or auxiliary system bus PTs if the auxiliary transformer is also energized, can be used for in-service readings to verify proper rotation, phase angles, and voltage magnitudes.

7.3 Motor Electrical Maintenance

7.3.1 Purpose

The purpose of this guideline is to provide suggested procedures and schedules for the inspection, care, and maintenance of switchgear-fed induction and synchronous motors (100 HP and larger).

7.3.2 Electrical Protection

Operation of motor protective relays should not be taken lightly. The following investigation steps should be completed before allowing restarts of motors after protective relay operations.

7.3.2.1 *Instantaneous Phase Overcurrent Tripping (50)*

Instantaneous phase overcurrent relay minimum trip points should be set well above surge and locked rotor values for the motor. An instantaneous target and corresponding trip indicates that either the electrical protection malfunctioned or was set improperly, or a permanent ground (solidly grounded system only), phase-to-phase, or three-phase electrical fault exists in the cable or motor. Contactor-fed motors are not equipped with instantaneous protection and will blow fuses to clear short circuit conditions. The motor should not be reenergized to avoid overstressing the plant electrical system with a second fault until the cause of the relay operation can be determined and repaired or reasonable testing to prove the electrical integrity of the motor/cables has been completed. The following investigative steps are recommended:

- Test the electrical protection for proper operation. The instantaneous trip elements are normally set for 250% of nameplate voltage locked-rotor amps.
- Physically inspect the outside of the motor, cables, and connections for evidence of electrical failure (i.e., odor and smoke damage).
- Perform a 1000-V megger test of the motor and cables from the switchgear cubicle and take a polarization index (PI). Measure the three phase-to-phase resistances and the three phase-to-phase impedances.

NOTE: Particular care must be exercised when testing motors from switchgear cubicles, especially when the bus stabs or primary disconnects are energized.

If the foregoing testing does not indicate a problem, motors and associated cables rated 2 kV and higher should be overvoltage- or hipot-tested at the recommended routine value for motors.

If an electrical failure is found, perform a routine inspection of the associated circuit breaker. Also, the motor should be isolated from the cables to determine if the failure is in the motor or in the cables. Motors with a neutral connection brought up to the terminal box should have the neutral opened to allow each phase to be tested separately.

After the foregoing investigative steps are completed and no cause for the relay operation can be found, attempt one restart of the motor. If successful, the motor can be returned to normal service. If unsuccessful, repeat the foregoing steps and install a portable fault recorder to monitor the electrical parameters before attempting another start of the motor.

7.3.2.2 Time-Phase Overcurrent Tripping (51)

Time overcurrent targets and corresponding trips are usually caused by mechanical problems in the motor or driven equipment, malfunctioning or improperly set protective relays, control anomalies, or loss of one phase of the supply.

If the motor trips on time overcurrent during *starting* (rotor normally thermally stressed), then perform the following actions:

- Verify that the motor was properly unloaded during the starting cycle.
- Visually inspect the bearings and lubricating systems for both the motor and the driven equipment.
- Rotate the motor and driven equipment (where practical) to verify mechanical freedom.

If no mechanical cause for the relay operation can be determined, complete the following electrical tests:

- Test the electrical protection for proper operation. Normally, the relays are timed for 5 seconds longer than a normal starting time at nameplate voltage locked-rotor amps.
- Perform a 1000-V megger test of the motor and cables from the switchgear cubicle. Measure the three phase-to-phase resistances and the three phase-to-phase impedances.

If the reason for the trip cannot be determined after completing the foregoing steps, after a minimum 1-hour cooling period; restart the motor for test with an operations or maintenance person at a safe distance to gauge the rotational capability.

If the motor trips on time overcurrent while *running* (stator normally thermally stressed), and no problems with the driven load or control system are identified, then do the following:

- Test electrical protection for proper operation. Normally, conventional overcurrent relays are set for 125%–140% of full-load amps.
- Test the control system for proper operation.
- Perform a 1000-V megger test of the motor and cables together from the switchgear cubicle. Measure the three phase-to-phase resistances and the three phase-to-phase impedances.
- Review bearing temperatures (where possible).
- Visually inspect the motor, cable connections, and driven equipment.

If no cause for the relay operation can be found after completing the foregoing items, after a minimum 1-hour cooling period, restart the motor for test after ascertaining a three-phase supply is present at the motor's source (no blown fuses), and closely monitor the running amperes.

7.3.2.3 Feeder Ground Tripping (51G)

Feeder ground targets and corresponding tripping should not occur unless there is a protective relay malfunction, setting errors, CT problems, or a permanent cable or motor insulation single phase to ground failure. The motor should not be reenergized until the cause of the failure can be determined and repaired or reasonable testing to prove electrical insulation integrity has been completed. The following investigative steps are recommended:

- Test the electrical protection for proper operation.

    NOTE: Limited current residual ground schemes (around 1000 amps for 100% ground fault) are sensitive to CT saturation, connection, and shorted-turn problems. Accordingly, the CTs and associated circuitry should be resistance- and saturation-tested to prove circuit integrity. To avoid false residual ground tripping from momentary CT saturation during starting conditions, instantaneous tripping should be disabled and the relays are normally timed to operate in around 0.3 seconds for a 100% ground condition. Designs using a single CT that wraps around all three phases (zero-sequence CT) are not prone to false tripping from saturation and other CT unbalance problems, and testing the CT is not required.

- Physically inspect the outside of the motor, cables, and connections for evidence of electrical failure (i.e., odor and smoke damage).
- Perform a 1000-V megger test of the motor and cables from the switchgear cubicle and take a PI. Measure the three phase-to-phase resistances and the three phase-to-phase impedances.

If the foregoing testing does not indicate a problem, motors and associated cables rated 2 kV and higher should be overvoltage- or hipot-tested at the recommended routine value before reenergization.

If no cause for the relay operation can be found after completing the foregoing steps, attempt a restart of the motor. If successful, the motor should be returned to normal service.

7.3.3 Routine Testing

The following testing is recommended when motor feeder circuit breakers are removed from their respective cubicles during circuit breaker and bus routine maintenance:

- Perform a 1000-V megger test from the switchgear cubicle to determine the health of the motor/cable insulation systems (if surge capacitors, arrestors, or both are installed, they can be left connected during the megger test).
- Perform three phase to phase resistance tests to determine the condition of the electrical connections; both internal and external (should be within 5% of each other). A poor electrical connection can cause unbalanced voltages and shorten the available life of the motor. Motors need to be derated for voltage unbalances as small as 1%.
- Perform three phase-to-phase impedance tests to ascertain the health of the stator winding and rotor bars (should be within 10% of each other). If there is an unbalance, rotate the rotor to see if it affects the readings (possible squirrel cage rotor problems).

In addition to the foregoing testing, during major turbine overhauls (HP for cross-compound units), motors rated 2 kV and higher and associated cables should be overvoltage (hipot)–tested to predict an undefined future life of the insulation system.

NOTE: If a rewind is required, it is suggested that the new winding be insulated with class H insulation to improve life expectancy. In coal-fired plants where fly ash abrasion of insulation is a problem, an RTV-like substance can be applied to the end winding surfaces, helping to reduce fly ash abrasion failures. RTV coating of end windings can also be used to mitigate salt contamination and tracking in open-ventilation medium-voltage motors in close proximity to coastal areas.

7.3.4 Internal Inspections

Where the failure of a motor would result in the loss or restriction of generation, the motor should be thoroughly inspected internally and frequently enough to ensure continuous service. At least one motor in each group of motors (boiler feed pump, circulating water pump, induced draft fan, forced draft fan, condensate pump, etc.) should be inspected during major turbine overhauls. When a motor is found to be in poor condition, all other similar motors in the group should be inspected. A major inspection of a motor should include the following checks:

- Remove the upper-end bells and internal dust shields and inspect the stator and rotor as far as practical. Verify that the iron-core ventilation passages are clear of contamination (particularly important in coal-fired plants).
- Check the air gap using a feeler gauge (in four positions, if possible).
  Maximum allowable eccentricity in both the horizontal and vertical directions is 10%.
  % Eccentricity = $200 (R_1-R_2)/(R_1 + R_2)$
  R_1 = largest air gap distance and R_2 = smallest air gap distance
- Inspect rotor bars and fan blades for cracks, if practical.
- Check the stator end turn areas for proper blocking and support. Undesirable end turn movement can cause black greasing if oil is in the environment or yellow dusting in a dry condition.
- Check for filler strip and wedge migration.
- Inspect the motor connection box for anomalies. Pay particular attention to the condition of the pigtail leads.
- Check the bearings for wear and replace the oil.
- Check and clean or replace the air or water cooler filters.
- Check coupling alignment. Laser alignment systems have proven to be excellent for verifying motor/coupling alignment.
- Inspect heaters and heater wiring.

When the inspection reveals excessive contamination, deterioration, movement, or fouling of the windings, the motor should be disconnected and sent out for cleaning and refurbishment.

Anytime medium-voltage motors (2 kV and higher) with shielded cables are disconnected, the single shield ground location should be changed from the switchgear location to the motor connection box. A thorough Electrical Power Research Institute (EPRI) study disclosed that high-frequency switching transients during second-pole closing can get as high as five times the voltage, which can stress or fail the turn-to-turn insulation in motor windings. Grounding the cable shield, at the motor end only, for motors that are not equipped with surge capacitors reduces switching transients by more than 50%.

Motors rated 2 kV and higher, which are sent out for rewinding or cleaning and refurbishment, should be hipot-tested in the service shop prior to delivery to the station. All motors should be megger, PI, resistance, and impedance tested immediately upon return to the station from a repair facility. The megger, PI, resistance, and impedance testing should be repeated from their respective cubicles when the motors are connected to the feeder cables. All test values should be recorded to permit comparison to measurements in the future.

7.3.5 On-line and Off-line Routine Inspections

Based on local environmental conditions, a periodic (at least annual) maintenance routine should be established for the following:

- Cleaning or replacement of air filters and connection box desiccant breathers (if applicable).
- Bearing lubrication system maintenance. Follow the manufacturer's recommendations for greasing bearings; over-greasing may cause a bearing failure.
- Motor heaters (which are normally energized when the motor is not running) should be equipped with ammeters or light-emitting diodes (LEDs) that indicate heater current flow is normal. The heaters and circuitry should be inspected for proper operation during the scheduled routine.

7.3.6 Motor Monitoring and Diagnostics

- Thermographic inspection of leads, frame, and bearings is usually economical for larger motors.
- Motors 1500 horsepower (HP) and larger are normally equipped with embedded resistance temperature detectors (RTDs) in their stator windings for monitoring stator temperatures. These devices can be used to drive recorder, DCS systems, and newer digital protective relays. In most plants, they are not being utilized but are always accessible in external connection boxes for measuring stator temperatures if an issue needs to be resolved on the temporary overload capability of the motor where the unit output is limited because of motor loading or to assess the effectiveness of the cooling system or cleanliness of the cooling passages. Generally, RTDs are slow to respond and not considered effective for severe events (i.e., motor stall or locked rotor conditions).
- Current spectrum analysis can be performed routinely for detecting incipient broken rotor bars and cracked short-circuiting rings. However, it may be more economical to perform the testing only on motors that exhibit vibration problems or have a history of squirrel-cage rotor problems.

- Either routine or continuous partial discharge monitoring of stator windings is available for motors rated 4 kV and higher. Normally, the cost of these devices can be justified only for motors that are problematic or part of critical plant processes.

7.4 Switchgear Circuit Breaker Maintenance

7.4.1 Purpose

The purpose of this guideline is to provide suggested procedures and schedules for the inspection, care, and maintenance of switchgear circuit breakers. Proper maintenance of switchgear circuit breakers is essential to obtaining reliable service and performance. Circuit breaker failures are a potential hazard to personnel and other plant equipment and could result in the loss of generation capability.

NOTE: This guideline does not apply to SF6, oil, and freestanding generator or switchyard circuit breakers.

7.4.2 General—Switchgear Circuit Breakers (200 V to 15 kV)

Modern switchgear circuit breakers have sensitive mechanisms with critical tolerances and internal forces and loadings that must be maintained in close adjustment to operate properly. Deviation in adjustment could result in improper operation or electrical failure. It is, therefore, imperative that the manufacturer's recommended settings and instructions are understood and followed. All personnel associated with maintaining circuit breakers and their associated cubicles should be well acquainted with the manufacturers' instructions and recommendations pertaining to both the circuit breakers and the cubicles.

NOTE: Particular care must be exercised when working on breakers that rely on stored energy systems (spring, air, or hydraulic), and appropriate safety practices must be followed.

Because the insulating properties of circuit breakers are adversely affected by the presence of moisture and contamination, all possible precautions must be taken to prevent their intrusion. Spare breakers, as well as other breakers that need to be stored outside their designated cubicles, should be stored in suitable heater-equipped enclosures to prevent the intrusion of moisture and contamination. For the safety of plant operators who stand in front of circuit breakers during live rack-in operations, it is essential that the breaker insulation systems prevent the occurrence of insulation flashovers and explosions. Switchgear rooms must be maintained in a clean condition. Cubicle doors and access panels must be kept closed to mitigate moisture and contaminant intrusion. Fans and dampers for positive-pressure switchgear rooms should be properly maintained. Ceiling water leaks should be repaired quickly, and water should not be used to hose down areas near switchgear locations.

No work should be attempted on racked-in energized circuit breakers at any time. All testing and adjustments should be performed with the breaker in the fully racked-out or test position. Additionally, no work should be attempted on the control circuits of racked-in circuit breakers for off-line units, where an inadvertent breaker closure would energize an at-rest generator. The foregoing applies to generator circuit breakers and to unit auxiliary transformer low-side breakers that can backfeed and energize the generator.

Adjustment of the main contacts and primary disconnects for source and tie breakers is particularly critical. Current flow through feeder breakers is usually well below the continuous rating. However, it is not uncommon for source and tie breakers to carry close to rated current; consequently, they are much more prone to contact and primary disconnect overheating that can lead to arcing grounds and short circuit failures.

7.4.3 Inspection and Testing Frequencies

Switchgear circuit breakers and cubicles should be mechanically inspected and electrically tested at the following intervals or events:

- Periodically, at 2–5-year intervals
- During unit overhauls
- Before placing new or modified breakers into service
- Before energizing breakers that have been out of service for over 12 months
- After an interruption of electrical short circuits, other than a ground fault in a resistance-grounded system
- After 1000 close-open operations following the last inspection

7.4.4 Mechanical Inspection

During the mechanical inspection process, the following items should be completed:

- Remove arc-chutes to facilitate the inspection.
- Use vacuum, dry air, or hand wipe to remove dust and other contaminants from the breaker and arc-chutes (do not use air pressure on arc-chutes that contain asbestos).
- Inspect silver-plated parts and replate as required, following manufacturer's recommendations.
- Inspect, adjust, clean, and replace main and arcing contacts as needed. Ensure the contacts have the proper "wipe," alignment, and synchronism and are adjusted as specified by the manufacturer.

Electrical Maintenance Guidelines 317

NOTE: A stamp impression of the contacts can be made using very thin tissue paper to ensure that the contact surfaces make evenly.

- Clean disconnect or finger clusters, and control fingers, check spring pressure, and lubricate lightly with approved grease to enhance proper engagement. Check all associated screws and bolts for tightness.
- Perform an overall inspection, looking for loose wiring and connectors or components, heating, cracked insulation, corrosion, and anomalies. Complete repairs as required.
- Check for tracking and corona activity, in particular in those areas close to the interface between insulated and noninsulated conductors.
- Close the breaker manually a number of times and compare variations in mechanical force needed and audible sounds. Variances should be investigated.
- Check for proper operation and condition of auxiliary contacts and relays.
- Vacuum-breaker bottles can leak. To ascertain proper vacuum integrity, some manufacturers recommend carrying out a "pull test" to measure the vacuum by the force required to pull apart the contacts.
- Lubricate moving parts as recommended by the manufacturer or in accordance with experience with a particular type of breaker.
- Prove that the mechanical trip push-button will trip a closed circuit breaker.

7.4.5 Electrical Testing

In general, acceptable measurement values for circuit breaker testing will depend on the model and type of circuit breaker. Experience, manufacturer's information, and review of previous test records will indicate a practical range of acceptable readings for a particular circuit breaker. Corrective action will need to be taken when test measurements are outside of that range. The following electrical testing should be completed to assess the condition of the circuit breaker and to determine if further maintenance is required:

- Before installing the arc-chutes, perform a micro-ohm test of each phase from disconnect stab (or cluster) to disconnect stab (or cluster) with the breaker closed, after manually opening and closing the breaker at least three times. If the breaker is not racked into a test device and is equipped with clusters, the micro-ohm readings should be taken from the mating surfaces that the clusters are installed on and not from the cluster itself. If the readings are acceptable, install the arc-chutes for further testing; otherwise, isolate and correct the problem.

- Perform an overvoltage test of the breaker. All circuit breakers should be tested with a 1000-V megger. If the megger test results are acceptable, an AC hipot (overvoltage) test should be performed on 2 kV and higher-voltage-rated breakers. Considering the severe consequence of an insulation breakdown, personnel safety, the low cost for a low capacity 30 kV AC test set, and that breaker insulation failures are easy to repair, the AC hipot testing of all medium-voltage (2 kV and higher) switchgear circuit breakers is recommended.

The following are the routine AC overvoltage test values for switchgear circuit breakers of various voltage ratings:

- 2.4–5 kV breakers   14.25 kV AC
- 7.2 kV breakers   19.5 kV AC
- 13.8 kV breakers   27.0 kV AC

The breaker should be able to hold the test voltage for 1 minute.

As illustrated in Figures 7.1 through 7.3, three tests are required to properly megger or overvoltage stress the circuit breaker:

- Open-pole insulation
- Phase to ground insulation
- Phase-to-phase insulation

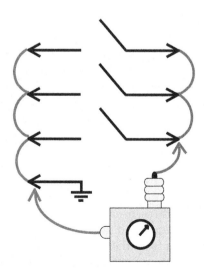

FIGURE 7.1
Open-pole insulation test.

Electrical Maintenance Guidelines

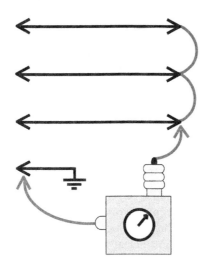

FIGURE 7.2
Phase-to-ground insulation test.

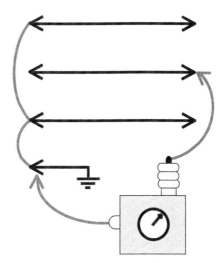

FIGURE 7.3
Phase-to-phase insulation test.

The following describes each test:

- *Open-pole insulation*: With the breaker open and A, B, and C phases jumpered together on both sides of the circuit breaker, connect the *hot* lead to the movable-contact side, and the test set *return* lead and breaker frame-ground to the stationary-contact side. Minimum insulation resistance for this test is 3 megohms per rated phase-to-phase kilovolt.

- *Phase-to-ground insulation*: With the breaker closed and all leads removed on the stationary side, connect the hot lead to the movable contact side, and the return to the breaker frame ground to test the phase-to-ground insulation. Minimum insulation resistance for this test is 3 megohms per rated phase-to-phase kilovolt.
- *Phase-to-phase insulation*: With the breaker closed and A and C phases jumpered together, connect the hot lead to B phase and the return and breaker frame-ground to A and C phases to test the phase-to-phase insulation. Minimum insulation resistance for this test is 6 megohms per rated phase-to-phase kilovolt.

Hipot leakage currents should be recorded during the foregoing testing and investigated when out of acceptable ranges or when there is a significant unbalance between phases. If the breaker does not pass the open-pole megger or hipot test, the arc-chutes may require cleaning. Care must be taken when cleaning arc-chutes that contain asbestos, especially if the material is friable. Arc-chutes that do not contain asbestos can normally be glass-bead blasted. Arc interruption by-products can be removed by hand from asbestos containing arc-chutes, with nonmetallic grit sandpaper if care is taken to avoid the asbestos material. Ceramic surfaces can also be wiped clean with an approved solvent.

NOTE: Applicable local regulations must be followed when handling asbestos materials or when performing maintenance activities with equipment containing asbestos or suspected of containing asbestos.

Vacuum breakers: The normal method for testing vacuum-bottle integrity is to overvoltage or hipot the bottles. Manufacturers' recommendations, intervals, and precautions should be followed. At a minimum, vacuum bottle hipot tests should be performed during unit overhauls or sooner if warranted. The tester should be aware of the possibility of x-ray radiation from vacuum bottles under test.

7.4.6 Operational Tests

Connect the breaker to a test stand and perform the following tests:

- *A 70% of rated-voltage close test*: Failure of the breaker to close may indicate the need for mechanism cleaning, lubrication, and adjustment.
- *A 50% of rated-voltage trip test*: Failure to trip may indicate the need to clean, smooth, lubricate, and adjust the latch mechanism.
- *Full-voltage close and open timing tests*: Using a high-speed cycle counter, measure the close and open time for three consecutive operations. If the average of the three timing tests falls outside the acceptable range, the bearings may need cleaning, lubrication, or replacement.
- Test the trip-free and safety interlocks for proper operation and correct any deficiencies.

Electrical Maintenance Guidelines 321

7.4.7 Cubicle Inspection

During the cubicle inspection process, the following items should be completed:

- Examine the bottom of the cubicle for parts that may have fallen from the breaker. The bottom of each cubicle should be maintained clean and free of any foreign objects to facilitate the detection of fallen parts.
- Verify that the mechanical safety interlocks and stops are intact.
- Check that the cubicle heaters (where applicable) are functioning properly.
- Verify that the rack-in mechanism is aligned correctly.
- Lubricate racking mechanism (jacking screws and bearings) according to station experience or manufacturers' recommendations. Check brush length of associated motor (when applicable).
- Perform an overall inspection looking for loose wiring or components and anomalies. Complete repairs as required.
- Verify that the shutter mechanism functions properly.
- The primary disconnects should be inspected for signs of overheating, cracked insulation, cleanliness, and misalignment.

NOTE: Normally, the bus side will be energized; hence, the proper safety measures must be followed.

7.4.8 Rack-In Inspection

The following systems checks should be completed prior to returning the breaker to service:

- With the breaker racked into test position, turn the DC on, witness a close-open operation, and verify that the cubicle door status lamps and mechanical semaphores are functioning properly.
- Arrange to have the open breaker racked into operating position and verify proper cell depth and alignment.
- During a bus outage, try to rack-in a closed breaker and prove that the safety interlocks trip the breaker open. Also during a bus outage, try to rack out an open breaker and prove that the safety interlocks will trip the breaker open.
- When the foregoing inspection process is satisfactory to the participating electrician or technician, an adhesive label should be attached to the front of the breaker that indicates the date of inspection and the name of the responsible person.

7.4.9 Generator DC Field Breakers

Generator field breakers should be inspected according to the same guidelines as detailed for other switchgear circuit breakers. In addition, particular attention should be given to the mechanical adjustments and condition of the discharge resistor insertion contacts. Ohmic measurements of field breaker discharge resistors should be taken during the inspection process to verify proper values. Considering the importance of generator field breakers, prior to returning racked-out generator field breaker to service, and allowing for possible troubleshooting time, a close-open test should be performed from the control room, with the breaker racked in (without excitation power), to ensure that the breaker is functioning properly.

7.5 Insulation Testing of Electrical Apparatus

7.5.1 Purpose

The purpose of this guideline is to provide suggested criteria and intervals for the insulation testing of electrical apparatus. This guideline is not intended to be a complete procedure on insulation testing, and personnel performing the tests are expected to be familiar with safety and other detailed aspects of high-voltage insulation testing. Testing is performed to verify equipment integrity and to provide a measure of confidence that the insulation will prove reliable until the next major outage.

NOTE: Although some overvoltage test currents are below the accepted lethal levels, the stored charge energy in the insulation capacitance can be lethal. The stored charge must be discharged or drained before the device can be considered dead. The charge is stored by dielectric capacitance between the object under test and the surrounding ground. The rule of thumb is, at minimum, to discharge by connecting a low resistance from the conductors of concern and ground and maintain the connection for at least five times longer than the overvoltage test duration.

7.5.2 Apparatus 460 V and Higher

To determine the integrity of 460-V and higher-voltage AC apparatus, a 500- or 1000-V DC megger test of the insulation should be performed periodically. The frequency of testing should be determined by site-specific environmental conditions and experience with the particular equipment. In general, the maximum interval between tests should not exceed 3 years.

The following presents a minimum megohm criterion, based on operating conditions, for all AC electrical apparatuses 460 V and higher that should be met before energizing the equipment. The following test criterion assumes

Electrical Maintenance Guidelines 323

that the temperature of the apparatus under testing is 25°C. This approach will be conservative for measurements carried out at temperatures higher than 25°C.

Institute of Electrical and Electronics Engineers (IEEE) and other industry standards or recommendations indicate that the minimum insulation megohms to ground for motors is 1 megohm per rated phase-phase kilovolt, plus 1 megohm at 40°C. The insulation megohms are determined by testing and need to be corrected for any temperature deviations from the reference temperature of 40°C. The actual correction factors are different depending on the class or type of insulation. In general, the higher the temperature, the lower the insulation resistance is because more free electrons are available for conduction. One can roughly calculate the minimum megohms as 3 times higher at 25°C (typical ambient testing temperature) or 3 megohms per kV for a three-phase test; motors are normally tested to ground without breaking internal phase connections, as presented in Figure 7.4. In the interest of simplicity, for a single-phase test, one would expect the insulation resistance to be three times higher, or 9 megohms per kilovolt, which is conservative since there is also a phase-to-phase insulation resistance that reduces the actual factor, especially if the untested phases are grounded.

If you are testing too much equipment at the same time, you may need to disconnect or isolate apparatus units in order to meet the minimum requirements. Although each type of electrical apparatus unit (buses, cables, circuit breakers, generators, and transformers) has its own minimum megohms, the

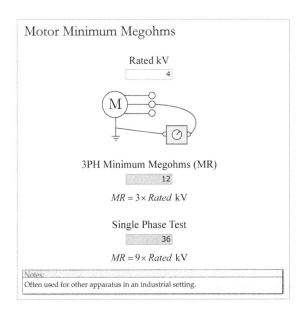

FIGURE 7.4
Motor insulation resistance.

industry in general uses the motor standard for the routine insulation testing of all apparatus. Experience has shown that 3 megohms per kilovolt will not be a problem for energizing any apparatus unless there are other contributing factors. The general practice is to use a 500- or 1000-V DC test instrument. Here, 1000 V is basically an overvoltage- or hipot-test for low-voltage equipment (480-V apparatus), and consideration should be given to using a 500-V test during routine or short outages to reduce possibilities of extending the outage for cleaning and dryout, and a 1000-V test for overhauls and longer outages where a higher-proof test for an undefined future life is desired.

Particularly concerting is the field practice of using 1 megohm per kilovolt (eliminating the plus 1 kV) and not correcting for temperature differences. The industry has reported failures and explosions when energizing electrical apparatus at 1 megohms per kilovolt.

NOTE: IEEE Standard 43, Testing Insulation Resistance of Rotating Machinery, indicates that when megger-testing one phase at a time (with the other two phases grounded), the required megohms should be multiplied by 2 unless guards are used, in which case the readings should be multiplied by 3. This guideline opted for the simpler and more conservative approach of multiplying by 3 in both cases, or 9 megohms per kilovolt, at ambient temperatures.

This criterion also applies to a complete system or parts of a system at the apparatus level. In other words, it is acceptable to isolate apparatus units to meet the requirement individually; that is, a motor can be isolated from the cables, and if the cables and motor meet the requirement separately, they can be reconnected and energized.

Electrical apparatuses that do not meet the aforementioned minimum megohm requirements should be cleaned, dried out, or refurbished prior to energizing. On some apparatuses, a 10-minute polarization test (PI) will help to determine if the low readings are due to moisture or other contamination. The PI test divides the 10-minute measurement by the 1-minute measurement. Because the test voltage is DC, the capacitive charging current decreases as the test duration continues, resulting in higher megohm values. In general, a PI value of 2 is considered acceptable in most cases, unless the apparatus has a history of lower PI values.

7.5.3 DC High Potential Testing

General: Generators and motors with 2 kV and higher voltage should be routinely overvoltage-tested during major turbine overhaul outages (HP turbine for cross-compound units). Normally, for convenience, cables that feed motors are included in the overvoltage test (motor routine values are not considered high enough to unduly stress cable insulation). Otherwise, the routine overvoltage testing of cables is generally not recommended.

It is impossible to tell before testing if the apparatus insulation will fail during the overvoltage (hipot) testing. Accordingly, time to procure material

Electrical Maintenance Guidelines 325

and repair or replace equipment must be provided when scheduling the test. However, the routine test values provided for motors and generators in this document are the minimum values (125% instead of 150% maximum) recommended by IEEE. These values are considered searching enough to provide a measure of confidence that the insulation will not fail before the next major outage, and yet they do not stress the insulation system enough to force an undesirable premature failure. The higher test values provided for new or refurbished apparatus assumes that the manufacturer or repair agency is financially responsible for a test failure (under warranty) and may not be performed if that is not the case.

A 1000-V megger polarization test of 2 or greater would normally be required before proceeding with a hipot test on generator and motor stator windings. The polarization test or PI (i.e., the ratio of the 10-minute to the 1-minute insulation resistance readings) is an indication of the fitness of the winding for the overvoltage test. A low PI (less than 2) may indicate that cleaning, dryout, or repair is required before the hipot can be performed. However, some insulation systems in good condition will not provide a PI of 2 or greater. Where that is the case, and the PI measurement is in agreement with historical measurements for the particular insulation system, you can proceed with the overvoltage testing.

A minimum insulation resistance using a 1000-V megger of 9 megohms per rated kilovolt is required when one phase is tested in isolation from the other phases (which is preferred for generators and cables) and 3 megohms per rated kilovolt when all three phases are tested together (motors). Apparatus with megger readings below minimum should not be hipot-tested. Apparatuses not meeting the recommended minimum megohm values should be dried out, cleaned, or refurbished prior to overvoltage testing.

NOTE: Overvoltage testing should be performed only by technicians or engineers who are familiar with the required safety and test procedures and are properly trained to perform the testing. Due to the high cost of test failures, the overvoltage testing of 10-MVA and higher generators should be witnessed by supervisors or engineers who also have experience in overvoltage testing.

Motors and generators are normally tested at the start of the outage to allow time to complete any required repairs during the outage window. Additionally, it is desirable to perform the testing while the apparatus is in a dry condition (before disassembly or in a cold standby condition). Generators are normally tested completely assembled and under hydrogen pressure (where applicable). On generators equipped with inner water-cooled stator coils, the water is normally evacuated by pulling a vacuum to facilitate megger testing for an acceptable PI and minimum insulation resistance values before proceeding with the overvoltage testing.

As illustrated in Figures 7.5 and 7.6, wye-connected generators are normally disconnected at the output and neutral ends. Each end of the winding is connected with copper bonding wire to avoid transient voltages if the test

326 Electrical Calculations and Guidelines

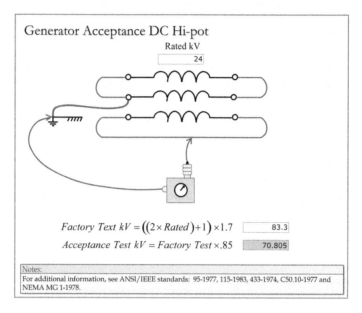

FIGURE 7.5
Generator DC acceptance test.

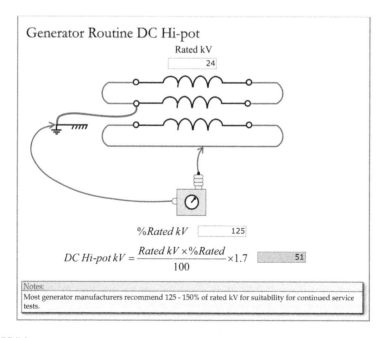

FIGURE 7.6
Generator routine test.

Electrical Maintenance Guidelines 327

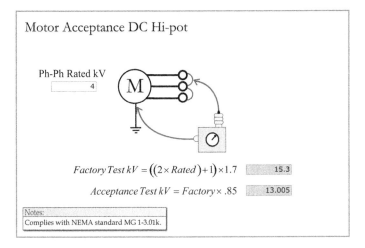

FIGURE 7.7
Motor DC acceptance test.

is suddenly interrupted, and each phase is individually tested to ground, with the other two phases grounded. All three phases for motors without neutral leads brought up to the termination box are normally tested simultaneously to ground. Motors and associated cables are usually tested together from the switchgear cubicle with all three phases bonded together, as shown in Figures 7.7 and 7.8. Cables should be separated from the motor only when the readings of the combined motor-cable system are not satisfactory.

Generators are normally tested with a rate of voltage rise or steps of approximately 2 kV DC, and motors are tested with a rate of rise or steps of approximately 1 kV DC. The operator should record the voltage and current

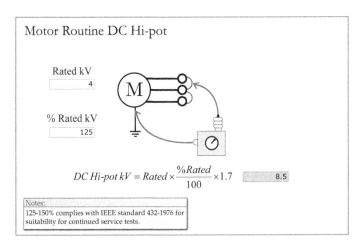

FIGURE 7.8
Motor DC routine test.

readings (after stabilization) for each step. The DC current should be closely monitored, and if the current starts to rise in an uncharacteristic manner, indicating that insulation breakdown is imminent, the test should be aborted by steadily reducing the voltage to zero. A successful test is concluded when the recommended voltage value is reached and the current is stable for 1 minute. At the conclusion of the test, the voltage should be steadily reduced to zero to avoid sudden changes and the development of potentially damaging transient voltages.

7.5.4 Generator Rotor Winding Overvoltage Testing

Insulation systems for large generator, cylindrical, rotor field windings are normally DC overvoltage- or hipot-tested at acceptance values after shipment to the site or following onsite rewinds, when another party is responsible for the cost of repair should it fail, and also to prove that the apparatus is fit for service. Figure 7.9 provides an illustration of the test and the suggested DC hipot test values. Most manufacturers suggest testing at 80% of the factory test value with both ends of the winding shorted together to mitigate transient voltages in case the test is suddenly interrupted. As you can see in the figure, a 500-V field winding would be acceptance-tested at 6.8 kV DC. Insulation systems for large generator, cylindrical, rotor field windings are normally DC overvoltage- or hipot-tested while still under warranty after completing 1 year of service, when another party is responsible for the cost of repair should it fail, and also to prove that the apparatus is fit for continued service.

Figure 7.10 provides an illustration of the test and the suggested DC hipot test values. Most manufacturers suggest testing at 60% of the factory test value with both ends of the winding shorted together to mitigate transient

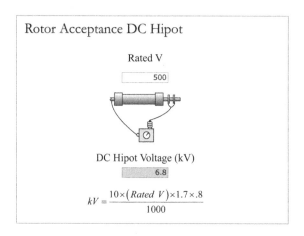

FIGURE 7.9
Generator rotor acceptance DC overvoltage testing.

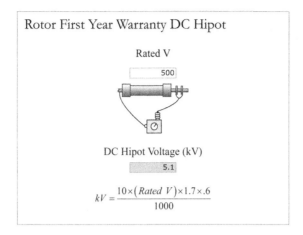

FIGURE 7.10
Generator rotor first-year warranty DC overvoltage testing.

voltages in case the test is suddenly interrupted. As you can see in the figure, a 500-V field winding would be tested at 5.1 kV DC after completing 1 year of service.

Cylindrical rotor generator field windings should be routinely tested during unit overhauls. Normally, a PI of 2.0 is required (measure of moisture or contamination) and is calculated by dividing the 10-minute megohm reading by the 1-minute measurement. On brush machines, the exposed insulation for the collector or slip rings usually have a hard buildup of oil/carbon from the bearings and brushes, and extensive cleaning of that area may be required before getting a satisfactory PI.

Figure 7.11 provides an illustration of the test and suggests using a 1000-V megger or insulation test instrument. As you can see in the figure, the

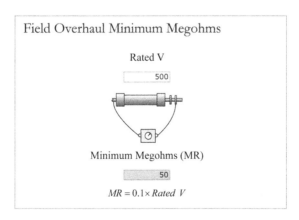

FIGURE 7.11
Generator rotor routine overhaul insulation testing.

minimum megohm value is 50, which is higher than a standard nonoverhaul requirement, which is only 9 megohms per kilovolt, or 4.5 megohms for a 500-V field. This can be performed with a 500-V megger or test instrument.

7.5.5 Generator-Neutral Buses or Cables

Generator-neutral bus and/or cable to the grounding transformer should be disconnected at each end and DC hipot-tested during overhauls at the routine value for the generator stator.

7.5.6 Cables 5 kV and Higher

Cables 5 kV and higher should not be routinely tested at cable hipot values during overhauls unless the integrity of the cable is suspect. Because of high test voltages and the limitations of the switchgear and associated CTs, the cables must be disconnected at both ends to perform a full, routine hipot test.

New cables should be hipot- or partially discharge-tested before and after installation in order to fix warranty responsibility and to ensure that the cables were not damaged during shipment or installation.

The hipot values for cable testing are dependent on the type of insulation material, insulation thickness, and the voltage class. The variations are too numerous to cover in this guideline. Normally, the after-installation test is performed at 80% of the factory test value, and routine tests are performed at 60% of the factory value. Please refer to the IEEE, EPRI, Insulated Cable Engineer Association (ICEA), Association of Edison Illuminating Companies (AEIC), and other appropriate standards for test values and details before hipot-testing cables.

Where the cable manufacturer does not recommend hipot testing, partial discharge testing should be performed instead. Cable insulation systems are chemically complex and represent many variations or families within the same generic type—that is, cross-link poly (CLP) and ethylene propylene rubber (EPR). Cable hipot values may shorten the life of some cable insulation systems by causing molecular changes, voids, or both in the insulation. Motor hipot levels are significantly lower than cable hipot values; consequently, they are considered safe for all cable insulation systems.

7.6 Bus and MCC Maintenance

7.6.1 Purpose

The purpose of this guideline is to provide suggested procedures and schedules for the inspection, care, and electrical maintenance of buses and MCCs.

Electrical Maintenance Guidelines 331

It is intended to include all major electrical buses; that is, generator-isolated phase and neutral buses, transformer outdoor buses, auxiliary power switchgear buses, MCC buses, battery distribution buses, and critical control power buses.

7.6.2 Bus Inspections

Where outages permit, electrical buses should be inspected at 2–3-year intervals. Generator- and transformer-isolated phase buses are normally inspected during unit overhauls.

The buses should be checked for moisture, contamination, overheating, excessive oxidization, discoloration, electrical tracking, flexible shunt or braid erosion, bird nests, rodents, proper sealing, and any other anomaly. The primary disconnects for each position (bus and load side) should be included in the inspection for switchgear buses. Deficiencies should be corrected during the inspection process. Accessible bus insulators should be cleaned and bus bolts checked for tightness. The bolts should be torqued to values that are compatible with the softest material involved, not necessarily to the value recommended for the particular bolt. Flexible braids and shunts that are eroded in a manner that would reduce their current-carrying capability should be replaced. Where necessary, silver-mating surfaces should be replated. Aluminum oxide can cause high-resistance connections, and treatment with an approved deoxide grease is necessary to prevent failures of aluminum connections. The oxide can form very rapidly, and the grease should be applied immediately after cleaning. Normally, the mating surface is brushed or file-carded after the grease is applied to ensure that any oxide formations that may have formed during the process are broken loose.

7.6.3 Bus Testing

The buses should be megger-tested during the inspection or overhaul process. The minimum megohms for a single-phase test (one phase at a time) is 9 megohms per rated phase-to-phase kilovolt; that is, each phase of a 4-kV bus should have at least 36 megohms to ground, with the other two phases grounded.

In addition to megger-testing, switchgear and MCC buses should be micro-ohm-tested during the bus inspection to prove the integrity of the associated bus connections and breaker and bucket primary disconnects. This can be accomplished by grounding all three phases at the source and measuring the micro-ohms from the source to each phase in each breaker cubicle or MCC position. The micro-ohms for phases located within the same position or cubicle should be within 5% of each other, and the measurements should increase slightly for positions that are farther from the source. Additionally, to take advantage of the accessibility (breakers out of the cubicles), the switchgear and loads should be megger-, resistance-, and impedance-tested during the bus inspection and testing.

NOTE: During bus outages, try to rack in and rack out closed breakers and prove that the cell safety mechanical interlocks actually trip the breakers. Care must be taken to ensure that the breakers are deenergized on both sides.

7.6.4 MCC Position Inspections

At 2- to 5-year intervals, MCC positions should be physically inspected. The MCC positions should be inspected for loose components, contamination, moisture, overheating, and connection tightness. The positions should be cleaned and the main contacts cleaned and dressed as required. Anomalies should be corrected during the inspection process.

NOTE: The MCC position may be energized at the power disconnect device (i.e., the molded case breaker or fused disconnect). No maintenance should be attempted on the energized portion of the MCC position.

7.6.5 MCC Position Testing

MCC loads should be megger-, resistance-, and impedance-tested during the position inspection process. Additionally, motor operated valves and dampers should be electrically operated through their range (where possible) to ensure proper operation.

7.7 Protective Relay Testing

7.7.1 Purpose

The purpose of this guideline is to provide suggested procedures and intervals for the testing of protective relaying. Proper relay operation is essential to mitigate equipment damage and hazards to personnel, as well as to reduce the duration of forced outages.

7.7.2 General

Routine tests on protective relays and associated equipment are normally made in accordance with standards, test manuals, and manufacturers' instructions. Only properly trained engineers and technicians should be allowed to maintain protective relay systems. All testing activities should be coordinated with control room personnel.

7.7.3 Testing Schedule (440 V to 765 kV)

Protective relay routine testing is normally performed at 2–5-year intervals or plant overhaul outages, but the gap between tests should not exceed 6 years.

7.7.4 Relay Routine Tests

When possible, the following steps should be completed during the routine testing:

- Where possible, megger and inspect CTs and PTs and associated wiring. Electronic relays may need to be isolated during the megger test.

  NOTE: Care must be taken to ensure that the CTs under testing are completely deenergized. CTs that are wrapped around an energized bus or connection, even though there is no current going through the bus or connection, must be treated as energized.

- Perform a thorough mechanical check of the protective relays, switches, wiring terminals, and auxiliary relays and timers. The inspection should include looking for metal chips or filings and other foreign objects in the area of the magnetic gaps. Contacts should be cleaned and dressed as required. Dirt and contamination should be removed with a soft brush or by low-pressure dry air or a vacuum.

- Complete an electrical calibration of the protective relay (normally with the relay in its case) using in-service taps and adjustments to ensure that it will operate properly (within 5%) over the intended range. Of particular importance is to verify the timing at the current level that is specified by the protection engineer responsible for the settings.

- Prove that the tripping and alarm circuits will perform their intended operations.

  NOTE: High- and medium-voltage circuit breakers, fuel trips, and turbine steam valve operations should be kept to the minimum required during each relay routine and trip test. One trip operation for each apparatus function is required; subsequent tests from different (86) or direct trip relays should be proved by drawing light bulb current on the particular trip wire or open trip cutout switch instead of actually retripping the equipment.

- Make a final in-service test to determine that the protective group is being supplied the proper currents and potentials as required for the various elements after the relay group is returned to service.

7.7.5 Primary Overall Test of CTs

Primary injection tests from each phase and CT group, with all secondary current devices included, should be performed to ensure proper operating values and performance under the following conditions:

- During the initial installation or replacement of CTs
- Anytime a CT's ratio is changed or there is a major wiring change in the secondary circuit
- If there is any reason to suspect faulty operation of the CTs

NOTE: Secondary tests (light bulb or resistive current) are acceptable for proving minor changes to the secondary wiring and for proving major changes where primary injection tests are not practical because the CTs are not accessible.

7.7.6 Documentation

Each station should maintain files on the protection philosophies, calculations, relay settings, and coordination of protective relaying for low-voltage, switchgear-fed loads and higher-voltage electrical systems.

Relay data cards or labels should be maintained at the relay and show the date of the last routine test, the name of the responsible technician or engineer, the name of the circuit, the instrument transformer ratios, and the relay settings in secondary values. Where a separate relay is provided for each phase, the data card/label should be attached only to the A-phase relay. The settings for multifunction digital relays do not need to be included on the relay card/label.

7.7.7 Multifunction Digital Relay Concerns

With multifunction digital relays, there are two areas of concern:

- When testing digital relays, there can be element or function interference, and it may be desirable to take elements or functions out of service temporarily to facilitate the testing. In this case, documented procedures should be established (i.e., check sum numbers, software comparison routines, etc.) to ensure that all desired alarming and tripping functions are placed back into service following the testing.
- The relay settings provided should be presented in such a manner that the commissioning and routine testing engineer or technician has absolutely no doubt about the intentions of the protection engineer responsible for issuing the settings. The information must be presented in a way that is not ambiguous or subject to interpretation.

Electrical Maintenance Guidelines 335

7.8 Battery Inspection and Maintenance

7.8.1 Purpose

The purpose of this guideline is to provide suggested procedures and intervals for the maintenance of plant battery banks. Batteries must be maintained correctly to prevent unnecessary outages and to ensure that backup emergency power is available for the turbine emergency DC oil pumps and other important loads. Proper battery maintenance is vital to ensure that control and shutdown systems function properly to mitigate damage to equipment and risk to plant personnel. This guideline covers lead acid batteries in detail; for other types of batteries, follow the manufacturer's recommendations.

7.8.2 General

Approved maintenance and safety procedures should be in place before any work is performed on station batteries. Routine tests on batteries and associated equipment are normally done in accordance with the standards and procedures outlined in the manufacturer's instructions and should also take into account any specific requirements or code of practices in the jurisdictional location where the batteries are installed. Additional references that can be consulted are IEEE Standard 450, "Recommended Practice for Maintenance, Testing, and Replacement of Large Load Storage Batteries for Generating Stations," and EPRI's *Power Plant Electrical Reference Series*, volume 9.

Room temperature can affect battery performance, and in some cases, it also may alter the set point of the charger voltage. When taking measurements, always record the temperature of the room where the battery bank is located. Batteries normally have reduced capacity at lower temperatures because the electrolyte resistance increases. At higher temperatures, they have increased capacity, but lower life expectancy. EPRI recommends a battery room temperature range between 60°F and 90°F, with an average temperature of 77°F. Batteries are to be kept clean and dry on the outside, and all necessary precautions should be taken to prevent the intrusion of foreign matter into the cells. All cell connections should be kept tight and free from corrosion.

All battery areas should be adequately ventilated, and the air should exhaust to the outside and not circulated to other indoor spaces. A concentration equal to or greater than 4% hydrogen is considered dangerous. Care should be taken to prevent pockets of hydrogen near the ceiling. If the ventilation system is out of service and work needs to be performed in the battery room, the area should be treated as hazardous, for both its possible lack of oxygen and its possible explosive nature. The room should be properly ventilated before proceeding with any work.

Only distilled or approved demineralized water shall be added to maintain the electrolyte level as near as practical to the marked liquid level lines.

Do not fill above the upper- or maximum-level lines. All water intended for battery maintenance should be tested for acceptable purity before use. Acid should never be added to or removed from a cell without specific instructions from the manufacturer.

Appropriately trained and qualified station personnel would normally perform weekly, monthly, quarterly, and annual maintenance inspections. All measurements should be properly documented, and any discrepancies found should be reported to the appropriate supervisor. The voltage chosen as the float voltage has an effect on the stored ampere-hour capacity of the battery. In general, as the float voltage is reduced, so is the stored ampere-hour capacity. Overall terminal voltages of batteries will vary depending on the number of cells. The typical number of cells for generating station applications is 60.

7.8.3 Floating Charges

There are various types of batteries; this guideline covers some of the more popular ones. All should be maintained, operated, and set up in accordance with standards and procedures outlined in the manufacturer's instructions. The guidance presented here provides data that are generally acceptable for lead acid batteries:

- Lead antimony batteries are normally to be kept on floating charge, at an average voltage of 2.19 V per cell but not more than 2.23 or less than 2.15 V per cell, per EPRI guidelines (except as specified differently by the manufacturer).
- Lead calcium batteries are normally to be kept on floating charge, at an average voltage of 2.21 V per cell but not more than 2.25 or less than 2.17 V per cell, per EPRI guidelines (except as specified differently by the manufacturer).

7.8.4 Inspection Schedules

The following intervals are recommended for inspecting battery banks:

- *Daily*: Operations should perform visual inspections of all battery banks, chargers, and associated ground detectors during each shift. Maintenance orders should be generated for any discrepancies found. This stipulation may be relaxed if appropriate and reliable alarm systems are in place.
- *Weekly*: Station electricians or technicians should visually inspect the chargers and battery banks once a week and take an overall voltage reading with a calibrated voltmeter. Anomalies found should be corrected during the inspection process.

Electrical Maintenance Guidelines

- *Monthly*: Station electricians or technicians should inspect the chargers and batteries, check the electrolyte levels, and add distilled water as needed every month. Any discrepancies should be corrected during the inspection process.
- *Quarterly*: Once a quarter, the station electricians should do the following:
    - Measure the specific gravity of each cell, starting at number 1, and record the reading.
    - Raise battery-charging voltage as high as circuit conditions will permit, but not to exceed 2.33 V per cell for a check charge. Allow 10 minutes for the voltage to stabilize and record the following readings:
        - Overall voltage at the beginning of the readings.
        - Individual cell readings, starting at cell number 1.
        - The overall voltage at the end of the readings should be as follows:
            - If the lowest of the individual antimony cell voltage is within 0.05 V of the average of all cells and the cells are gassing freely, the check charge is complete and the charger should be returned to the normal floating charge.
            - If the lowest of the individual calcium cell voltage is within 0.20 V of the average of all cells and the cells are gassing freely, the check charge is complete and the charger should be returned to the normal floating charge.
            - If one or more antimony cells indicate 0.05 V below the average of all cells, or if one or more calcium cells indicate 0.20 V below the average of all cells, the check charge is to be carried on to an equalizing charge. The equalizing charge should be set to 2.33 V per cell multiplied by the number of cells, and held for a period of 8 hours.
            - If the equalize charge does not bring the individual cell voltages to acceptable limits, consult with the manufacturer. If the cell voltages are within acceptable limits, return the charger to the normal floating charge.
- *Annually*: A thorough examination of the battery intercell connections should be performed once a year. Any connection with oxidation or corrosion should be cleaned using the approved cleaning methods given by the battery manufacturer. Interconnection bolts should be checked for proper tightness using approved insulated wrenches. Connections should be tightened to the manufacturer's recommended torque specifications using a torque wrench with the appropriate insulation and scale.

- To test that the batteries still have an acceptable ampere-hour capacity, load discharge tests should be performed during convenient outages using either an appropriate load bank or the turbine emergency DC oil pump load. The duration of the test should be sufficient to prove that the batteries have sufficient capacity to prevent damage to turbine/generator bearings during coastdown conditions. A reasonably high discharge rate should be applied to the batteries with the charger switched off (*this discharge, at a minimum, should correspond to the actual load expected during an emergency condition, and care should be taken not to exceed the minimum voltage values specified by the battery manufacturer*). The actual battery discharge voltage/time characteristic should be compared with that obtained from manufacturer for the same discharge conditions. Any deviation from the curves should be discussed with the manufacturer. The test discharge rate used must be within the specified battery/equipment limits and properly account for temperature values.
- Impedance testing can also be performed if the manufacturer recommends it. A test instrument designed for that purpose should be used.

7.8.5 Safety Precautions

Extreme care must be exercised when performing maintenance on batteries, as they contain either acid or caustic solutions. Before working on batteries, note the location of the nearest eyewash station and safety shower. The following recommendations are provided to minimize hazards to personnel:

- Wear approved protective clothing, rubber gloves, and eye protection when adding liquid to a battery or performing any other activity when there is a possibility of coming into contact with cell electrolyte.
- If acid or caustic materials or solutions make contact with the eyes or skin, flush with a copious amount of water and seek first aid. Provide in the battery room, or have at hand, suitable eyewash and acid splash dilution facilities that are regularly inspected.
- Exercise care when handling acid or caustic materials and solutions.
- If spillage occurs, it should be cleaned up immediately. Neutralizing materials should be available in the battery room to neutralize any spilled active material. Usually, 1 pound of baking soda is mixed with 1 gallon of water to create a suitable neutralizing mixture.
- Smoking, open flames, or heat that could cause the ignition of hydrogen gas is generally prohibited in a battery room. Activities that either may or can cause flames, sparks, or heat should be prohibited

Electrical Maintenance Guidelines 339

unless a competent or qualified person has performed a risk assessment and all the recommendations of the assessment have been applied. Approved eye protection should be worn when performing any activity that could cause a battery cell to explode.
- Care must be taken not to use materials that may generate static electricity arcing (e.g., polythene plastic sheeting).
- When working in a battery room, ensure that the exit remains unblocked and unlocked, and that all doors have panic bolts on the inside. Doors should be locked when no work is being carried out in a battery room in order to prevent unauthorized access.
- The use of noninsulated metallic tools on or around the exposed battery terminals or intercell connections is prohibited.
- The use of oil, solvent, detergent, or ammonia solution on or around the battery cases is prohibited. These solutions may cause permanent damage to the special, high-impact plastic case materials.
- Be careful when carrying metallic articles, such as watches, jewelry, pens, glasses frames, or rulers, when working on batteries. All metallic items not required for performing the assigned activity should not be taken into the battery room.

7.8.6 Operation and Troubleshooting

- *Low voltage*: Critical protective relay DC circuit voltage levels should be monitored to ensure that the protection can operate properly. Whenever the DC voltage level is at or below 95% of nominal, plant operators, electricians, and technicians should immediately investigate and resolve the problem or remove the protected equipment from service.
- *High voltage*: Whenever the DC voltage level is found to be above the equalize level 110% or normal, indicating a charger malfunction, plant operators, electricians, and technicians should immediately investigate and resolve the problem or remove the protected equipment from service.
- *DC grounds*: Critical protective relay DC circuits should be monitored for grounds to ensure that the protection can operate properly. Normally, ground detectors are set to alarm for a 10% ground. At the first indication of a DC ground, plant operators, electricians, and technicians should immediately investigate and resolve the problem or remove the protected equipment from service.

NOTE: It is not permissible to interrupt the DC to in-service circuit breakers or critical protective relays temporarily unless the responsible protection engineering organization determines that adequate backup protection

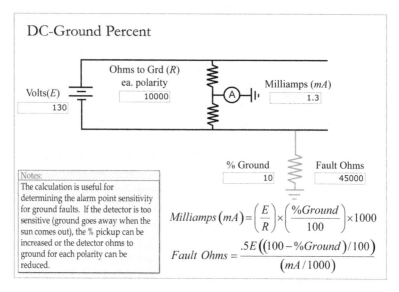

FIGURE 7.12
DC ground fault ohms.

is in service. Critical protective relays are those that are associated with the high-voltage bulk power systems, subtransmission systems, or main-unit generators.

Figure 7.12 presents a calculation for determining the DC ground fault resistance in a typical plant battery ground detector scheme. It shows two 10,000-ohm resistors that are connected to each polarity of a 130-V DC system with the common point connected to ground. A 10% ground or alarm actuation point is equal to 45,000 ohms to ground in this case.

7.9 Personnel Safety Grounds

7.9.1 Purpose

The purpose of this guideline is to provide suggested procedures and schedules for the application, inspection, care, and maintenance of personnel safety grounds. It applies to switchgear (draw-out circuit breakers and contactors) that feed electrical loads 480 volts and higher. Typical short circuit currents for low-voltage (480-V) switchgear systems are in the same range as the higher medium-voltage systems (15,000–45,000 amps). This guideline does *not* apply to low-voltage systems and loads that are fed from MCCs or distribution panels. The short circuit currents for these systems are usually lower than those available from switchgear-fed auxiliary power buses and feeders, and the smaller terminations do not permit the clamping of personal ground with enough capacity to protect workers from inadvertent energizations.

7.9.2 General

Personal grounds are applied on all three phases for the following reasons:

- Prevents the development of dangerous, and even lethal, induced voltages where the employee or contractor is working.
- Provides a short circuit current path that will result in the automatic tripping of any breakers involved in inadvertent or accidental switching energizations by in-service protective relays.

To accomplish this, portable grounds and grounding devices must be able to handle the following:

- Mechanically withstand the short circuit electromechanical forces developed at the particular location.
- Have the thermal capacity to carry the short circuit current at the specific location.

Portable personal grounds are often made from flexible 2/0 conductors and clamps that have been approved for this purpose by vendor short circuit testing. Conductors of a size equal to 2/0 will fuse open in approximately 18 cycles, with short circuit values of around 35,000 amps. To be effective, the fuse time of the personal grounds must be longer than the protective relay clearing times (relay plus circuit breaker).

Personnel grounds must be connected directly to the station ground grid, where available, to reduce the circuit impedance and associated voltage rise during short circuit conditions. Additionally, all connections must be secure to reduce the contact resistance and corresponding heating of the connections and to ensure that the grounds can mechanically withstand the electromechanical forces when carrying short circuit currents.

Personal grounds should be located in between work locations and any possible sources of inadvertent energization. Where there is the possibility of induced voltages from neighboring energized conductors or circuits, the grounds must be located in the general proximity of the employee to prevent a voltage increase at that location from short circuit currents flowing in the adjacent energized circuits. It is not unusual for cables to share cable tray and duct banks with other energized circuits. For example, if motor cables are within 1 foot of another circuit and parallels the circuit for 900 feet, a 35,000-ampere ground or double line to ground fault in the neighboring circuit can induce 1–5 kV (depending on how the fault current returns to the source) into the out-of-service motor cables for the duration of the fault. Consequently, the end of the conductor that the worker is touching (i.e., the motor end) needs to be grounded to prevent a hazardous voltage rise.

7.9.3 Special Grounding Considerations

The following locations in generating stations usually require special consideration for safety grounding:

- *Low-voltage side of generator step-up transformers (GSUTs)*: In this case, applying grounds on the low-voltage side of the GSUT will usually not protect workers from an inadvertent high-voltage side energization even if the grounds are located in between the worker and the step-up transformer. The short circuit currents are normally too high for personal grounds; the clamps will not be able to withstand the electromechanical forces and the grounding cables will fuse open almost instantaneously from the high current values. Accordingly, high-voltage side grounds must be applied to protect workers on the low-voltage side of generator step up transformers where the current is more manageable. Grounds should also be applied on the low-voltage side to protect workers from an inadvertent energization from the unit auxiliary transformer.
- *Low- and medium-voltage switchgear buses*: The protective relaying fault clearing times for switchgear buses often exceeds the thermal capability of personal grounds. Figure 7.13 presents the calculation procedure for 2/0 personal grounds at typical generating station low- and medium-voltage bus short circuit levels. The actual short circuit current levels should be used and the protective relay time to isolate the fault should be determined to verify that the ground will not fuse open before the protection can operate. Double grounds or larger conductor sizes could be applied to increase the capability of the personal grounds, or a manufacturer's switchgear grounding device could be used to ground the bus.
- *Motor or feeder cables*: Where motor or feeder cables share duct banks or cable trays with other energized or in-service feeders, there is concern about induced voltages from the neighboring circuits. If one of the adjacent circuits develops a ground fault or double line to ground fault, the corresponding high currents can produce a strong magnetic flux that links with the deenergized motor or feeder cables. This flux can easily induce voltages into the cables in the range of 1–5 kV or even higher, depending on the magnitude of the current, the distance from the motor or feeder cables to the adjacent circuit, the distance of parallel, and the path of the return short circuit current. Figure 7.14 illustrates the problem; as you can see, grounding

Electrical Maintenance Guidelines

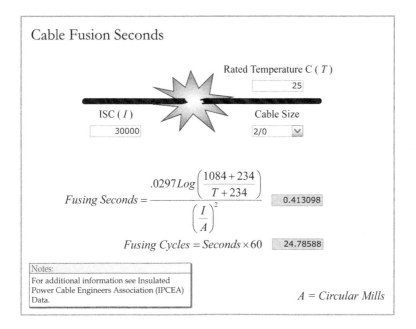

FIGURE 7.13
Cable short circuit fusion.

the switchgear end aggravates the hazard since the capacitive reactance impedance is bypassed, allowing more induced current to flow through the worker. Grounding the switchgear end also increases the hazard because the full amount of the induced voltage is referenced to ground at the opposite end of the cable from the worker. Safety personal grounds need to be near workers to protect them from induced voltages and not at a remote location. Grounding the motor end only prevents workers from exposure to induced voltages at that end but not at the switchgear end. A switchgear-grounding device could be used for work at the switchgear end. For work at the

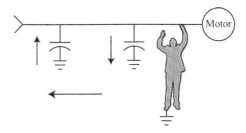

FIGURE 7.14
Induced voltage hazard.

motor end, the preferred method is to *not* ground the switchgear end of the cables. The cables can then be disconnected, using properly rated and approved medium-voltage gloves (as if it was energized), and personal grounds can be applied to the cables at the motor end immediately after they are disconnected. A reverse procedure can be used to reconnect the cables.

The following list provides induced voltages per 100 feet of parallel, with an adjacent energized circuit at various short circuit current levels. It is assumed that the energized circuit is within 1 foot of the conductors where work is to be performed and the return path for the short circuit current is 100 feet away. If the circuit separation is increased to 10 feet, or if the short circuit return path is reduced to 10 feet, the induced voltages will be reduced by half. Conversely, if the circuit separation is reduced to 0.1 feet, or if the short circuit return path is increased to 1000 feet, the induced voltages will increase by a factor of 150%.

- 50,000 amps = 530 V
- 40,000 amps = 424 V
- 30,000 amps = 318 V
- 20,000 amps = 212 V
- 10,000 amps = 106 V
- 5000 amps = 53 V

7.9.4 Maintenance

- *Personal grounds*: Personal grounds should be inspected at 2-year intervals. Particular attention should be given to the clamps and terminations on each end of the cable. Cable clamping bolts should be checked for proper torque.
- *Switchgear grounding devices*: Grounding devices should be inspected every 2 years. Primary disconnect finger clusters should be inspected for proper spring tension, all mating surfaces should be inspected for acceptable silver plating, and all bolts involved in current carrying connections should be checked for proper torque values. Proper alignment of the poles and carriage should also be verified to ensure that racking in the grounding device will not damage the cubicle primary disconnects or racking mechanism.
- *Switchgear ground and test devices*: Ground and test devices should be inspected at 2-year intervals. In addition to the maintenance items delineated for grounding devices, ground and test devices should be thoroughly cleaned because approximately half the device becomes energized under normal use.

- *Ground disconnects*: Ground disconnects used for personnel safety should be inspected on 2- to 3-year intervals. Disconnects should be inspected for proper spring tension, all mating surfaces should be inspected for acceptable silver plating (particular attention should be given to hinged areas), and all bolts involved in current carrying connections should be checked for proper torque values. Proper alignment and engagement of the poles should also be verified during this inspection.

7.9.5 Electrical Testing

- *Personal grounds*: During the inspection process, personal grounds should be micro-ohm tested to verify proper current-carrying capability. A small amount of increased resistance during fault conditions will force a premature failure of the personal ground. For example, an increased resistance of 1 milliohms can cause an extra 100 kW of heating during fault conditions. The micro-ohm readings should be documented and compared to prior readings and other personal grounds for reasonableness. Differences should be investigated and resolved.
- *Switchgear grounding devices*: Grounding devices should be micro-ohm tested, during the inspection process, to verify proper current carrying capability. The micro-ohm readings should be documented and compared to prior readings for reasonableness. Differences should be investigated and resolved.
- *Switchgear ground and test devices*: In addition to micro-ohm tests as specified for switchgear grounding devices, ground and test devices should also be megger- and hipot-tested during the inspection process.
- *Ground disconnects*: During the inspection process, ground disconnects should be micro-ohm tested to verify proper current carrying capability. The micro-ohm readings should be documented and compared to prior readings for reasonableness. Differences should be investigated and resolved.

7.10 Generator Automatic Voltage Regulators and Power System Stabilizers

7.10.1 Purpose

The purpose of this guideline is to provide suggested procedures and schedules for the testing, inspection, care, and electrical maintenance of automatic voltage regulators (AVRs) and power system stabilizers (PSSs).

7.10.2 AVRs

AVRs control generator excitation to maintain a constant terminal voltage as the machine is loaded. In general, var flows from a higher voltage to a lower voltage point on the electrical system. Therefore, it is important to maintain the desired generator output terminal voltage to control electrical system var flows.

7.10.3 PSSs

PSSs are installed on larger generators to dampen out intertie watt oscillations. Each generator has a local mode natural frequency at which it wants to oscillate at, and the cumulative effect of operating a number of geographically dispersed generators in a network can create an intertie watt oscillation in the neighborhood of 1.0 cycle per second. The watt oscillations at system intertie points are undesirable, and the stabilizers act on the excitation system to dampen out the oscillations. On older units, the PSS is mounted in separate enclosures, and on newer excitation systems they may be incorporated into the same enclosure as AVR.

7.10.4 Certification Tests

Accurate models of generators that feed the bulk power electrical system (100 kV and higher) and their associated controls are necessary for realistic simulations of the electrical grid. Baseline testing and periodic performance validation are required by regional transmission authorities to ensure that the dynamic models and databases that are used in the grid simulations to represent plant excitation systems are accurate and up to date. The testing involves off-line tests as well as full load dynamic testing and is normally performed by an excitation engineering specialist.

Each bulk power-generating unit should have an excitation system model that was certified by dynamic testing. The certification document and associated testing reports should be maintained in the station file system.

Recertification testing of bulk power generators is normally performed during major unit overhauls at approximately 5-year intervals to ensure that the excitation model has not changed. New generation or units with modifications that could affect the excitation system model should be tested within 180 days of a start or restart date.

In more recent years, NERC is requiring that the minimum excitation limiter (MEL) or reactive ampere limit in the AVR be set by excitation engineers to prevent excursions into the loss of field mho circle. Ideally, a 20% margin is used; that is the MEL will start its control function well before the loss of field relay can operate. For NERC requirements, the actual MEL curve should be determined by testing the AVR. This is really a specialty area involving the particular MEL setting parameters, conversions to relay ohms, control

Electrical Maintenance Guidelines

response time delays, machine steady state stability limits, slip frequencies, and core end iron overheating. Figure 7.15 illustrates the loss of field mho circle, four generator capability curve points, and an ideal MEL control curve, all plotted on the same base using secondary or relay ohmic values. The figure shows four capability curve output boxes C1–C4 in ohmic values; the inputs for each output box are MW and MVAR at the maximum capability curve values. The points on the generator capability curve in Figure 7.15 are

- 1.0 PF
- 0.95 PF
- 0.70 PF
- 0.00 PF

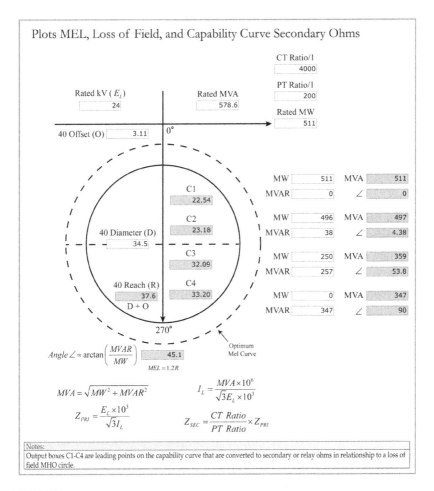

FIGURE 7.15
Loss-of-field, MEL, and capability curve ohmic plots.

The maximum watt-only point is outside the loss of field mho circle. The 0.95 PF leading point is close to the AVR stability limit and is also outside the loss-of-field mho circle, but further var increases could actuate the loss of field relay if it loses synchronization and slips poles. The 0.7 leading PF point is where there is a particular concern about overheating the core iron end packets and is also outside the loss of field mho circle. The maximum leading var-only area is inside the loss-of-field mho circle and would not require pole slipping to actuate the loss-of-field protection (assuming that the MEL control function does not sufficiently limit the leading vars into the generator).

7.10.5 Routine Tests

The following tasks should be completed prior to the full load dynamic testing:

- Where possible, megger and inspect CTs and PTs and associated wiring.
- Perform a thorough mechanical check of the switches, wiring terminals, and auxiliary relays and timers. Contacts should be cleaned as required. Dirt and contamination should be removed with a soft brush or by low-pressure dry air, or vacuum.
- Electrical testing should be performed to prove that the voltage regulator is controlling properly and volts/Hz and the reactive ampere limiter or MEL are functioning properly. The PSS should be checked for proper tuning.
- Prove that tripping and alarm circuits will perform their intended operations.
- Make a final in-service test with the unit on-line to prove that the voltage regulator and PSSs are being supplied the proper currents and potentials, and, by observation, determine that the equipment appears to be functioning properly.

7.10.6 Generating Station Responsibilities

- Each generating station should maintain files on their excitation system certification and routine tests. The associated station drawings must be maintained accurately. Files shall also be maintained on the investigation of the AVR and PSS failures.
- The individual generating station failure investigation reports should be sent to the generator excitation engineering support group in a timely manner for further analysis and possible upgrades to mitigate reoccurrence.

- Routine test cards or labels shall be visibly mounted on voltage regulator and PSS equipment. The cards or labels should show the date of the last routine test and the name of the responsible technician or engineer.

7.10.7 Excitation Engineering Responsibilities

- The excitation engineering group is responsible for reviewing and improving the design and calculating settings for generating station AVR and PSS systems.
- The excitation engineering group should also maintain files on key electrical drawings, instruction manuals, reference books, settings and associated calculations, and their analysis of excitation system failures.
- The excitation engineering group is responsible for developing routine test procedures for proving that voltage regulators and PSSs are functioning according to the intended design.
- The excitation engineering group will either perform the routine testing of AVR and PSS systems or will train station forces to perform the tests and provide testing and troubleshooting support as required.
- The excitation engineering group is responsible for performing dynamic testing for certification and recertification purposes.

Bibliography

Baker, T., *EE Helper Power Engineering Software Program* (Laguna Niguel, CA: Sumatron, Inc., 2002).
Baker, T., *Electrical Maintenance Guidelines* (Laguna Niguel, CA: Sumatron, Inc., 2003).
Baker, T., *Generator Rotor Shorted Turn Analyzer Instruction Manual* (Laguna Niguel, CA: Sumatron, Inc., 2004).
EPRI, *Power Plant Electrical Reference Series*, Vols. 1–16 (Palo Alto, CA: Stone & Webster Engineering Corporation, 1987).
Kerszenbaum I. and Maugham, C., *Utilization of Repetitive Surge Oscillograph (RSO) in the Detection of Rotor Shorted-Turns in Large Turbine-Driven Generators* (Annapolis, MD: IEEE Electrical Insulation Conference, 2011).
Klempner, G. and Kerszenbaum, I., *Handbook of Large Turbo-Generator Operation and Maintenance* (Piscataway, NJ: IEEE Press, 2008).
Westinghouse Electric Corporation, *Applied Protective Relaying* (Newark, NJ: Silent Sentinels, 1976).

8
Managing a Technical Workforce

Countless operations around the globe are feeling the impact of an ongoing skilled workforce crisis. Combined with the types of advanced digital logic technologies that are employed today, it presents a serious challenge for generating station management. This chapter will explore strategies for both the staffing and management of a technical workforce at large generating stations, and focuses on the following three crafts or journeymen positions:

- Electricians
- Instrument technicians
- Protective relay technicians

8.1 Strategies

There are two main strategies used at most generating stations:

- Minimize staffing and contract much of the work
- More self-sufficient staffing, with little or no reliance on contracting

More self-sufficient area maintenance can provide the following economic advantages over a contracting approach if managed properly:

- Provides a more efficient workforce since it is a form of specialization of labor.
- Allows for faster turnaround times following equipment or unit tripping due to improved onsite expertise.
- Improved utilization of employee strengths.
- Greater employee job satisfaction.
- Enhanced control of plant maintenance, resulting in improved personnel safety and equipment availability.
- Reduction in training hours.
- Better employee understanding of management expectations.

8.2 Methodology

On a 5-day-a-week basis, over the course of 1 year, the following loading seems to be practical:

- 2 days for routine work
- 1 day for troubleshooting
- 2 days of float

The annual routine man-hours can be identified for each work element that the craft is responsible for, and assumes that no problems are uncovered during that process. If there is enough routine work for 2 days per week (104 days annually) on average, that much equipment will generate approximately 1 day per week (52 days annually) for troubleshooting purposes. The additional 2 days of float per week (104 days on an annual basis) are allocated for vacations, holidays, management meetings for safety, organizational reasons, and unforeseen maintenance.

Some people in the industry will advocate a super-technician position that can effectively handle the three foregoing journeymen positions. The disadvantages of that approach are there are no super-people—even the best of us are just mortals. From a practical point of view, based on my many years of experience, people are not capable of handling all three responsibilities effectively. The super-technician is normally paid more, but those who were previously electricians, for instance, still perform mostly as electricians but end up getting paid a higher wage.

The mainstream responsibilities for the three crafts, from the least- to the highest-paid, are

- Electricians—Electrical apparatus maintenance
- Instrument technicians—Physical measurements of pressure, flow, and temperatures, as well as associated closed-loop process control
- Protective relay technicians—Protective relaying, electrical measurements, and generator excitation

However, there are a number of gray areas involving programmable logic computers (PLCs) and other equipment that could be assigned to any of the journeymen positions in order to balance the workload effectively. In example, there may not be enough electrical protection work to justify a full-time protective relay technician. In that case, the technician also could be assigned responsibility for turbine governor control, turbine supervisory instrumentation, main unit tripping, and the management of electrical drawing revisions, as appropriate, in order to achieve a reasonably balanced loading.

8.3 Area Maintenance

The author is reminded of a television commercial years ago for a large supermarket chain in California: "It may be Mr. Albertson's supermarket," a worker proudly exclaims, "but the produce section belongs to me." Work assignments based on areas of responsibility provide the following advantages:

- Pride of ownership.
- Improved specialization of labor.
- More focused training is permitted.
- Employees can easily understand management expectations.
- Employee performance over the course of a year will be visible and obvious.
- Assignments can be based more on employee strengths.
- Naturally fosters friendly competition.
- More interest is paid to housekeeping.
- Additional interest in organizing to reduce future time spent on different tasks.
- Allows supervision to focus on the big picture by shifting many of the day-to-day responsibilities to the journeymen.

Consideration should be given to providing employees with business cards that identify their area of responsibility (e.g., "Unit 1 Electrician") to foster ownership. Workers will naturally take more interest in housekeeping for their area of responsibility and look for ways to reduce the man-hours required for different tasks. Operations should be able to contact the responsible journeyman directly, without going through management. Journeymen should be encouraged to order and organize the parts and supplies needed for their area of responsibility and to work directly with operations for electrical clearances or tag-out procedures.

The areas of responsibility could be treated as separate business units. For example, if there is an outage on Unit 1, the Unit 2 journeyman could assist the Unit 1 outage and charge for their time accordingly. Annual appraisals could be centered on completing the routine work for each person's area of responsibility, troubleshooting strengths, housekeeping, and managing the area effectively by reducing costs.

The vast majority of routine technical work can be accomplished during short unplanned outages and does not require long overhaul outages. For example, protective relay testing can be performed during any outage and not necessarily overhaul outages and still meet North American Electric

Reliability Corporation (NERC) testing requirements. Some of the work can be completed when the unit is on-line and may not require a complete unit outage.

The work should be organized in a manner that reduces time wasters. For example, each journeyman should be able to locate vendor information, drawings, tools, test instruments, and spare parts for their area of responsibility in a few minutes. When they have free time, they should be working on optimizing the organization of the foregoing items and fabricating test or maintenance fixtures that will reduce future maintenance hours. Lockable job boxes can be placed where they will be needed. Test equipment also can be configured in a manner that reduces setup and takedown hours.

8.4 Safety

In addition to the economic advantages of area maintenance as outlined here, large gas, oil, and nuclear generating stations have many hazards, and consequently station personnel rely on others for their safety. Obviously, operating mistakes can have a large impact on employee safety, but on the technical maintenance side, station personnel depend on electricians to maintain the electrical equipment properly to prevent electrical explosions; counts on instrument technicians to maintain the controls, trips, and logic permissives for the safe operation of large turbines and boilers; and rely on protective relay technicians for the automatic isolation of hazardous electrical short circuits and other anomalies by the protective relay systems. Therefore, it is imperative that the equipment is maintained properly, that employees understand what is expected of them, and have formal responsibilities. In a technically complex generating station, a safe environment can best be achieved through some form of specialization of labor.

Bibliography

Baker, T., *Assign Areas to Maximize Power-Plant Skills* (Barrington, IL: Applied Technology Publication, June 2015).

Index

A

AC, *see* Alternating current
Acceptance DC Hipot, 252–253
Acetylene (C_2H_2), 306
ACSR cable, *see* Aluminum conductor steel-reinforced cable
Admittance method, 26, 76, 100
AEIC, *see* Association of Edison Illuminating Companies
AGC, *see* Automatic generator control
AIEE, *see* American Institute of Electrical Engineers
Air gap torque, 181
Alarm-only ground
 detector schemes, 40
 schemes, 291–292
Alternating current (AC), 8, 106, 180, 299
 factory test, 236
 Hipot testing, 232–233
 network model, 90
 power systems, 8
 resistance value, 11, 12
 RMS values, 11
 theory, 2
Aluminum bus ampacity, 63
Aluminum conductor steel-reinforced cable (ACSR cable), 11, 66, 231
 3-foot spacing, 68
 12-foot spacing, 69
Aluminum oxide, 331
American Institute of Electrical Engineers (AIEE), 91
American National Standards Institute (ANSI), 219, 275–276
Amps squared R (I^2R), 11, 114
ANSI, *see* American National Standards Institute
Antimotoring relays, 275
Applied volts/Hz ratio, 173
Association of Edison Illuminating Companies (AEIC), 330
Asymmetrical currents, 116

Automatic bus transfer schemes, 293
Automatic generator control (AGC), 268
Automatic voltage regulators (AVRs), 74, 173, 267, 346
 electrical maintenance, 345
 instability, 270
Auxiliary power transformers, 281
Auxiliary sudden pressure, 280–281
Auxiliary system
 arc voltages, 102
 bus PTs, 308–309
Auxiliary system protection; *see also* Generator protection
 bus tie overcurrent, 121–125
 bus transfer schemes, 151–155
 CT, 107–112
 high-impedance grounding, 131–147
 MCCs source overcurrent, 118–121
 motor overcurrent, 112–118
 overcurrent schematic, 106–107
 residual ground protection, 129–131
 software program, 101–102
 switchgear overcurrent coordination, 103–106
 transformer current differential protection, 148–151
 transformer high-speed, sudden-pressure protection, 147–148
 transformer primary side overcurrent, 126–129
 transformer secondary side overcurrent, 125
Auxiliary system switchgear operation, 288
 alarm-only ground schemes, 291–292
 high-side source transformer overcurrent protection, 290
 load feeder ground protection, 290
 load feeder overcurrent protection, 289–290
 operator inspections, 288–289
 protection, 289

355

Index

Auxiliary system switchgear operation (*Continued*)
  purpose, 288
  source and tie overcurrent protection, 290
  source and tie residual ground protection, 290–291
  source transformer neutral ground protection, 291
  switchgear bus transfers, 292–293
Auxiliary transformer differential, 279
AVRs, *see* Automatic voltage regulators

B

Backup impedance, 168, 276–277; *see also* Cable impedance
  CTs, 170
  FLA, 168
  generator backup impedance settings, 169
  Mho circle reach points, 171
  offset mho circle reach points, 171, 172
  out-of-step slip cycle, 169
Balanced conditions, 55
Basic impulse level (BIL), 208
Battery
  banks inspection, 336
  floating charges, 336
  general, 335–336
  inspection and maintenance, 335
  inspection schedules, 336–338
  operation and troubleshooting, 339–340
  purpose, 335
  safety precautions, 338–339
  types, 336
BES, *see* Bulk power electrical system
Bidirectional current flow, 8
BIL, *see* Basic impulse level
Bohr model of atom, 2–3, 5, 8
Boost vars, 74
B-phase angle, 192
Breaker-and-a-half configuration, 161, 162
Breaker failure, 197–198, 281
Buck vars, 74

Bulk power electrical system (BES), 32, 163
Bus(es), 63, 226
  delta bus transfer currents, 122, 123
  expansion and contraction, 226
  ground detectors, 273–274
  horizontal bus bar forces, 227, 228
  inspections, 331
  magnetic flux, 227
  maintenance, 330
  parallel currents, 124
  purpose, 330–331
  roughly estimating bus transfer motor currents, 122
  testing, 331–332
  tie overcurrent, 121, 122, 123–125
  tie protection oversights, 125
  transfer elementary, 153
  transfer schemes, 151–155
  voltage drop, 70, 71
Bushings, 307–308

C

Cable impedance, 100; *see also* Backup impedance
  data, 63–64
  ohmic value, 90
Cable insulation
  ratings, 132
  systems, 330
Cable(s), 228
  ACSR cable, 231
  5 kV and higher, 330
  line loss, 230–231
  short circuit fusion seconds, 228–230
  short circuit withstand seconds, 228, 229
California Public Utilities Commission hearings (CPUC hearings), 245
Capacitance, 13, 21, 22
Capacitive reactance, 20–21
Capacitive resonance, 25
Capacitive vars, 230
Capacitive voltage rise, 72
Capacitors, 13, 15
Carbon dioxide (CO_2), 306
Carbon monoxide (CO), 306
Certification tests, 346–348

Index 357

CHP, *see* Combined heat and power
Circuit breaker
    duty, 233–235
    switching, 256
Circuits, 13–16
Close-in short circuit scenarios, 163–164
CLP, *see* Cross-link poly
CMOS, *see* Complementary metal oxide semiconductor
Collapsing delta, 72, 73
Combined heat and power (CHP), 31
Combustion turbine (CT), 32, 33
Complementary metal oxide semiconductor (CMOS), 247
Conductance, 25
Conductor
    buses, 63
    insulated cable, 63–66
    overhead ACSR cables, 66–69
    parameters, 62
    strands, 209
Contact transitions, 248
Conventional generator stator ground scheme protection, 207
Conventional impedance relays, 167
Conventional theory, 6, 7, 11
Conversions, 55
    MVA, 56–58
    ohmic, 55–56
Copper
    bus ampacity, 63
    end turns, 184
    losses, 262, 263
    ohms to degrees Celsius, 240
Corrected voltage base, 78–79
Correction factors, 251
Coulomb (Q), 13
C-phase
    angle, 192
    winding, 45
CPUC hearings, *see* California Public Utilities Commission hearings
Critical clearing time, 164
Cross-link poly (CLP), 330
CT, *see* Combustion turbine; Current transformer
Cubicle
    inspection, 321
    labeling, 289
Current, 6–11, 21, 74
    differential protection, 148, 149
    flow, 7
    spectrum analysis, 314
Current transformer (CT), 103, 105–106, 107, 162, 170, 272, 308; *see also* Delta–wye transformer
    burden, 108–109
    delta lead and lag connections, 53
    open circuit, 108
    primary overall test, 334
    ratio, 52
    reflected ohms, 108
    safety ground, 107–108
    saturation, 109–112
    wye connection, 51
Cylindrical generator two-pole rotor slip frequency currents, 184
Cylindrical rotor generator
    field windings, 238, 329
    stator coil indirectly cooled construction, 208
Cylindrical rotor shorted turns, 242
    field winding, 243
    RSO test connections, 243
    RSO test waveforms, 244

D

DC, *see* Direct current
DCSs, *see* Distributed control systems
Degrees Celsius to copper ohms, 241
Delta
    bus transfer currents, 122, 123
    configuration, 39, 55
    delta-connected secondary windings, 138
    delta-connected source transformers, 291
    delta–delta connection, 48
    delta–delta transformer, 84–85
    delta-winding connection, 38–39
Delta-wye
    connection, 45–46
    delta–wye–zigzag transformer configuration, 46, 49
    electromechanical transformer differential schemes, 52
    grounded transformer, 95

Delta–wye transformer
    phase-to-phase faults, 87
    phase-to-phase vectors, 88
    short circuit, 87
Demystifying ground fault
        calculations, 93–96
DGA, *see* Dissolved gas analysis
Dielectric breakdown test, 307
Differential relays, 272
Digital motor protection, 117
Direct axis
    saturated subtransient per-unit
        reactance, 96–97
    saturated transient per-unit
        reactance, 97
    synchronous per-unit reactance, 97
Direct current (DC), 1, 111, 167, 232,
        266, 297
    field current, 173
    grounds, 339
    high potential testing, 324–328
    high-voltage, 282
    low-voltage, 281–282
    network model, 90–91
Direct DC field breaker, 98
Disconnect, 183
Dissolved gas analysis (DGA), 305–307
Distributed control systems (DCSs),
        240, 267
Double breaker double bus, 162, 163
Drop pickup transfers, 293
Dynamic instability, 213

E

Eddy current effects, 3
EE Helper Power Engineering software
        program, 10
Electrical apparatus
    apparatus 460 V and higher, 322–324
    buses, 226–228
    cable, 228–231
    calculations, 225
    DC high potential testing, 324–328
    generator-neutral buses or
        cables, 330
    generator rotor winding overvoltage
        testing, 328–330
    generators, 235–244

insulation testing, 322
metering, 244–250
motors, 250–261
purpose, 322
switchgear circuit breakers, 231–235
transformers, 261–264
Electrical engineers, 55
Electrical faults, 213, 289
Electrical maintenance guidelines, 295
    battery inspection and maintenance,
        335–340
    bus and MCC maintenance, 330–332
    generator AVRs and PSSs, 345–349
    generator electrical maintenance,
        295–302
    insulation testing of electrical
        apparatus, 322–330
    motor electrical maintenance,
        309–315
    personnel safety grounds, 340–345
    protective relay testing, 332–334
    switchgear circuit breaker
        maintenance, 315–322
    transformer electrical maintenance,
        302–309
Electrical operating guidelines, 265
    operation of auxiliary system
        switchgear, 288–293
    operation of large electric motors,
        285–288
    operation of large generators, 265–282
    operation of large power
        transformers, 282–285
Electrical power engineers, 55
Electrical Power Research Institute
        (EPRI), 313, 330
Electrical protection, 309; *see also*
        Generator protection
    feeder ground tripping, 311–312
    instantaneous phase overcurrent
        tripping, 309–310
    time-phase overcurrent tripping,
        310–311
Electrical resistance, 11
Electrical studies
    conductor parameters, 62–69
    conversions, 55–58
    ohmic short circuit calculations,
        75–76

Index 359

per-unit short circuit calculations, 84–100
per-unit system, 76–84
power transfer calculations, 72–74
study accuracy, 70
transformer tap optimization, 58–62
two-generator system, 74–75
voltage studies, 70–72
Electrical systems, 30–32, 164
Electrical testing, 317–320, 345
Electrical theory, 2
Electricians, 352
Electric Power Research Institute project (EPRI project), 243
Electromagnetic waves, 8
Electrons orbit, 7
Electrostatic voltage transfer, 285, 306
Emergency cooling and loading, 284–285
EPR, *see* Ethylene propylene rubber
EPRI, *see* Electrical Power Research Institute
EPRI project, *see* Electric Power Research Institute project
Equivalent symmetrical conductor spacing, 68
Ethane (C_2H_6), 306
Ethylene (C_2H_4), 306
Ethylene propylene rubber (EPR), 330
Excitation engineering responsibilities, 349
Expansion and contraction, 204, 226
Exponential transmission losses, 246

F

Fault
clearing time, 166
current distribution, 165
Federal Energy Regulatory Commission guidelines (FERC guidelines), 245, 265, 295
Feeder
cables, 342
differential, 279
ground tripping, 288, 311–312
FERC guidelines, *see* Federal Energy Regulatory Commission guidelines

Ferromagnetic materials, 6
Field
current open circuit testing, 242, 302
grounds, 269–270
temperature measurements, 240
winding, 243
FLA, *see* Full load amps
Fleming's right-hand rule, 4
Floating charges, 336
Flux, 4
probe analysis, 242, 301
Flywheel effect, 27
Force (F), 4
Frequency, 6–11, 9
Full load amps (FLA), 82, 112, 168
Full-load watt losses, 262
Full-voltage close and open timing tests, 320

G

General Electric (GE), 176
Generating station
electrical configurations, 32–34
transformers, 41
Generator DC
acceptance test, 326
field breakers, 322
Generator electrical maintenance, 295; *see also* Motor electrical maintenance; Transformer electrical maintenance
overhauls, 299–301
purpose, 295–296
routine online slip-ring brush-rigging inspections, 296–298
routine unit outages, 298–299
vibration, 301–302
Generator protection, 157, 272
auxiliary and main transformer sudden pressure, 280–281
backup impedance, 276–277
bus ground detectors, 273–274
DC high-voltage, 282
DC low-voltage, 281–282
differential, 272
feeder differential, 279
generator relay data, 157–160

Generator protection (*Continued*)
  high-voltage switchyard
    configurations, 161–163
  high-voltage switchyard protection
    concerns, 163–168
  inadvertent energization, 278
  main and auxiliary transformer
    differential, 279
  negative phase sequence, 275–276
  out of step, 277
  overall unit differential, 279–280
  overexcitation, 274–275
  overfrequency, 277
  pole flashover, 278–279
  reverse power, 275
  stator ground, 272–273
  sync check, 278
  underfrequency, 277
  unit switchyard disconnect position
    switch, 280
  voltage restraint overcurrent,
    276–277
Generator protective functions, 161, 168;
    see also Generator protection
  backup impedance, 168–172
  breaker failure, 197–198
  disconnect, 183
  generator deexcitation, 190–191
  generator differential, 222
  generator third-harmonic
    monitoring, 211–213
  GSUT instantaneous neutral
    overcurrent ground fault,
    198–199
  GSUT neutral overcurrent breaker
    pole flashover, 199–201
  horizontal reactance
    supervision, 218
  inadvertent energization, 194–197
  isolated phase bus ground detector,
    202–205
  lockout relay, 221–222
  loss of field, 183–190
  loss of potential, 205
  negative phase sequence, 191–194
  out of step, 213–217
  overfrequency, 217–220
  overvoltage, 201
  reverse power, 182–183

  stator ground, 206–209
  subharmonic injection schemes,
    209–210
  Sync Check, 178–181
  transformer GSUT ground bank
    neutral overcurrent, 201
  underfrequency, 217–220
  volts/Hz, 172–178
Generator relay data, 157
  generator data, 158
  primary to relay ohms, 160
  relay to primary ohms, 160
  system data, 159
  transformer data, 158
Generator rotor
  acceptance DC Hipot, 237–238
  first-year warranty DC Hipot,
    238, 239
  winding overvoltage testing,
    328–330
Generator routine test, 326
Generator(s), 11, 31, 235, 325, 327, 345;
    see also Motors; Transformer(s)
  AVRs, 345, 346
  breaker transformers, 204
  capability curve, 184, 185
  certification tests, 346–348
  copper ohms to degrees Celsius, 240
  cylindrical rotor shorted turns,
    242–244
  data, 158
  deexcitation, 97–98, 190–191
  Degrees Celsius to copper ohms, 241
  differential, 222
  excitation engineering
    responsibilities, 349
  generating station responsibilities,
    348–349
  generator-neutral buses or cables, 330
  isolated phase bus breaker, 201
  neutral grounding transformer
    sizing, 140
  PSSs, 345, 346
  purpose, 345
  rotor acceptance DC Hipot, 237–238
  rotor first-year warranty DC Hipot,
    238, 239
  rotor routine overhaul insulation
    testing, 238–239

Index 361

routine tests, 348
stator acceptance DC Hipot, 235–236
stator routine DC Hipot, 236–237
temperature, 240–242
three-phase short circuits, 96–97
volts/Hz, 174
X/R ratio, 244, 245
Generator stator
  acceptance DC Hipot, 235–236
  routine DC Hipot, 236–237
Generator step-up transformer (GSUT), 32, 76, 123, 162, 166, 245, 274, 342
  ground bank current, 202
  instantaneous neutral overcurrent ground fault, 198–199
  neutral overcurrent breaker pole flashover, 199–201
  short circuit currents, 198
  transformers, 41
Generator third-harmonic monitoring, 211–213
  circuit depiction, 211
  generator third harmonic at no load, 212
Geometrical mean distance (GMD), 67
Geometrical mean radius (GMR), 62
GMD, *see* Geometrical mean distance
GMR, *see* Geometrical mean radius
Grain-oriented steel, 262
Graphic modules, 10
Ground detection schemes, 131
Ground disconnects, 345
Grounded wye–broken delta grounding, 142–146
  neutral ohms, 146
Ground fault calculations, demystifying, 93–96
Ground faults, 209
GSUT, *see* Generator step-up transformer

H

Hall-effect
  devices, 241
  sensor, 242
Heaters, 286
Henry (H), 12
Hertz (Hz), 9
High-impedance
  detection oversights, 146–147
  differential schemes, 162
  grounded wye–broken delta grounding, 142–146
  grounding schemes, 131, 132, 206
  induced voltages, 132–134
  neutral grounding, 138–142
  primary to secondary capacitive coupling, 138
  transient voltage mitigation, 134–137
  unlimited ground fault currents, 131–132
High pressure turbo-generators (HP turbo-generators), 98, 299
High-side source transformer overcurrent protection, 290
High voltage, 339
High-voltage ground faults (HV ground faults), 199
High-voltage switchyard configurations, 161
  breaker-and-a-half configuration, 161, 162
  double breaker double bus, 162, 163
  ring bus configuration, 161
High-voltage switchyard protection concerns, 163, 168
  close-in short circuit scenarios, 163–164
  fault current distribution, 165
  GSUT, 166
  infeed ohms, 165, 166
  transmission faults, 164
  transmission line backup impedance zones, 164
High-voltage transmission, 1
Hipot leakage currents, 320
Hipot values for cable testing, 330
Horizontal bus bar forces, 227
Horizontal reactance supervision, 218
Horsepower (HP), 31, 59, 102, 253, 314
HP turbo-generators, *see* High pressure turbo-generators
HV ground faults, *see* High-voltage ground faults
Hydrogen (H$_2$), 306
Hysteresis loss, 6

I

I^2T calculation, 140–141
ICEA, *see* Insulated Cable Engineer Association
IEEE, *see* Institute of Electrical and Electronics Engineers
Impedance
  loss, 262
  swing, 186, 187
  zone 1 impedance, 280–281
Inadvertent energization, 278
  of at-rest generating units, 194
  current, 196
  negative phase sequence reactance, 197
Independent System Operator (ISO), 230, 245
Induced voltages, 132–134
Inductance, 12–13, 23
  spacing impact on, 67
Induction motors, 286
Inductive reactance, 20–21
Inductive resonance, 25
Infeed ohms, 165, 166
In-phase connection, 46–47
Inspection and testing frequencies, 316
Inspection schedules, 336–338
Instantaneous overcurrent elements, 128
Instantaneous phase overcurrent tripping, 286–287, 309–310
Instantaneous trip (IT), 116
Institute of Electrical and Electronics Engineers (IEEE), 250, 323, 330
Instrument technicians, 352
Instrument transformer connections, 49–53
Insulated cable, 63–66
Insulated Cable Engineer Association (ICEA), 330
Insulation systems, 252, 328
Insulation testing of electrical apparatus, 322
  apparatus 460 V and higher, 322–324
  cables 5 kV and higher, 330
  DC high potential testing, 324–328
  generator-neutral buses or cables, 330
  generator rotor winding overvoltage testing, 328–330
  purpose, 322
Insulators, 307–308
Internal inspections, 305–306, 312–314
Interrupting rating, 233
Iron losses, 261
ISC, *see* Three-phase short circuit current
ISO, *see* Independent System Operator
Isolated phase bus ground detector, 202
  generator breaker and step-up transformers, 204
  neutral instability, 203
  single potential transformer ground detection, 203
  two-transformer ground detection scheme, 205
IT, *see* Instantaneous trip

K

Kilovolt-amps (kVA), 36, 253
  expressions, 41
Kilovolts (kV), 1, 102, 157
Kilowatt-hour (kWh), 248

L

Lagging
  area, 185
  connection, 46, 47
Large electric motors operation, 285
  feeder ground tripping, 288
  heaters, 286
  instantaneous phase overcurrent tripping, 286–287
  operator inspections, 285–286
  protection, 286
  purpose, 285
  starting duty, 286
  time-phase overcurrent tripping, 287
Large generators, 244
  field grounds, 269–270
  generator protection, 272–282
  moisture intrusion, 270–271
  on-line operation, 268
  purpose, 265–266

Index

routine operator inspections, 271–272
shutdown operation, 267
startup operation, 266–267
system separation, 269
voltage regulators, 270
Large power transformers operation, 282
  emergency cooling and loading, 284–285
  oil pump operation, 285
  operator inspections, 282–283
  purpose, 282
  sudden pressure relays, 283
  transformer differential or sudden pressure relay operations, 284
Lead antimony batteries, 336
Lead calcium batteries, 336
Leading area, 185
Leakage air-gap flux, 185
LEDs, see Light-emitting diodes
Lenz's law, 4
Light-emitting diodes (LEDs), 286, 314
Line current, 34
Line voltage drop, 70–71
Load feeder ground protection, 290
Load feeder overcurrent protection, 289–290
Load tap changers, 305
Locked rotor amps (LRA), 82, 114, 116, 253–254
Lockout relay, 221–222
Loss of excitation, 274
Loss of field, 183, 347
  cylindrical generator two-pole rotor slip frequency currents, 184
  generator capability curve, 184, 185
  generator capability curve, loss-of-field, and Mel plots, 189
  impedance swing, 186, 187
  lagging area, 185
  MEL, 188
  mho circle, 186, 190
  relays, 183
  relay settings, 188
  two-zone loss of excitation scheme, 187, 188
  underexcitation, 186
Loss of potential, 205
Loss-of-synchronization protection, 277

Low pressure (LP), 182, 267, 299
  turbo-generators, 98
Low voltage, 339
  480-volt systems, 102
  busway, 64
  side of GSUTs, 342
  switchgear buses, 342
Low-voltage cable
  in magnetic conduit, 64
  in nonmagnetic conduit, 65
LP, see Low pressure
L/R, see Resistor–inductor
LRA, see Locked rotor amps
Lubricate racking mechanism, 321

M

Magnetic
  field, 7
  flux, 227
  flux shielding, 3
  material, 5
  phasors, 5
Magnetism, 2–6
Maintenance and operating errors, 157
Manual or motor-operated (MOD), 280
MCCs, see Motor control centers
Mechanical inspection, 316–317
Medium-voltage
  auxiliary power bus faults, 167
  switchgear buses, 342
Megavars, see Megavolt ampere reactive consumption (MVAR consumption)
Megavolt ampere reactive consumption (MVAR consumption), 72, 247
Megavolt-amps (MVA), 39, 55, 101, 157, 227
Megawatts (MW), 74, 101, 157
Megger-testing, 331
MEL, see Minimum excitation limiter
Metering, 244
  ISO guidelines, 245
  Process Systems, 246
  theory, 247
  three-element, 247
  two-element, 247, 248
  watt demand, 248–249
  watts, 249, 250

Methane (CH$_4$), 306
mho, 167
Mho circle reach points, 171
Microfarad (μF), 18
Millihenry (mH), 19
Minimum excitation limiter (MEL), 188, 346, 347
Minimum seconds per scope revolutions, 180
Minimum trip (MT), 103
MOD, *see* Manual or motor-operated; Motor-operated disconnect
Modified wye–wye potential transformer grounding, 207
Moisture intrusion, 270–271
Momentary rating, 233–234
Motion (M), 4
Motor control centers (MCCs), 102, 112, 118
  feeder, 252
  maintenance, 330
  position inspections, 332
  position testing, 332
  purpose, 330–331
  source feeder protection oversights, 121
  source overcurrent, 118–121
Motor electrical maintenance, 309; *see also* Generator electrical maintenance; transformer electrical maintenance
  electrical protection, 309–312
  internal inspections, 312–314
  motor monitoring and diagnostics, 314–315
  online and offline routine inspections, 314
  purpose, 309
  routine testing, 312
Motor-operated disconnect (MOD), 161–162
Motor overcurrent, 112
  digital motor protection, 117
  motor overcurrent oversights, 117–118
  rotor overcurrent protection, 114–115
  short circuit protection, 116
  stator overcurrent protection, 112–114

Motor(s), 250, 325; *see also* Generator(s); Transformer(s)
  acceptance DC Hipot, 252–253
  cables, 342
  contribution, 98–100
  DC acceptance test, 327
  DC routine test, 327
  form wound coils, 256
  insulation resistance, 250–252
  locked rotor amps, 253–254
  monitoring and diagnostics, 314–315
  motor HP derating for supply voltage unbalances, 255
  overcurrent oversights, 117–118
  reliability, 258
  routine DC Hipot, 253
  switching transients, 256–257
  symmetrical short circuit contributions, 99
  unbalanced voltages, 254–255
  voltage drop, 258–261
  X/R ratio, 255–256
MT, *see* Minimum trip
Multifunction digital relay, 334
MVA, *see* Megavolt-amps
MVAR consumption, *see* Megavolt ampere reactive consumption
MW, *see* Megawatts

N

National Electrical Code (NEC), 63, 65, 265, 295
National Electric Manufacturers Association (NEMA), 117, 252
NDE, *see* Nondestructive examination
NEC, *see* National Electrical Code
Negative phase sequence, 91, 191, 275–276
  120-Hz reverse rotation-induced currents, 191
  magnitude, 193
  Per ANSI/IEEE Standard C50. 13 generators, 192
  positive phase sequence magnitude, 193
  relay settings, 195
  withstand, 195
  zero phase sequence magnitude, 194

Index 365

NEMA, *see* National Electric Manufacturers Association
NERC, *see* North American Electric Reliability Corporation
Neutral grounding, 138–142
　detector fault ohms, 141
Neutral instability, 144, 203
Nitrogen-blanketed spare transformers, 308
No-load tap changers, 305
Nondestructive examination (NDE), 300
Non-network isolated electrical systems, 32
North American Electric Reliability Corporation (NERC), 188, 265, 295, 353–354
No transformer, 75–76

O

OA, *see* Oil and air
Occupational Safety and Health Administration (OSHA), 132, 265, 295
OD, *see* Outside diameter
OEM, *see* Original equipment manufacturer
Offline routine inspections, 314
Offset mho circle reach points, 171, 172
Ohmic approach, 90
Ohmic short circuit calculations, 75–76
Ohmic values, 55–56
Ohms (Ω), 11
Ohm's Law, 6, 55, 122
Oil and air (OA), 78, 127
Oil-filling, 305
Oil pump operation, 285
Oil sudden-pressure relays, 308
One cycle sine wave, 9
On-line operation, 268
Online routine inspections, 314
Online slip-ring brush-rigging inspections, 296–298
Open delta connection, 50
Open-pole insulation, 318, 319
Open-pole test, 233
Operational tests, 320
Operator inspections, 271–272

Optimum tap, 59
Orbiting electrons potential energy, 7
Original equipment manufacturer (OEM), 219, 298
OSHA, *see* Occupational Safety and Health Administration
Out-of-phase synchronizing operations, 178
Out of step, 213, 277
　blinder calculations, 216
　protection, 213
　single blinder relay settings, 215
　single blinder scheme, 214
　slip cycle, 169
　swing impedance, 214
　vertical blinder impedance points, 217
Outside diameter (OD), 300
Overall differential, 162
Overall unit differential, 279–280
Overcurrent
　bus tie, 121–125
　MCCs source, 118–121
　motor, 112–118
　schematic, 106–107
　transformer secondary side, 125–126
Overexcitation, 274–275
Overfrequency, 217–220, 277
　withstands, 219
Overhauls, 299–301
Overhead ACSR cables, 66–69
Overvoltage, 201

P

Parallel circuits, 14
Parallel impedance, 24–27, 26
Paralleling two sources, 292
Parallel reactance, 15
Parallel resistors, 14
Parallel resonant circuits, 27
Parallel sources, 76
Parts per million (PPM), 307
Per ANSI/IEEE Standard C50.13 generators, 192
Permanent magnet, 6
Personal grounds, 344, 345
Personnel safety grounds, 340
　cable short circuit fusion, 343

Personnel safety grounds (*Continued*)
  electrical testing, 345
  general, 341–342
  induced voltage hazard, 343
  maintenance, 344–345
  purpose, 340–341
  special grounding considerations, 342–344
Per-unit short circuit calculations, 84
  demystifying ground fault calculations, 93–96
  generator de-excitation, 97–98
  generator three-phase short circuits, 96–97
  motor contribution, 98–100
  sequence impedances, 91
  three-winding transformer short circuits, 86–89
  transformer ground fault procedures, 91–93
  transformer ohmic short circuit calculations, 89–91
  transformer short circuits, 84
  transformer three-phase and phase-to-phase fault procedures, 84–86
Per-unit system, 76
  basic formulas, 77–78
  converting amps to per-unit R and X, 81
  converting amps to per-unit Z, 82–84
  converting per-unit to ohms, 82, 83
  converting per-unit Z to amps, 79–80
  corrected voltage base, 78–79
  MVA base, 82
  transformer, 76–77
PF, *see* Power factor
Phase-ground faults, 213
Phase-ground test, 233
Phase-neutral voltage, 40
Phase–phase fault critical clearing times, 213
Phase–phase test, 233
Phase rotation sequence, 43–44
Phase to ground insulation, 318, 319, 320

Phase-to-phase insulation, 318, 319, 320
Phase-to-phase insulation resistance, 251
Phasing test, 308–309
Phasors, 22
Photons, 6
Photon striking, 7
PI, *see* Polarization index
PLC, *see* Programmer logic controller
PLCs, *see* Programmable logic computers
Polarization index (PI), 238, 309
Polarization test, 324, 325
Pole flashover, 278–279
Pole flashover current, 199, 200
Pole hydrogenerators, 208
Polyvinyl chloride (PVC), 298
Portable personal grounds, 341
Positive phase sequence, 91
Positive phase sequence magnitude, 193
Potential transformer (PTs), 49–50, 170, 308
Power circuits, 23
Power factor (PF), 42–43, 230, 304
Power system stabilizers (PSSs), 269, 345, 346
Power transfer calculations, 72–74
Power transformer
  connections, 45–49
  losses, 261–263
  X/R ratio, 263–264
PPM, *see* Parts per million
Primary capacitive coupling to secondary capacitive coupling, 138
Primary impedance, 28
Primary to relay ohms, 160
Primary voltage, 27
Process systems, 246
Programmable logic computers (PLCs), 352
Programmer logic controller (PLC), 102
Protection engineering, 101, 102
Protective relay technicians, 352
Protective relay testing, 332
  documentation, 334
  general, 332
  multifunction digital relay, 334

Index

primary overall test of CTs, 334
purpose, 332
relay routine tests, 333
schedule, 333
PSSs, *see* Power system stabilizers
PTs, *see* Potential transformer
PVC, *see* Polyvinyl chloride
Pyrolytic growth avoidance in tap changers, 304–305

R

Rack-in inspection, 321
Radial systems, 32
RATs, *see* Reserve auxiliary transformers
RC, *see* Resistor–capacitor
Reactance, 20–21
Reactive ampere, 346
Reliability, 258
  theory, 8
Repetitive surge oscilloscope testing (RSO testing), 242, 302
  test connections, 243
  test waveforms, 244
Reserve auxiliary transformers (RATs), 34, 41, 55, 165
Reset-set flip-flop logic (RS flip-flop logic), 248
Residual ground
  protection, 129–131
  schemes, 131
Residual voltage relays, 152–153
Resistance, 11–12, 16, 23
Resistance temperature detectors (RTDs), 117, 240, 314
Resistance-test, 304
Resistor–capacitor (RC), 17
Resistor–inductor (L/R), 18, 19
Resonance, 23
Resonant frequency, 24
Revenue meters, 249
Reverse power, 182–183, 275
Revolutions per minute (RPM), 100, 184
Ring bus configuration, 161
RMS, *see* Root mean square
Room temperature, 335
Root mean square (RMS), 11, 236

Rotor
  overcurrent protection, 114–115
  routine overhaul insulation testing, 238–239
Routine DC Hipot, 253
RPM, *see* Revolutions per minute
RS flip-flop logic, *see* Reset-set flip-flop logic
RSO testing, *see* Repetitive surge oscilloscope testing
RTDs, *see* Resistance temperature detectors
Rule of thumb, 322

S

Safety, 354
  precautions, 338–339
Secondary capacitive coupling, Primary capacitive coupling to, 138
Secondary voltage, 27
  magnitude, 28
Self-magnetic flux, 12
Self-sufficient area maintenance, 351
Sequence impedance method, 91, 94, 96
Series capacitive impedance, 22
Series circuits, 13–14, 23
Series current lagging circuit, 167
Series impedance, 21–24
Series inductive impedance, 23
Series microfarads, 16
Series resonant impedance, 24
Short circuit, 29, 75, 77, 78
  fusion seconds, 228–230
  protection, 116
  withstand seconds, 228, 229
Shorted turns, 242, 301
Shutdown operation, 267
Silicone greases, 307
Simple transformer, 27
Single blinder scheme, 214, 215
Single conductor, 4
Single-element volts/Hz tripping scheme, 176, 177
Single-phase system, 2
Single potential transformer ground detection, 203
Skin effect phenomenon, 8, 192
Slip frequency currents, 184

Slope, 149
Source overcurrent protection, 290
Source residual ground protection, 290–291
Source transformer neutral ground protection, 291
Spacing impact on inductance, 67
Spare
　breakers, 315
　transformer maintenance, 308
SSTs, *see* Station service transformer
ST, *see* Steam turbine
Startup operation, 266–267
Station maintenance, 296
Station responsibilities generating, 348–349
Station service transformer (SSTs), 32
　transformer, 41
Stator
　iron damage, 209
　overcurrent protection, 112–114
Stator ground, 206, 272–273
　conventional generator stator ground scheme protection, 207
　cylindrical rotor generator stator coil indirectly cooled construction, 208
　high-impedance grounding, 206
　modified wye–wye potential transformer grounding, 207
　stator iron damage, 209
Steady-state instability, 184
Steam turbine (ST), 32, 33
Step-up Transformer, 204
Stray
　capacitance, 138
　losses, 262
Subharmonic injection schemes, 209–210
Subtransmission systems, 32
Sudden pressure
　protection, 147–148
　relays, 283, 284, 308
Susceptance, 26
Switchgear
　automatic bus transfer schemes, 293
　bus transfers, 292
　drop pickup transfers, 293
　ground and test devices, 344, 345
　grounding devices, 344, 345
　and MCC buses, 331
　overcurrent coordination, 103–106
　paralleling two sources, 292
Switchgear circuit breaker maintenance, 315–316
　cubicle inspection, 321
　electrical testing, 317–320
　generator DC field breakers, 322
　inspection and testing frequencies, 316
　mechanical inspection, 316–317
　operational tests, 320
　purpose, 315
　rack-in inspection, 321
Switchgear circuit breakers, 231
　AC Hipot Testing, 232–233
　ANSI/IEEE remote source, 235
　circuit breaker duty, 233–235
　problems, 232
Switching transients, 256–257
Switch time transfer schemes, 293
Sync check, 178, 278
　AC, 180
　maximum symmetrical synchronizing current, 181
　out-of-phase synchronizing operations, 178
　relay function, 179
　relay settings, 179
Synchronous
　generators, 184, 274
　reactance, 97
System data, 159

T

Tank circuit, 27
Technical workforce management, 351
　area maintenance, 353–354
　methodology, 352–353
　safety, 354
　strategies, 351
Thermal modeling, 117
Thermal stability testing, 242, 301
Three-element metering, 247
Three-element volts/Hz protection scheme, 176

Index 369

Three-phase
  balanced apparent power formula, 56
  basics, 34
  fault, 213
  ferro-resonance, *see* Neutral instability
  ferroresonance, 145
  Ohm's law, 75
  power, 1, 2
Three phase-phase voltage combinations, 34
Three-phase short circuit current, 55, 75, 79, 81, 90, 92, 103, 118, 196, 290
  generator, 96–97
Three-phase system, 1
  advantages, 2
  balanced conditions, 39–40
  capacitance, 13
  circuits, 13–16
  delta-winding connection, 38–39
  electrical systems, 30–32
  generating station electrical configurations, 32–34
  high-impedance, 40–41
  inductance, 12–13
  instrument transformer connections, 49–53
  line current, 34–35
  magnetism, 3–6
  parallel impedance, 24–27
  PF, 42–43
  phase rotation sequence, 43–44
  phase to neutral voltage relationships, 45
  power, 37–38
  power transformer connections, 45–49
  reactance, 20–21
  resistance, 11–12
  series impedance, 21–24
  theory, 2–3
  time constants, 16–19
  total voltage drop, 35–36
  transformers, 27–30
  voltage, current, and frequency, 6–11
  voltage squared to VA, 36–37
Three-winding transformer short circuits, 86–89

Three-wire pulses, 248
Tie overcurrent protection, 290
Tie residual ground protection, 290–291
Time constants, 16–19
Time-phase overcurrent tripping, 287, 310–311
Transformer differential relays, 126, 130, 150, 151, 222, 279
  operations, 284
  taps, 150
Transformer electrical maintenance, 302; *see also* Generator electrical maintenance; Motor electrical maintenance
  avoiding pyrolytic growth in tap changers, 304–305
  DGA, 306–307
  dielectric breakdown test, 307
  electrostatic voltage transfer, 306
  inspections, 302–304
  insulators and bushings, 307–308
  internal inspection, 305–306
  phasing test, 308–309
  purpose, 302
  spare transformer maintenance, 308
  sudden-pressure relays, 308
  transformer testing, 304
Transformer(s), 11, 27–30, 261; *see also* Generators; Motors
  backup protection, 126
  current differential protection, 148–151
  data, 158
  differential, 284
  electromechanical magnetic force, 127
  gas, 308
  ground fault procedures, 91–93
  GSUT ground bank neutral overcurrent, 201
  high-speed, sudden-pressure protection, 147–148
  loss model, 261
  ohmic short circuit calculations, 89–91
  overcurrent, 281
  power transformer losses, 261–263
  power transformer X/R ratio, 263–264

Transformer(s) (*Continued*)
   primary side overcurrent, 126
   protection oversights, 128–129
   secondary side overcurrent, 125
   short circuits, 84
   sudden pressure, 280–281
   tap optimization, 58–62
   testing, 304
   three-phase and phase-to-phase fault procedures, 84–86
   volts/Hz, 175
   winding tap correction, 79
Transient voltage mitigation, 134–137
Transmission
   faults, 164
   line backup impedance zones, 164
Tripping, 215, 221
Troubleshooting, operation and, 339–340
Turbine
   manufacturers, 219
   turbine/generator, 178
Two-element metering, 247, 248
Two-generator system, 74–75
Two-lamp method, 44
Two-transformer ground detection scheme, 205
Two-zone loss of excitation scheme, 187, 188

U

UATs, *see* Unit auxiliary Transformer(s)
Unbalanced voltages, 254–255
Underexcitation, 186
Underfrequency, 217–220, 277
Undisturbed generator bus PTs, 308–309
Unit auxiliary Transformer(s) (UATs), 32, 41, 55, 97, 98, 166, 267
Unit breaker pole flashover, 200
Unit differential, 162
Unit outages, 298–299
Unit switchyard disconnect position switch, 280
Unity power factor, 42
Unloaded transmission lines, 230–231
Utility configuration, 32

V

Vacuum breakers, 320
Vacuum-drying, 305
VARs, *see* Volt-amps-reactive
Vector, 22
Vertical impedance blinders, 214, 217
Vibration, 301–302
Voltage, 6–11, 24
   bus voltage drop, 70, 71
   capacitive voltage rise, 72
   collapsing delta, 72, 73
   drop, 258, 261
   line voltage drop, 70–71
   motor 4/0, 259
   motor 750, 260
   regulators, 270
   restraint overcurrent, 276–277
   studies, 70
Volt-amps (VA), 30, 74, 139, 159
Volt-amps-reactive (VARs), 41
Voltage transformer (VT), 50
volts/Hz, *see* Volts per hertz
Volts/Hz, 60–61
Volts per hertz (volts/Hz), 172
   AVR, 173
   excitation, 173
   generator, 174
   lower impedance, 178
   settings, 177
   single-element volts/Hz tripping scheme, 176, 177
   three-element volts/Hz protection scheme, 176
   transformer, 175
   Westinghouse transformer volts/Hz withstand curve, 174, 175, 176
VT, *see* Voltage transformer

W

Watt demand, 248–249
Watts (W), 37, 249, 250
Wave traps, 27
Western Electricity Coordinating Council (WECC), 220
Westinghouse transformer volts/Hz withstand curve, 174, 175, 176

Index

Worst-case angle, 180
Wye configurations, 35
Wye-connected generators, 325–326
Wye–delta
　configuration, 45
　conversion, 40
　phase-to-phase faults, 88
　phase-to-phase vectors, 88
　transformer, 86, 105
Wye–wye
　potential transformers, 206
　transformer bank, 46, 48

X

X/R ratio, 244, 245, 255–256

Z

Zero phase sequence, 91
　component, 192
　magnitude, 194
Zone 1 impedance relays, 166, 167, 280–281
　switchyard, 221–222